VOLUME II. APPLICATIONS

Developed from selected papers presented at the First International Conference on Separations Science and Technology, New York, NY, U.S.A., April 15-17, 1986.

Edited by

C. JUDSON KING
University of California
Berkeley, California, U.S.A.

and

JAMES D. NAVRATIL
Colorado School of Mines
Golden, Colorado, U.S.A.

LITARVAN LITERATURE

DENVER VIENNA

For cataloging in publication data use main entry under title: Chemical Separations. Volume II. Applications.

"Developed from selected papers presented at the First International Conference on Separations Science and Technology, New York, NY, April 15-17, 1986."
Bibliography:p.
Includes Index.
1. Separation (Technology) I. King, C. Judson
II. Navratil, James D., 1941-

LC 86-81868
ISBN 0-937557-02-1 (V. I)
ISBN 0-937557-03-X (V. II)
ISBN 0-937557-04-8 (Set)

This book represents either original work or information obtained from highly regarded sources. Reprinted material is quoted with permission, and sources are indicated. Although the data and information are regarded as reliable, the editors and publisher cannot assume responsibility for the validity of the information or its use. Furthermore, the papers contained herein were refereed by at least one competent expert in the field. To expedite publication, the authors' revised typescripts were, for the most part, directly reproduced; we hope this will not inconvenience the reader.

International Standard Book Number 0-937557-03-X (Volume II)
Library of Congress Card Number 86-81868
Printed in the United States of America

CONTENTS

FUEL AND PETROCHEMICAL APPLICATIONS

METALLURGICAL APPLICATIONS

NUCLEAR APPLICATIONS

APPENDIX

INDEX

PREFACE

The First International Conference on Separation Science and Technology was held April 15-17, 1986, in New York, in conjunction with the 191st National Meeting of the American Chemical Society. The Conference was sponsored by the Subdivision of Separations Science and Technology, Industrial and Engineering Chemistry Division, American Chemical Society with financial support from the U.S. National Science Foundation and the Donors of The Petroleum Research Fund (administered by ACS).

This volume contains selected papers based on plenary and poster presentations made at the Conference. The papers cover some of the applications of separation methods used in the biological, fuel, metallurgical, nuclear, petrochemical and waste treatment fields.

C. Judson King
Berkeley, California

James D. Navratil
Golden, Colorado

BIOLOGICAL APPLICATIONS

LARGE-SCALE PURIFICATION OF BIOPOLYMERS

Jan-Christer Janson
Pharmacia Biotechnology
Research Department and
Biochemical Separation Center
Uppsala University Biomedical Center
Uppsala, Sweden

The key problems in downstream processing of pharmaceutical products originating from biological sources have always been the same. However, they have come in the lime-light in an unprecedented way by the boom of modern biotechnology at the beginning of this decade. These problems are primarily connected with terms like purity, process hygiene, recovery, throughput, reproducibility, practicality and cost. Among these, this paper will mainly deal with the concept of purity and purification and only occasionally touch upon some of the other problems. For the viability of a commercial process they are, however, as important.

INTRODUCTION

For many of the probably more than two hundred biotechnology companies in the world, currently developing production processes for polypeptides based on rDNA-technology, it is becoming more and more evident that the product purification is the bottleneck in their activities and that skill in this art is the key to success or failure in their business. To many readers this probably sounds like an old truth, however, it was not until a couple of years ago that companies which had been in the field for many years begun understanding the true meaning of the concept of purity with regard to products intended for repeated injection in human beings. Even if it is difficult to get official statements from company representatives other than that the product should be depleted of contaminants detectable with available assay techniques, it seems as if a contamination level of foreign immunogens (including neo-determinant structures) not exceeding 5 ppm is generally regarded as immunologically tolerable.

Thus, the first and prime goal of downstream processing is that of getting a final product possessing the highest possible purity. And, when we are talking in terms of contamination levels below 5 ppm, it does not matter very much which techniques are used to remove the bulk of the contaminants early in the purification process, or how the product is concentrated. What really matters is the choice and optimization of the techniques and methods used to remove the final crucial trace contaminants. In most cases these are close to identical with the desired product itself and could in fact very well be the same molecule which for various reasons has been modified, either during the biosynthesis or through proteolytic degradation during the earlier stages of the purification process. For the economy of the process it is essential to achieve maximum recovery of the final product. This is the second goal of downstream processing. Or, to quote Peter Dunnill (1): "Downstream processing is largely a matter of not losing more of the desired product than is absolutely necessary. Therefore, the reasons for such disappearance, physical loss, physical damage and irreversible chemical change, are of central concern". Dunnill concludes that there is now evidence that globular proteins are not sensitive to shear forces, at least not in the absence of gas-liquid interfaces. Thus, there is little risk for physical damage of the products. On the other hand, there is much evidence that irreversible chemical change, i.e. proteolytic degradations is one of the most important causes of product losses. E. coli contains at least eight different proteolytic enzymes and, as a consequence, the half life for human interferon alpha in crude E. coli homogenates is approximately 12 hours at room temperature. Also, proteolytic activity in crude extracts shortens the life of affinity purification adsorbants based on immobilized proteins such as mono- and polyclonal antibodies.

The traditional way of avoiding losses by proteolysis is to keep the temperature low and to work fast. In addition, some workers add protease inhibitors such as 0.5-1mM phenylmethylsulphonylfluoride (PMSF) and 2-5mM EDTA. In large scale work this might be impractical and less efficient. Recently it was shown that proteolytic degradation in crude microbial homogenates can be reduced considerably by adsorption of the proteases to hydrophobic gel media (2).

PURIFICATION BY CHROMATOGRAPHY

The basic strategy in protein purification is to work fast, to combine techniques and to use simple techniques which achieve both purification and concentration (precipitation, adsorption, etc.) early in the purification process. There is a clear need for fast, initial work-up procedures, preferably applicable directly to the unclarified homogenate. Here, affinity partitioning in aqueous polymer 2-phase systems (3,4) should have considerable potential, as should affinity adsorption to magnetizable microparticles. The latter technique, however, still seems to suffer from lack of suitable high gradient magnetic separators.

The final purification of a protein product can in general only be achieved by a combination of several separation techniques and here various types of liquid chromatography are essential elements. In Table 1 are listed the most common types of chromatography together with their principal mode of separation. Common to all chromatographic methods is the requirement of particle free, low viscosity sample solutions. This is achieved either by centrifugation or by cross-flow micro filtration, followed if necessary by adequate dilution.

Table 1. Liquid Chromatography in Biotechnology

Separation Principle	Type of chromatography
Size and shape	Gel filtration
Net charge	Ion exchange chromatography
Isoelectric point	Chromatofocusing
Hydrophobicity	Hydrophobic interaction chromatography
	Reversed phase chromatography
Biological function	Affinity chromatography
Antigenicity	Immunosorption
Carbohydrate content	Lectin affinity chromatography
Content of free -SH	Chemisorption ("Covalent chrom.")
Metal binding	Immobilized metal affinity chrom.
Miscellaneous	Hydroxyapatite chromatography
	Dye affinity chromatography

When discussing the scaling-up of protein chromatography, it is convenient to distinguish between isocratic, or starting buffer chromatography (e.g. gel filtration and chromatofocusing) and non-isocratic or gradient elution chromatography (e.g. ion exchange chromatography and affinity chromatography). Thus, in gel filtration, scaling-up means keeping the ratio between the sample zone width and the number of chromatographic cascades in the column constant. In chromatofocusing it means keeping the ratio between the sample mass (i.e. dry weight proteins) and the number of chromatographic cascades constant. As a measure of the number of cascades in a column, the plate number, N, can be used. Or, simpler, the column length if the same batch of column packing material is used and the same packing quality is obtained in the process scale column as in the laboratory trial column. If any of the physical characteristics such as particle diameter or particle size distribution is changed when going to a larger column, then this has to be compensated for by changing the column length.

Remember also that doubling the length of a chromatographic column, all other parameters kept constant, will only increase the resolution by a factor of $\sqrt{2}$.

In various types of adsorption chromatography (non-isocratic chromatography or gradient elution chromatography) such as ion exchange chromatography and affinity chromatography, scaling-up means keeping the ratio of the sample mass (i.e. dry weight protein) to the mass of adsorbent constant. Thus, at equal packing densities of the adsorbent in the process column and in the laboratory trial column, the scale factor for the sample loading (identical samples) is the ratio between the column volumes. For the sample loading step and the product elution step, the scale factor for the volume flow rate is the ratio between the cross-sections of the columns. In the gradient elution step, the scale factor for the gradient slope, i.e. the change in ionic strength or change in any other displacer concentration per column volume, is the ratio between the column volumes.

Continuous vs. stepwise Gradient Elution Methods

In a continuous gradient the concentration of the eluting buffer is increased gradually, either by mixing the starting buffer with a continuous feed of the high (limit) concentration buffer or by using a programmable switch-valve followed by an on line static or dynamic mixer. In either case fairly reproducible gradients are obtained in which the protein sample components elute in the reverse order of their binding strengths. The main advantage of this method is the higher degree of purification obtained together with the often high reproducibility in the peak composition. The disadvantages lay with the low flow-rates and the dilution of the separated components. Another consequence is the handling of fairly large volumes of buffer solutions which is both time consuming and less practical in large scale work.

In the stepwise elution procedure, the concentration of the eluting agent (salt,etc.) is increased abruptly by switching over from one buffer solution to another. The main advantage of this technique is its simplicity and speed of operation. The proteins are normally eluted in sharp peaks at high concentrations. The method lends itself ideally to automation and consumes comparatively little buffer. The main disadvantage is the lower degree of purification. Each peak is a mixture of several components and the same component is often found in more than one peak. Higher purification factors may be obtained by fine tuning the operating conditions, however, this could result in very narrow windows for the pH and salt concentration intervals and thus with a risk for less satisfactory reproducibility.

Column or stirred-batch adsorption followed by stepwise elution has become an important initial concentration and partial purification step in many modern large scale protein purification processes. This is particularily true for ion exchangers. Anion exchangers especially seem to bind both lipopolysaccharides (bacterial endotoxins) and nucleic acids more strongly than they bind most proteins, allowing the former to be removed early in the process. Often a second ion exchange chromatography step, using the same ion exchanger but under more favourable conditions and applied immediately after the concentration adsorption step, has been shown to contribute significantly to an increased specific activity of the protein product. By skilful design of this intermediate purification step, it should be possible to remove most of the residual pyrogens, DNA and proteolytic activity. This strategy can provide a suitable starting material which will contribute to an increased life span of more sophisticated (and considerably more expensive) separation media such as immunosorbents based on immobilized monoclonal antibodies.

New Gel Media For Large Scale Chromatography

The classical problem (5) in scaling-up protein chromatography used to be the low rigidity of the then available adequately porous, hydrophilic and inert column packing materials based on cross-linked dextran, agarose and polyacrylamide. This problem was partly compensated for by special column design, notably the "stacked" column concept (6,7), Fig. 1.

Figure 1. Three stacks of KS 370/15 columns. This column type was developed around 1970 to enable the use of semi-rigid polysaccharide and polyacrylamide based gel materials for large-scale chromatography of proteins (6).

Today, the development of new rigid gel materials based on cross-linked agarose, cross-linked cellulose and various resins made of synthetic organic polymers, has provided the large-scale research workers and the biochemical process engineers with a large degree of freedom in chosing column type and also in varying the operating conditions. In Table 2 are listed some modern gel materials for large-scale chromatography, their composition and manufacturers.

When comparing different gel material from a process chromatographers view-point it is essential to take into consideration two aspects which both contribute to an economically successful

Table 2. Modern gel media for large-scale chromatography

Trade name	Composition	Manufacturer
Cellufine	Cellulose (cross-linked)	Chisso Corp. Japan
Sephacryl HR	Poly(N,N-bisacrylamide) + allyldextran	Pharmacia AB, Sweden
Sepharose FF	Agarose (cross-linked)	Pharmacia AB, Sweden
Toyopearl	Vinyl polymer with hydroxyl groups	Toyo Soda K.K, Japan
Trisacryl LS	Poly(N-acryloyl-2-amino-2-hydroxy-metyl-1,3-propan diol)	Reactifs IBF, France

performance of a column. The first is the life span of the column packing material and the other is the throughput obtained under optimum operating conditions. The life span achieved with any bed material is a consequence of the fouling effect of the samples frequently applied to the column and the chemical stability of the matrix enabling regular and efficient cleaning. Cleaning in place (C.I.P) is a concept which is synonymic with good process hygiene and which means that the process chromatography system is regularly (and often automatically) washed with a suitable agent to remove adsorbed and precipitated material such as lipids, nucleic acids and proteins. The preferred agent is a 0.2-0.5 M NaOH-solution which has a good cleaning effect against both lipids (saponification) and precipitated protein. This NaOH-solution also sterilizes the complete column system, destroys pyrogens and has no adverse affect on the products subsequently separated on the column.

The second aspect of prime importance in process chromatography is the throughput, i.e. the mass of the adequately purified and collected fraction of the desired compound that can be recovered in a certain period of time. Thus one should not too much concentrate one's interest towards the flow-rates attainable in large columns as such, because these are strongly influenced not only by the matrix rigidity, but also by the average particle diameter and the particle size distribution profile of the column packing material. Rather one should pay more attention to the dynamic capacities achieved at different flow-rates for the compound under investigation. This information is normally not available from the suppliers of gel materials but is easily collected by small scale laboratory trials. How this is done by studying the shape of the breakthrough curves in ion exchange chromatography and affinity chromatography has recently been reported by Chase (8,9,10). The differences observed between different gel media are significant and reflect the importance of both the micro- and macrostructures, i.e. both the polymer type and the pore distribution of the gel matrix. Cross-linked agarose (11) seems to possess particularily favourable characteristics as a base matrix for large-scale ion exchange and affinity adsorbents respectively (12,13), Fig. 2.

Figure 2. Stainless steel column (80 cm diameter, total volume 75 l) developed for large-scale ion exchange chromatography of proteins. Packed with DEAE-Sepharose Fast Flow, a cross-linked agarose based ion exchanger, this column can accommodate 135 metric tons of human plasma per year (9 cycles per day, 5 days per week, 40 weeks per year), producing approximately 4.5 tons of partially purified human serum albumin.

Thus, it seems as if we are in a situation today where the kinetic characteristic of the interactions between the proteins to be separated and the column packing material is the main rate limiting parameter to take into consideration during the optimization work. Invariably, the process development work starts on a laboratory scale. It is then essential that the same type of column packing material is used for the small scale optimization of the operating conditions that will later be used in the production scale. It is also of considerable advantage for a successful and reproducible process development that two other parameters are kept constant throughout the scaling-up work. These are the column length and the sample composition.

As was discussed earlier, the scale factor for the sample loading is the ratio between the column volumes. However, in order to get the desired linearity in the scaling-up phase, it is also necessary to keep the chromatographic path length for the various components of the sample constant. This is achieved by keeping the column length constant and scaling-up then is achieved by increasing the column diameter.

Today column design is of such a high level of refinement that increased column diameter should not give rise to any significant increase in band broadening provided the column bed is properly packed. The stacked column concept will still be advantageous in isocratic chromatography with softer gels. It is also easier to get a short column packed to a high and homogeneous particle density, i.e. a small void volume (V_o), giving a higher relative plate number, N, than a longer column. From a separation efficiency point of view, it is thus probably more advantageous to pack three 30 cm bed height columns and connect them in series than to pack one 90 cm long column.

The second parameter which should be kept constant is the sample composition. Thus the laboratory trials should be performed with process scale samples, i.e. lysates or extracts should have been processed under industrially adequate conditions. It is a well known fact that process changes up-stream will always influence the outcome of the recovery steps down-stream. Also the absolute concentration of the proteins in the sample should not be changed during the scaling-up procedure. In gel filtration chromatography, an increase in viscosity of the samples applied to the column could give rise to a sharp decline in performance. In ion exchange chromatography, the binding capacities of most adsorbents are dependent on the concentration of the proteins in the samples (9).

PURIFICATION USING IMMOBILIZED MONOCLONAL ANTIBODIES

Ever since the pioneering work by Secher and Burke (14), immunoaffinity chromatography on immobilized monoclonal antibodies has been the most powerful technique available for the purification of human leucocyte interferons; particularly, for those produced by recombinant DNA technology (15,16). In normal human leucocytes there are 15-20 different interferons ("buffy coat" interferons) and as it seems as if there are certain clinical advantages to use the natural buffy coat interferon mixture, adsorbents based on polyclonal antibody preparations are frequently used for their isolation and purification (17).

Among other low volume, high value proteins which have attracted attention for isolation and purification by immunosorption to immobilized monoclonal antibodies, should be mentioned blood coagulation factor VIII (18), urokinase (19), pregnancy specific α_1-glycoprotein (20), superoxide dismutase (21), interleukin-2 (22) and human tissue-type plasminogen activator (23).

With monoclonals, it is in principle possible to choose from a variety of specificities and affinities in order to tailor make the optimum immunoaffinity adsorbent for a particular antigen. The user may decide to utilize a common antigenic determinant, if available, to a group of related proteins (e.g. interferons) if this is clinically advantageous in chemotherapy, or to utilize an antigenic determinant unique to a single molecular species. One attractive feature of monoclonals is the possibility to select cell lines producing antibodies with very favourable binding constants allowing mild conditions for desorption (i.e. a moderate change in pH) and thus increasing the life span of the adsorbent.

By definition monoclonals exhibit single-epitope attachment to their antigens. This, together with a high degree of homogeneity in the complex formation should lead to comparatively favourable kinetics for sorption and desorption. Provided the carrier gel material has adequate macroporosity and a low substitution with antibodies, the over-all rate constants for adsorption and desorption should allow high flow-rates during both sample application and antigen elution, the limit set only by the rigidity and particle diameter of the carrier.

For a deeper understanding of the problems and opportunities connected with scaling-up of immunoaffinity separations the reader is recommended to study recent papers by Chase (24,25), Fowell and Chase (26) and Calton (27).

In addition to the high and controllable specificities and affinities of adsorbents based on monoclonal antibodies, there are a number of other advantages which advocate their use in industrial scale processes. These are listed in Table 3.

Table 3. Advantages of adsorbents based on immobilized monoclonal antibodies:

- High and controllable specificities
  (single-step purification factors several thousand fold possible).
- Controllable affinities
  (low or moderate affinities, K (diss) = 10^{-6} - 10^{-9}, preserves antigen (yields > 90%) and increases life span of adsorbent to often several hundred cycles).
- High binding capacities
  (at least 10 fold higher as compared with adsorbents based on polyclonal antibody mixtures).
- Sharp desorption peaks
  (high concentration of antigen in eluate)
- High reproducibility from batch to batch of adsorbent
  (constant supply of highly uniform antibody).

One serious disadvantage is the high cost of monoclonal antibodies and thus of adsorbents based on these, restricting their largescale use at present to a few high value proteins such as the various lymphokines, some therapeutic enzymes and Factor VIII. However, the situation is likely to change when the production scale for monoclonals has increased from present 1000

1 to planned 10,000 l airlift fermenters and when the antibody yield has increased by improved media design (28).

The next disadvantage, which is shared with all adsorbents based on immobilized proteins, is the high risk of fouling and irreversible chemical denaturation, notably proteolytic degradation. General fouling by non-specific adsorption of various biopolymers and lipids is best prevented by inclusion of pre-column purification steps. Thus, also in order to prevent proteolytic attack, one should never apply crude extracts directly to columns packed with adsorbents based on monoclonal antibodies. Even if it is possible, at least in principle, to select hybridomas which produce antibodies with increased stability against proteolytic attack, there is always the risk of contaminating the antigens with trace amounts of proteases which will affect their storage stability and maybe create new immunogenic determinants (neo-determinant structures) and thus problems in patients after repeated administration. It is also necessary to prove the absence of antibody in the final product.

The inherent delicate nature of proteins makes sterilization of adsorbents based on these a major problem. For good process hygiene, a low count of microbes in the column is a necessity. In Table 4 is presented a procedure for chemical sterilization of lectin based affinity chromatography columns as recommended by Sparrman (29).

Table 4. Sterilization of affinity chromatography columns

1. Equilibrate with buffer containing 2% Hibitane digluconate and 0.2% benzyl alcohol.
2. Allow to stand for 24 hours.
3. Wash with sterile buffer.
4. Equilibrate with 2% Hibitane digluconate and 0.2% benzyl alcohol.
5. Allow to stand for 24 hours.
6. Wash extensively with sterile buffer.

(Successfully applied to the sterilization of columns containing lectins immobilized to CNBr-activated Sepharose).

PSEUDO-IMMUNO-AFFINITY PURIFICATION OF IN-VIVO FUSED PROTEIN-PROTEIN A COMPLEXES

By applying the techniques of genetic engineering, it is possible to facilitate the purification of rDNA products by the in-vivo synthesis of fusion proteins. By adding DNA sequences coding for one or two polypeptide domains exhibiting a unique affinity to a particular ligand, it is possible to taylor-make affinity chromatographic purification procedures with very high selectivity.

By fusing two IgG-binding domains of Staphylococcal "Protein A" to human insulin-like growth factor-1 (IGF-1), using Staphylococcus aureus as host, Uhlén and Nilsson could demonstrate the efficacy of this technique (30). An acid labile Asp-Pro cleavage site was introduced at the fusion point as well as a signal sequence to enable extracellular export of the complex through the bacterial envelope. The cells were separated from the 7 l culture medium by tangential cross-flow filtration through a Pellicon 0.22 u, 0.5 square meter microfilter and the particle free filtrate was pumped through a 200 ml column packed with a custom designed IgG-derivative of Sepharose Fast Flow (31). The flowrate was 7 l/h (linear flow-rate 12 cm/min) and the breakthrough loss of the fusion complex < 1%. After washing and elution with Glycine-HCl buffer pH 3.0, the desorbed complex was dialysed against 0.2M HAc and lyophilized with an overall yield of > 95% and a purity of > 90%. This process was recently scaled-up to a 1000 l fermenter. The particle free culture medium was pumped through a 2 l column of IgG-Sepharose Fast Flow at a linear flow-rate of 120 cm/h in 25 cycles of 35 l each. Due to the very favourable kinetics in this adsorption process (the molecular weight of the complex is 21,000 dalton), as much as 80% of the total column capacity could be utilized in each run with a breakthrough leakage of less than 1% (32).

PURIFICATION OF MONOCLONAL ANTIBODIES

There are at least four different chromatographic methods reported for the purification of monoclonals either from hybridoma suspension culture supernatants or from ascites fluids. These are based on: 1. Immunosorption to immobilized antibodies or antigens, 2. Pseudoimmunosorption to immobilized Staphylococcal Protein A (binds to the Fc-moiety of most IgG), 3. Hydrophobic interaction chromatography and 4. By a combination of ion exchange chromatography and gel filtration.

A rapid method for isolation of monoclonals from large volumes of hybridoma cell culture supernatants was recently reported by Malm (33, 34). The process was based on three consecutive chromatographic steps: 1. Desalting on Sephadex G-25, 2. Ion exchange chromatography on S-Sepharose Fast Flow and 3. Gel filtration on Sephacryl S-200 HR. The crucial step is the adsorption to the ion exchanger. This provides an approximately 100 fold concentration combined with a purification resulting in a 70-95% pure antibody with a recovery of more than 90%. The key feature is the high flow-rate possible during loading and washing of the S-Sepharose Fast Flow. A linear flow-rate of 300 cm/h. over a 3 cm bed height resulted in a breakthrough loss of less than 2% with a culture supernatant to gel bed volume ratio of 100 and approximately 5% loss with a ratio of 450. On displacement of the adsorbed monoclonals, a marked dependence of the elution volume of the desorbed peak on the flow-rate was observed. Thus, by reducing the elution flow-rate to 5 cm/h, the antibodies could be recovered in a volume corresponding to approximately 0.5 V_t . Systems can be designed for culture supernatants ranging in volume from 100 ml to hundreds of litres. In Table 5 is shown the process scheme for a sample volume of 50 l.

Since monoclonal antibodies are a heterogeneous group of proteins one would expect that the conditions for adsorption to and desorption from the ion exchanger have to be optimized with regard to pH and salt concentration in each case. However, the conditions given in Table 5 have been successfully applied to 10 different mouse monoclonals covering all 4 subclasses and with volumes ranging from 1 to 43 litres of hybridoma culture supernatant.

Table 5. Purification of Monoclonal Antibodies by Chromatography

1. Desalting of 50 litres hybridoma cell culture supernatant on Sephadex G-25 equilibrated in 20 mM citrate buffer pH 5.3 (3 runs on a 75 l column GF 04/60). Total process time 1.5h.
2. Ion exchange chromatography on S-Sepharose Fast Flow equilibrated in the citrate buffer. Column volume 300 ml (BioProcess Column BP 113). Elution buffer: 20 mM citrate buffer pH 5.0 + 140 mM NaCl. Adsorption and washing flow-rate: 300 cm/h. Elution flow-rate: 5 cm/h. Regeneration with 0.1M NaOH.
3. Gel filtration chromatography on Sephacryl S-200 HR equilibrated in 50 mM Tris-HCl pH 7.5 + 150 mM NaCl. Linear flow-rate through the 100 cm bed height column (BP113) was 7.5 cm/h.
4. Result: 95-99% pure monoclonal antibody
   85-95% total yield

   Process time for desalting and ion exchange chromatography were less than 6 h. Process time for gel filtration was approximately 13.5 h.

AUTOMATION OF CHROMATOGRAPHIC PROCESSES

Closed column chromatography is ideally suited to automation using microprocessor based control equipment (35). The system should comprise an adequate number of solenoid valves, air sensor, air trap, on-line filters, flow indicator, flow meter and flow stops on both ends of the column to prevent it from being run dry in case of power failure. Furthermore, there should be a variety of monitors to measure UV, conductancy and pH in the eluent. In Fig. 3 is shown a typical liquid flow scheme for a process chromatography system.

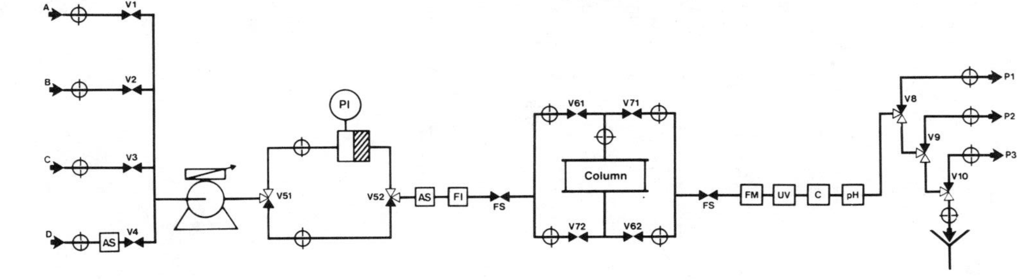

Fig. 3. Liquid flow scheme for a typical process chromatography system. Valves V51 and V52, V61 and V62, V71 and V72 are activated in pairs. The valve arrangement around the column enables reversed flow elution of proteins adsorbed to ion exchangers and affinity chromatography adsorbents. Lower loop V51-V52 only used during sample application. Pumps, valves, sensors, meters and monitors are connected to Process Chromatography Controller C-3 (Pharmacia AB).

REFERENCES

1. P. Dunnill, Process Biochem. 18:10 (1983) 9-13.

2. P. Hedman, and J.G. Gustafsson, Develop.Biol.Standard., Vol. 59 (1985) 31-39, S. Karger, Basel.

3. H. Hustedt, K.H. Kroner, U. Menge, and M-R. Kula, Trends Biotechnol., 3:6 (1985) 139-144.

4. G. Kopperschläger and G. Johansson, Anal. Biochem., 124 (1982) 117-124.

5. J-C. Janson and P. Dunnill in "Industrial Aspects of Chromatography", Ed.: Spencer, B., North Holland, Amsterdam, 1974, pp. 81-105.

6. J-C. Janson, J. Agr. Food Chem., 19 (1971) 581-588.

7. Column KS370/15, "The Stack", brochure from Pharmacia Fine Chemicals AB, Uppsala, Sweden, 1972.

8. H.A. Chase, J. Chrom., 297 (1984) 179-202.

9. H.A. Chase in "Ion-exchange Technology", eds. Naden, D. and Streat, M., Ellis Horwood Ltd., Chichester, England 1984, pp. 400-406.

10. H.A. Chase in "Discovery and Isolation of Microbial Products", ed., M.S. Verrall, Ellis Horwood Ltd., Chichester, England 1985, pp. 129-147.

11. Sepharose CL and Sepharose FF, Pharmacia AB, Uppsala,Sweden

12. J.M. Cooney, Biotechnology 2:1 (1984) 41-55.

13. J-C. Janson, Trends Biotechnol. 2:2 (1984) 1-8.

14. D.S. Secher and D.C. Burke, Nature 285 (1980) 446-450.

15. T. Staehelin, D.S. Hobbs, H-F. Kung, C-Y. Lai and S. Pestka, J. Biol.Chem., 256:18 (1981) 9750-9754.

16. B. Vaks, Y. Mory, J.U. Pederson and O. Horovitz, Biotechnology Letters, 6:10 (1984) 621-626.

17. H. Borg, Personal communication.

18. T.S. Zimmerman and C.A. Fulcher, US Patent No. 4,361,509 (1982).

19. D.A. Vetterlin and G.J. Calton, Thromb. Haemostasis 49 (1983) 24-28.

20. M. Heikinheimo, U.H. Stenman, B. Bang, M. Hurme, O. Makela and H. Bohn, J. Immunol. Methods 60:1-2 (1983) 25-31.

21. I. Bianchi, Eur. Pat. No.: 112299 (1985).

22. Y. Kiso, K. Okamoto, M. Makiyama, M. Shimokura, Y. Hirai, K. Kawai and H. Kikuishi, Pept. Chem., 22 (1985) 103-108.

23. M. Einarsson, J. Brandt and L. Kaplan, Biochim. Biophys. Acta, 830 (1985) 1-10.

24. H.A. Chase, Chemical Engineering Science, 39:7-8 (1984) 1099-1125.

25. H.A. Chase, J. Biotechnology, 1 (1984) 67-80.

26. S.L. Fowell and H.A. Chase, J. Biotechnology (1986), in press.

27. G.J. Calton, Meth. Enzymol., 104 (1984) 381-387.

28. M. Boss, Pract. Biotechnol., 7:2 (1986) 7-9.

29. M. Sparrman, Ph.D. thesis, Uppsala 1984.

30. M. Uhlén, and B. Nilsson, Proceedings of Biotech '85 Europe, Geneva, May 1985, pp 171-179.

31. Affinity Chemicals Workshop, Pharmacia AB, Uppsala,Sweden.

32. M. Uhlén, Personal communication.

33. B. Malm, Poster at the Biotech '85 USA exhibition in Washington DC, October 21-23, 1985.

34. B. Malm and J. Berglöf, to be published.

35. H. Johansson, in The Proceedings of Biotech 85 Europe, pp 193-202, Online Publications, Pinner, UK, 1985.

MEMBRANE TECHNOLOGY AND CHROMATOGRAPHY: APPLICATION FOR BIOTECHNOLOGY

MEMBRANE CHROMATOGRAPHY--IS A MARRIAGE OF SEPARATION TECHNIQUES ON THE HORIZON?

Doris C. Warren
Department of Chemistry
Houston Baptist University
7502 Fondren Road
Houston, Texas 77074

Biotechnology application demands on separation science is leading to a reexamination of separation techniques in order to solve complex problems. When the fields of protein and nucleic acid chemistry are considered, the vast majority of separation methods fall into three broad categories--those based on size differences; those based on differences in the electrical charge carried by the molecules; and those based on some specific biological or chemical property of the molecule under investigation. All three of these categories are addressed by the various chromatographic and membrane methods. However membrane materials are being redesigned with similar characteristics to traditional chromatographic column packing materials. New column designs are beginning to be assessed. On the horizon there appears to be coming a rethinking of technology and a combination of the best of both separation techniques into a new approach--membrane chromatography.

Biotechnology application demands on separation science appear to be leading to new ways to solve complex purification problems. Represented by numerous research projects, patents and now emerging products for the purification of biomolecules markets, the trend seems to be towards a marriage of two powerful purification techniques--membrane technology and chromatography into membrane chromatography.

Biotechnology, one of the most exciting areas of science today, is developing at a rapid pace because of the many applications of recombinant DNA research, microbial biosynthesis, cloning, and gene splicing. Developments are on a fast track from the research laboratory to application in many industrial areas such as in medical, agriculture, and industrial processes. Many problems related, not only at the research laboratory level, but also to moving from laboratory research to scaled-up operations, remain to be solved.

Biomolecules range in size from small organic compounds, peptides, and proteins to oligonucleotides (including large fragments of DNA). The growth in the field of biotechnology depends to a great extent upon the separation, detection, and characterization of these molecules. The approaches to solving the particular problems posed by the purification of biomolecules has led to the application of the whole arsenal of a wide variety of techniques which fall within the realm of both membrane technology and chromatography. Membrane technologies such as microfiltration (MF), ultrafiltration (UF), dialysis, and electrodialysis (ED) are being used. Chromatographic techniques such as gel filtration, ion exchange, hydrophobic interaction, reversed-phase, all types of polyacrylamide gel electrophoresis, and affinity chromatography have a unique role in the purification of biomolecules.

Current separation methods for the purification of high molecular weight biomolecules such as proteins, enzymes, and nucleic acids lean heavily on physical processes which cause the minimum disturbance to both the physical and chemical properties of the molecules and result in the maximum retention of any biological activity that the molecule may possess. When the fields of protein and nucleic acid chemistry are considered, the vast majority of separation methods fall into three broad categories--those based on size differences; those based on differences in the electrical charge carried by the molecules; and those based on some specific biological or chemical property of the molecule under investigation.

In both membrane technology and chromatography, components of a mixture are carried through or brought in contact with a stationary phase by the flow of a mobile phase (a gas or liquid). In chromatography, separations are based on the distribution of the components between the two phases. In membrane technology, separations are based on several different parameters.

The literature is full of methods for the purification of biomolecules by chromatographic methods and membrane technology. In practice more than one technique or a variety of steps are necessary in the design of a protein purification scheme. The membrane approach using microfiltration and ultrafiltration has traditionally been considered as separate and somewhat unrelated techniques to chromatographic methods such as ion exchange, gel filtration, hydrophobic interaction, affinity chromatography and a variety of gel polyacrylamide electrophoresis techniques (PAGE). We now see recurrent themes common to both types of separations and the divisions are no longer clear cut.

We speak of liquid chromatographic (LC) techniques in terms of the physical form of the stationary phase and the mechanisms by which the separations can be effected. The two types of LC are liquid-solid and liquid-liquid chromatography. The stationary phase may be found as a thin layer, i.e., thin layer chromatography (TLC); a slab or cylindrical rod of a gel material as in PAGE; or a column packed with a gel or spherical particles or beads which may have the surface coated with a liquid or with structures chemically bonded to the surface (ion exchange or affinity chromatography). The mechanisms by which separations are effected include adsorption, partition, ion exchange, exclusion, or some form of ligand interaction such as affinity.

In membrane technology, one thinks of materials such as cellulose acetate or polysulfone, fabricated into sheets and produced in disk, spiral-wound or hollow fiber configurations. The mechanisms used to effect separations can include separations based on size of the biomolecule (microfiltration and ultrafiltration, reverse osmosis)--all of which depend on the pore size of the membrane. The differences in the concentration of ions on each side of a membrane is used to separate the desired ion in dialysis; while electrodialysis separates ionic components by means of charge-specific membranes and the driving force of an electric current. Gas diffusion membranes separate one gas from a mixture of gases by means of their different permeation rates through the membranes. Finally, in supported liquid membranes a liquid, held by capillary action in the pores of a microporous membrane, chemically facilitates transport of certain solutes from one side of the doped microporous membrane to the other by mechanisms such as ion exchange.

When we take a new look at membrane and chromatographic techniques and organize them by application in the purification of biomolecules rather than by categories under technologies, we see that we are back to separations based upon the following categories:

- Differences in electrical charge carried by the molecules

- Difference in molecular size

- Specific biological or chemical property of the molecules

Ion exchange, various types of PAGE, and electrodialysis exploit electrical charge characteristics of the biomolecule to accomplish separations. Gel exclusion or permeation chromatography, sodium dodecylsulfate (SDS) PAGE, gradient gel electrophoresis and MF and UF are techniques that purify molecules based on their sizes. Techniques such as affinity chromatography, or affinity or immunoelectrophoresis are used to exploit specific biological or chemical properties of the biomolecules.

Since the surface structure of membrane materials may be chemically altered in much the same manner as materials used in packing chromatographic columns, we have recently seen the development of membrane products directed towards the biotechnology purification market. For example, membranes prepared from nylon, such as those produced by Cuno, can be made hydrophyllic or hydrophobic to suit the particular biological application.

Millipore's latest product, Immobilon©, is a membrane prepared from polyvinylidene, which has affinity groups attached to the surface. Proteins are covalently bound via amino linkages.

Membranes containing ion-exchange sites, such as Du Pont's Nafion©, are routinely used in the industrial preparation of chlorine and alkali. Gershoni at Yale recently obtained a patent (U.S. 4,512,896) on the transfer on macromolecules such as nucleic acid and proteins from a chromatographic substrate to an immobilizing cationic charge modified membrane matrix. A result is a nucleic acid or protein blotting product comprising a chromatographic matrix having the charge modified microporous membrane on a surface. The cationic charge modifying agent is bonded to substantially all of the wetted surfaces of a microporous membrane.

For several years membranes have been used as "blotting" agents to allievate problems inherent in the purification of proteins by electrophoretic methods. Widely used techniques have been developed using membranes in conjunction with PAGE. "Blotting" techniques-capillary and electrophoretic transfer to derivatized paper and membranes-have been developed to address problems associated with the _in situ_ detection and analysis of electrophoretically resolved molecules. Capillary transfer and electrophoretic transfer both require that the gel be placed in contact with the material to which the proteins or nucleic acids will be transferred. Charge modified nylon membranes have been used for this purpose.

It is now time to examine chromatographic methods and membrane technology; take the best features from both; and design new systems to purify biomolecules. This approach can lead to possible solutions of scale-up problems in the purification of these biomolecules. The place to begin in the development of _membrane chromatography_ is with the "column."

The column packing may take the physical form of membrane systems. Polymeric material in thin-film or sheet form can be arrayed as layers of discs, in a spiral wound, or hollow fiber configuration. The membrane material, as the bed, or "packing," will be altered taking advantage of the vast body of knowledge developed in liquid chromatography. Instead of using a membrane always only in its traditional role such as MF or UF, it will also be used as a modified-surface membrane packing material. Essentially the spherical particle packing material used in chromatography today is the concept which may be altered. Then the problems associated with packed spherical materials may be circumvented.

There are already new column designs in use and being proposed in scaled-up chromatographic applications. At the 1986 Pittsburgh Conference, Sepragen was showing its innovative Superflo columns. These columns have a unique radial-flow design. The eluant flows radially inward from an outer channel to an inner channel through a middle region containing the conventional chromatographic packing.

According to Sepragen, the combination of large cross-sectional area and small bed height allows high flow rates (high production) with low pressure drops. One advantage cited is for separations depending upon binding interactions (ion exchange, affinity). When the amount of sample increases, the length of the column and the flow rate can be increased proportionally. There is no need to reoptimize process conditions to yield the same resolution and separation time.

Cuno was recently granted a patent (U.S. 4,496,461) on a chromatographic column design for effecting chromatographic separation using as the stationary phase a swellable fibrous matrix that is in sheet form and has chromatographic functionality. The column design is such that the stationary phase is spirally wound around the longitudinal axis of the solid phase to form a plurality of layers around the axis. A spacer between each layer separates the layers, permits controlled swelling of the matrix, and enhances the distribution of sample flowing radially through the stationary phase. The enhancement occurs by channelling the sample flow through the matrix and substantially evenly dispersing the sample axially and circumferentially across the matrix, according to the patent.

Therefore, trends in the purification of biomolecules appear to be leading to the marriage of two separation techniques--liquid chromatography and membrane technology into membrane chromatography.

APPLICATION OF ULTRAFILTRATION TO THE PURIFICATION AND CONCENTRATION OF BOVINE INTESTINAL ALKALINE PHOSPHATASE (BIAP) ENZYME

C.S. Slater, P.P. Antonecchia, L.S. Mazzella
and H.C. Hollein

Manhattan College
Chemical Engineering Department
Riverdale, NY 10471

Ultrafiltration is an effective purification and concentration process for enzymes. These studies investigated the applicability of this technique for the purification of the enzyme bovine intestinal alkaline phosphatase (BIAP). Studies in a thin-channel system evaluated various membrane types for flux and separation efficiency. An Amicon YM100 membrane was shown to be more effective in purification efficiency and production rate than the other commercially available membranes tested. Enzyme activity was retained while the crude enzyme mixture was purged of the lower molecular weight protein impurities. Studies in a stirred cell demonstrated that increases in stirrer speed and temperature minimized gel layer build-up and improved the separation process. A dialysis run was utilized to stabilize flux decline.

INTRODUCTION

Processes to concentrate and purify enzymes are becoming increasingly important to the biochemical industry. Process scale-up for production of a variety of biochemicals warrants a thorough investigation of these processes and their optimization. In a production scheme employing fermentation and product recovery, it is the separation and recovery processes that are the most costly and complex.

One technique that shows great promise for both lab-scale and production-scale concentration and purification is ultrafiltration. Ultrafiltration (UF) processes have great versatility in biochemical processing. They are typically employed in downstream separation and recovery of fermentation products. UF can be utilized in isolating the desired product by separating the microbial cells or debris from the broth, in purifying the enzyme in the clarified liquor and in a possible finishing step involving formulation in a sterile pyrogen-free solution [1]. The dewatering of the broth and purification of the product are the major use of UF in an overall biochemical production process. Therefore, UF processess can be utilized in three types of separations. The first of these is to produce a permeate stream with the desired enzyme while retaining and removing larger macromolecules or other undesirable matter. The second would be to have the membrane concentrate the desired product. Thirdly, the membrane can act to concentrate the desired product while simultaneously removing lower molecular weight impurities. One of the most rigorous applications of UF is the concentration of one product in a protein mixture while purging other lower molecular weight constituents [2].

While UF is a sound process for both separation and concentration of enzymes, other processes play key roles in enzyme purification. Some of these processes include centrifugation, affinity and ion-exchange chromatography, crystallization, electrophoresis, evaporation, extraction, filtration, microfiltration and sedimentation. All of these have advantages and disadvantages. They are usually utilized in an extensive process sequence if high purity enzyme is the desired end product.

The objective of this study was the concentration and purification of the enzyme bovine intestinal alkaline phosphatase (BIAP). Several separation processes have been employed to purify alkaline phosphatase. Seargent and Stinson [3] present a process sequence of butanol extraction, acetone precipitation, ion-exchange and/or affinity chromatography with Sephadex and Sepharose type resins. Continuous paper electrophoresis [4] and preparative-scale polyacrylamide gel electrophoresis [5] have also been investigated.

The economics of UF is a major criteria in process feasibility [6]. The main economic advantage is a reduction in system design complexity and energy usage since UF processes can simultaneously concentrate and purify process streams. An energy savings is obtained since no phase change is required for separation and UF processes are normally operated at near ambient temperature. An operating savings is obtained since UF processes normally require no chemical reagents or catalysts for processing. Productivity can be increased due to highly selective commercially available or customed manufactured membranes. Gentle process treatment of the desired product can be accomplished by UF to significantly increase the yield.

Economic disadvantages also exist. The capital investment can be high if low flux rates for purification demand a large system design. Productivity losses can occur due to adsorption on the membrane or sheer inactivation of the enzyme. Maintenance costs for chemicals used in membrane cleaning and rejuvination or for eventual membrane replacement are other economic disadavantages.

UF processes stand-out as economically sound in comparision to other traditional separation techniques. O'Sullivan et al [7] and Breaton [1] compared the operating cost of UF for enzyme purification compared to evaporation and pre-coat rotary drum filtration. Their data indicate that UF is 45% and 35% less costly to operate than evaporation and pre-coat rotary drum filtration, respectively. The filter aid, e.g., diatomaceous earth used in filtration was a significant operating cost. Evaporation has higher operating costs because of increased energy usage.

EXPERIMENTAL

The experimental objectives were to concentrate and purify a crude grade of BIAP. The BIAP enzyme has a molecular weight of 140,000 [8]. Studies examine separation efficiency and flux for various membranes and module configurations.

The crude mixture of BIAP utilized was obtained from Sigma Chemical Co. and has an activity of 1.7 units/mg. One unit of enzyme activity is that quantity that will hydrolyze 1 umol of p-nitrophenyl phosphate per minute at 30°C and at a pH of 10.15 [9]. The reaction kinetics are determined utilizing a spectrophotometer at a wavelength of 405 nm. The majority of the impurities in the BIAP mixture are low molecular weight proteins, predominately albumin with a molecular weight of 69,000 [3]. The BIAP mixture was prepared in a 0.05 M phosphate buffer. The buffer had a pH range of 6.8 to 7.0 and was prepared by mixing equimolar solutions of 0.05M monobasic and dibasic phosphate until the desired pH is reached [10]. The buffer was found effective in earlier studies in BIAP purification [11].

Enzyme purification was defined in terms of several parameters. The purification factor is a measure of the enzyme activity per mg of retentate (concentrate) divided by the activity per mg of feed:

$$P.F. = \frac{(a_f/C_f)}{(a_o/C_o)} \tag{1}$$

where enzyme activity in the initial feed and final retentate are denoted as a_o and a_f, respectively. The total protein concentration in the initial feed and final retentate are represented by C_o and C_f, respectively. The maximum possible P.F. for a given run is equal to the volumetric concentration factor for that run, V_o/V_f. Enzyme activity yield, Y, is the percent enzyme activity remaining in the final retentate with respect to the initial feed:

$$Y = \frac{(a_f)(V_f)}{(a_o)(V_o)} \times 100 \ . \tag{2}$$

Rejection or retention of both enzyme activity and total protein are based on overall system separation [12].

The rejection or retention of enzyme activity, R_a, is defined as:

$$R_a = \frac{\ln(a_f/a_o)}{\ln(V_o/V_f)} \ . \tag{3}$$

The rejection of total protein, R_p, is defined as:

$$R_p = \frac{\ln(C_f/C_o)}{\ln(V_o/V_f)} \ . \tag{4}$$

The ultrafiltrate or permeate flux, heretofore referred to as flux, is the volumetric production rate of ultrafiltrate from the ultrafiltration system per unit membrane area per unit time. It is expressed as a volumetric flux using the units cm^3/cm^2-sec.

For effective operation the flux should be high and remain stable. The purification factor should be as close to the volumetric concentration as possible if the batch concentration mode is used. Enzyme yield should be as close to 100% as possible. Rejection of the enzyme should be as close to 1.0 as possible. This term is another mathematical treatment of enzyme yield. It is undesirable to have the rejection of total protein close to 1.0 because the system's function is to remove the protein impurities in the BIAP mixture. The protein that is retained should be the BIAP.

The experimental approach was to evaluate various membranes in a screening study utilizing an Amicon TCF10 (Amicon Corp., Danvers, MA) thin-channel ultrafiltration unit. The unit ran in a batch concentration mode to examine the separation and flux for the BIAP mixutre. Final studies were performed in a 2000 ml stirred cell run in both concentration and dialysis mode.

The TCF10 system circulates the process solution in a thin channel above the membrane surface. This allows for a recirculating stream to minimize macro-solute build-up on the membrane surface. The unit accepts 90 mm diameter circular membranes. The channel dimensions (W x L x H) are: 9.5 x 0.38 x 414 mm and the effective filtration area is 40 cm^2. The process solution's temperature can be controlled and the recirculation rate is 0 to 500 cm^3/min [13].

The membranes evaluated in the screening study were a PM30, a polysulfone membrane which has a nominal molecular weight cut-offs of 30,000. XM50 and XM300 membranes are composed of acrylic co-polymer and have respective nominal molecular weight cut-offs of 50,000 and 300,000. YM30 and YM100, cellulosic membranes, with molecular weight cut-offs of 30,000 and 100,000, respectively, were also evaluated.

The stirred cell system has a total capacity for batch operation of 2000 ml and can be arranged in a dialysis mode with a 10L dialysis tank. The pressure driving force across the membrane is provided by purified nitrogen. Temperature can be controlled through a heating/refrigeration intercooling system. The unit can accommodate 150 mm flat circular membranes. The effective mass-transfer area of the unit is 165 cm^2. A stirrer bar rotates with a range of 0 to 550 rpm above the membrane surface to circulate the process solution.

RESULTS AND DISCUSSION

The screening studies were done utilizing PM30, XM50, XM300, YM30 and YM100 membranes in a TCF10 thin-channel system. The BIAP feed concentration in all studies was 0.2 g/L. A cross-flow channel velocity of 300 cm/sec was utilized at an operating pressure of 2.07 bar (30 psi). Solution pH was 6.9 and temperature was 20°C. Studies were run in a batch concentration mode achieving a volumetric concentration ratio of up to 4.0 in some studies. The YM100 membrane appears to be the best for effective concentration and purification of the enzyme (Table 1).

Table 1

Separation and Flux Data for Membrane Screening Studies

Membrane	V_o/V_f	P.F.	Y(%)	R_a	R_p	$J_f \times 10^3$ (cm^3/cm^2-s)
PM30	4.0	1.1	99	0.99	0.93	4.62
XM30	3.33	1.0	99	0.975	0.92	0.86
XM300	4.0	1.0	85.5	0.89	0.89	2.45
YM30	2.85	1.1	95	0.95	0.86	3.81
YM100	3.33	1.3	84	0.86	0.63	7.20

All the low molecular weight membranes, i.e., PM30, YM30 and YM50 retain a high degree of enzyme activity but fail to allow the albumin to permeate. The enzyme yield for the PM30, YM30 and XM50 were 99, 99 and 95%, respectively. The rejection of enzyme activity for these membranes were 0.99, 0.95 and 0.975, respectively. This shows that the membranes effectively retain the BIAP. These membranes are also retaining the albumin and other impurities, as expected. Total protein rejection for the PM30, YM30 and XM50 were 0.93, 0.86 and 0.92, respectively. Therefore the purification factor for the runs ranged from 1.0 to 1.1.

The flux characteristics of the 30,000 MW membranes were quite different. The flux after a 4 fold concentration for the PM30 membrane was 4.62×10^{-3} cm^3/cm^2-sec (Figure 1). The initial value was 1.0×10^{-2} cm^3/cm^2-sec and dropped through the run. The figure indicates that it would have probably dropped further if the run were allowed to progress. The YM30 had a very stable flux which remained at 3.81×10^{-3} cm^3/cm^2-sec for virtually all the run. The XM50 membrane exhibited an extremely low flux with a final value of 8.6×10^{-4} cm^3/cm^2-sec after a 3.33 fold volumetric concentration (Figure 2). This flux only dropped slightly during the run.

The high molecular weight membrane, XM300, was expected to allow the enzyme and the protein impurities to permeate but did not do so. The results indicate that enzyme rejection was 0.89 and total protein rejection was also 0.89. This suggests that pore plugging and surface adsorption had occurred early in the run causing the membrane to retain the components of the crude BIAP mixture. This would also explain the extremely low fluxes for the XM50 membrane. Permeate flux for the XM300 dropped throughout the 4.0 fold volumetric concentration (Figure 2). The flux dropped about 50% to a final value of 2.45 x 10^{-3} cm^3/cm^2-sec.

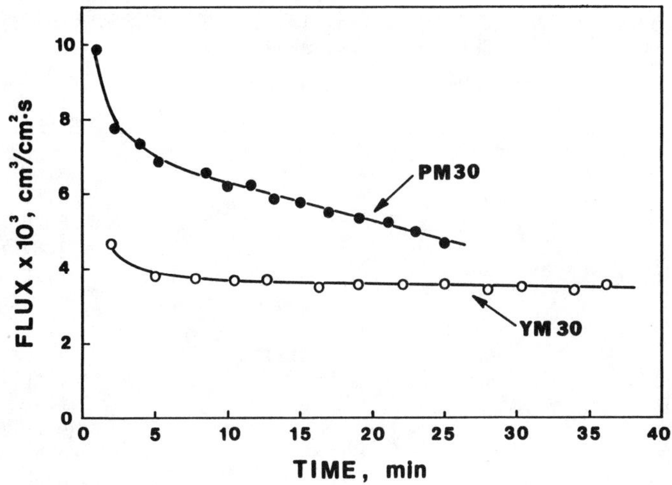

Figure 1. Ultrafiltrate Flux vs. Time for PM30 and YM30 Membranes.

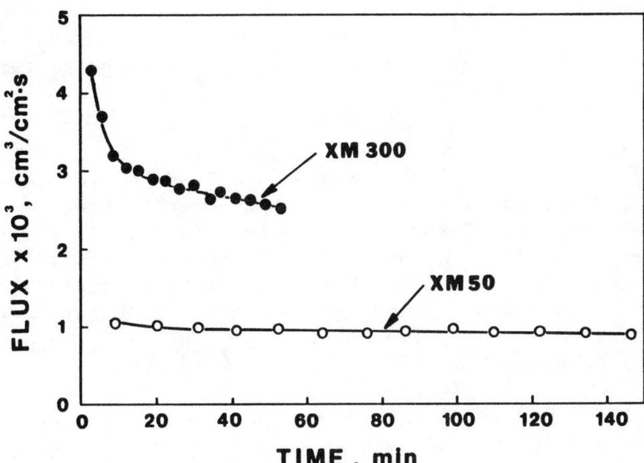

Figure 2. Ultrafiltrate Flux vs. Time for XM50 and XM300 Membranes.

The YM100 membrane appears to purify the crude BIAP and also has the best flux characteristics of the membranes studied. The purification factor for the two YM100 run was 1.3. Enzyme yield was 84% and rejection 0.86. Protein rejection was 0.63. The permeate flux at the end of the run ($V_o/V_f = 3.33$) was 7.2×10^{-3} cm^3/cm^2-sec (Figure 3). This flux is considerably higher than the other membranes that were evaluated.

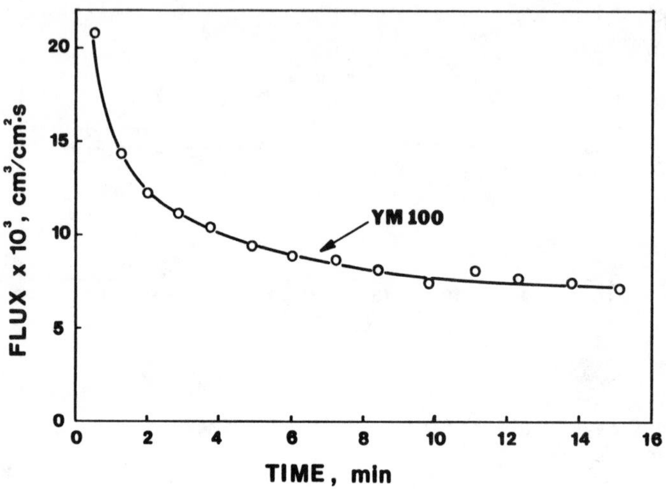

Figure 3. Ultrafiltrate Flux vs. Time for YM100 Membrane.

Stirred cell experimentation utilized YM100 membranes in a modified 2000 ml unit. Studies examined the unit run in a batch concentration mode. These studies were done with an initial BIAP concentration of 0.2 g/L in a 0.05 M phosphate buffer. The major process parameters varied in these studies were pressure, stirrer speed and temperature. A complete analysis of the stirrer speed and pressure effects is given by Slater et al [11] and will be summarized here.

Stirrer speed investigation was done at 0.34 bar (5 psi) and 10°C. The results of increasing the stirrer speed from 0 to 310 rpm are quite evident in both flux and separation data. The initial flux value for the 0 rpm study was 6.5×10^{-3} cm^3/cm^2-sec and decreased rapidly to 9.0 x 10^{-4} cm^3/cm^2-sec at the conclusion of the run (Figure 4). Rejection of the enzyme and total protein were low because they were compressed into the gel layer at the membrane surface; therefore, a low purification factor of 1.09 was obtained. As the stirrer speed was increased, flux and separation efficiency were enhanced. Stirrer speed was increased to 310 rpm at the same pressure and temperature to minimize the gel layer. Under these conditions the purification factor was increased to 1.50 out of a possible 2.0. Enzyme rejection was 0.90 and total protein rejection was 0.32. This corresponds to an enzyme yield of 94%. The final flux value was also enhanced to 2.1 x 10^{-3} cm^3/cm^2-sec (Figure 4).

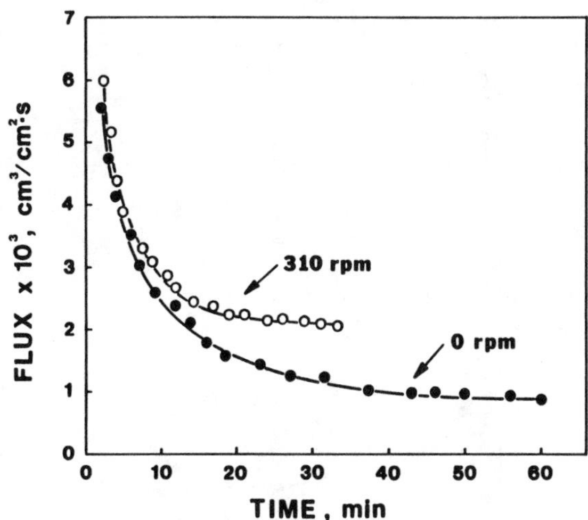

Figure 4. Ultrafiltrate Flux vs. Time for Studies at 0 and 310 rpm at 5 psi and 10°C.

At higher pressures at 10°C similar results are obtained. At 1.38 bar (20 psi) and 0 rpm stirrer speed the initial flux was 1.60 x 10^{-2} cm^3/cm^2-sec and was 8.0 x 10^{-4} cm^3/cm^2-sec at the end of the run, a 2.67 volumetric concentration (Figure 5). Enzyme and protein rejection were extremely low as they were in the 0.34 bar (5 psi) study. An extensive polarization layer had formed which adversely effected the separation efficiency. The purification factor was 1.08. Increase in stirrer speed improved the operational performance. At 310 rpm final flux was 2.0×10^{-3} cm^3/cm^2-sec and the enzyme yield was 98% (Figure 5). This purification factor had been increased to 1.44 which was an efficiency of 72%. The stirred cell studies indicate that the gel concentration layer is rapidly established at no stirrer speed or flow past the surface. This effect negates the intended use of the membrane for protein mixture fractionation. To overcome these effects the gel layer is minimized by high stirrer speed. The stirrer speed is high enough to minimize the gel layer formation but low enough to prevent shear inactivation of the enzyme. Once the gel layer is minimized the separation effectiveness is increased.

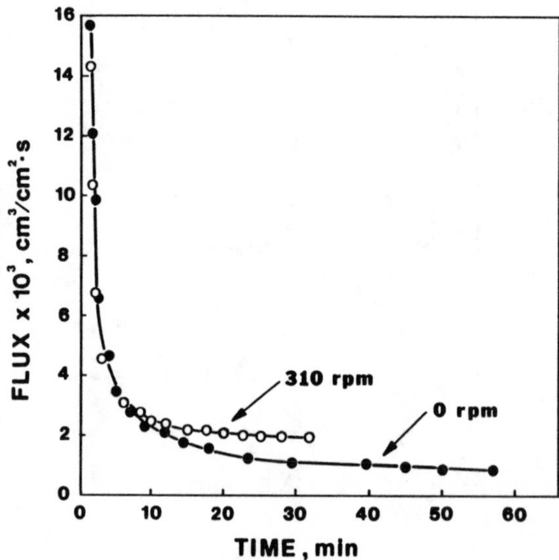

Figure 5. Ultrafiltrate Flux vs. Time for Studies at 0 and 310 rpm at 20 psi and $10^{\circ}C$.

Studies at increased temperature were performed. For long processing times high temperatures will lead to enzyme denaturation so this would not be practical in biochemical production. Therefore moderate increases in temperature were examined. Studies at $20^{\circ}C$ and 0.69 bar (10 psi) indicate that there seems to be some improvement in flux profiles between the $10^{o}C$ and $20^{o}C$ runs. With no cell agitation the flux after 40 minutes of operation at a volumetric concentration of 2.6 was 1.45×10^{-3} cm^3/-cm^2-sec (Figure 6). This can be compared to a value of 1.0×10^{-3} cm^3/cm^2-sec obtained at the same time at $10^{o}C$. This 45% increase in flux would be important in this worst case where the flux and separation efficiency were very poor at $10^{o}C$.

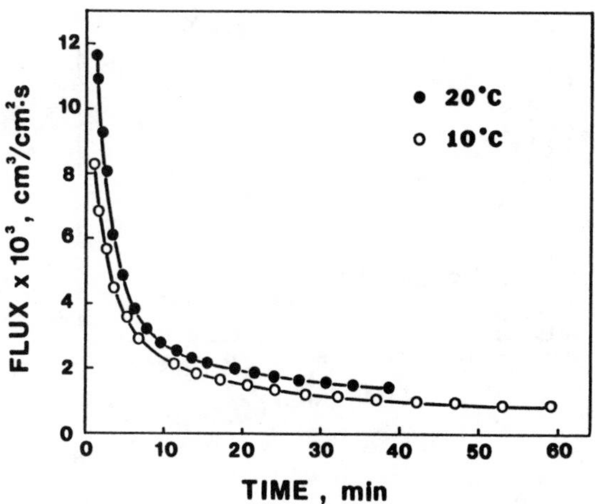

Figure 6. Ultrafiltrate Flux vs. Time for Studies at 10 and 20 ^{o}C at a Stirrer Speed of 0 rpm and 10 psi.

Increasing the stirrer speed at $20^{o}C$ improves the flux characteristics over no agitation in an analogous way to the $10^{o}C$ studies. The increase in temperature from 10 to 20 ^{o}C at a stirrer speed of 180 rpm and 0.69 bar (10 psi) also increases the flux. The flux after a volumetric concentration of 2.5 was 2.0×10^{-3} cm^3/cm^2-sec for the $20^{o}C$ run. This can be compared to a value of 1.25×10^{-3} cm^3/cm^2-sec after the same degree of concentration at $10^{o}C$ (Figure 7). These values correspond to a 60% improvement in flux. This increase in system output is observed throughout the duration of the run. This study demonstrates that if it is possible in an enzyme purification scheme to increase temperature it will improve production characteristics. It is important to understand that an optimization study should be done to determine the loss of enzyme activity at increased temperature vs. improved production.

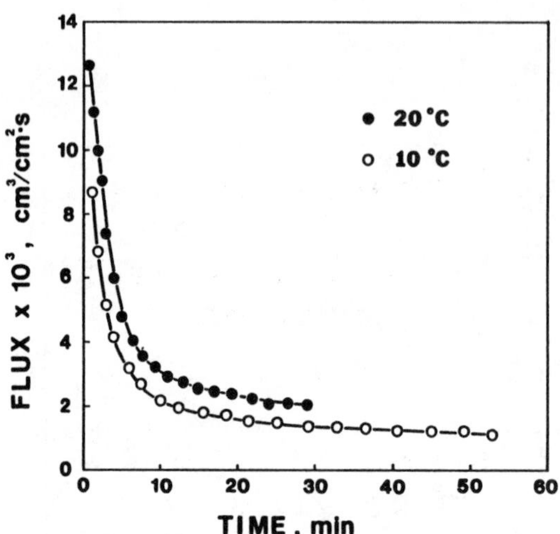

Figure 7. Ultrafiltrate Flux vs. Time for Studies at 10 and 20°C at a Stirrer Speed of 180 rpm and 10 psi.

Studies were run in a dialysis mode on the system. The solution parameters remained the same and the 0.05 M phosphate buffer was utilized as the dialysate. The initial BIAP feed concentration was 0.2 g/L and the process temperature was 10°C. A pressure of 1.38 bar (20 psi) and stirrer speed of 310 rpm were utilized since they were the optimal settings in the batch concentration runs. Under dialysis conditions the flux was greatly stablized at 2.0 x 10^{-3} cm^3/cm^2-sec. (Figure 8). The ratio of dialysis volume to feed volume was 5.3. The enzyme yield was 98% and the purification factor was 1.6.

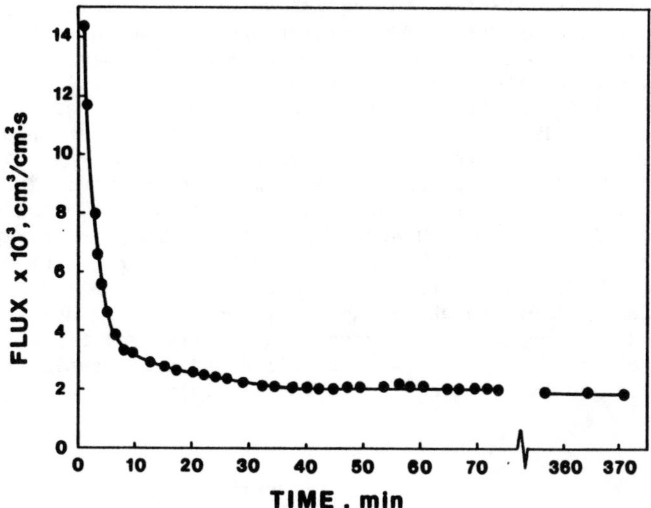

Figure 8. Ultrafiltrate Flux vs. Time for a Dialysis Run at 20 psi, 10°C and a Stirrer Speed of 210 rpm.

CONCLUSIONS

Bovine intestinal alkaline phosphatase (BIAP) enzyme can be concentrated and purified utilizing ultrafiltration. A membrane screening study in a thin-channel system evaluated PM30, YM30, XM50, YM100 and XM300 membranes for separation efficiency and flux. The YM100 membrane was effective in purging a crude BIAP mixture of undesired protein impurities. The smaller molecular weight cut-off membranes, i.e., PM30, YM30 and XM50 retained high degrees of enzyme activity, but failed to allow the protein impurities to permeate. Therefore, their utility was limited to a concentration role. The larger XM300 membrane failed because of severe protein surface adsorption which caused the membrane to retain all the solution components. The YM100 membrane retained a significant amount of enzyme activity and also allowed the protein impurities to permeate. It had a higher flux than the other membranes evaluated and flux degradation was not a major problem.

A stirred cell geometry was investigated with the YM100 membrane. Studies examined the variation of the process parameters of pressure, temperature and stirrer speed on operational effectiveness. Pressure was varied from 0.34 bar (5 psi) to 1.38 bar (20 psi), temperature from 10 to 20°C and stirrer speed from 0 to 310 rpm. Both increases in stirrer speed and temperature improved final flux characteristics. These process parameters are important in minimizing the gel-layer build-up for the BIAP separation process. Although increased pressure improved initial production values, this change did not translate to an increase in final flux or improved separation efficiency. Increasing the stirrer speed from 0 to 310 rpm increased the final flux by an average of 140% at the various pressures examined at 10°C. Purification factors and enzyme activity recovery were dramatically improved. A temperature increase of 10°C improved final flux by an average of 50%. Flux was stabilized and purification increased slightly by operating in a dialysis mode.

ACKNOWLEDGMENTS

Partial support for this work was provided by the National Science Foundation's College Science Instrumentation Program through grant #CSI-8551851.

LIST OF SYMBOLS

a	enzyme activity (units/mg)
C	protein concentration (g/cm^3)
J	ultrafiltrate flux (cm^3/cm^2-sec)
J_f	final ultrafiltrate flux (cm^3/cm^2-sec)
P.F.	purification factor
ΔP	applied pressure gradient (psi)
R_a	enzyme activity rejection
R_p	protein rejection
V	volume (cm^3)
Y	enzyme activity yield (%)
ω	stirrer/agitator speed (rpm)

subscripts

f	final retentate
o	initial feed

REFERENCES

1. N.C. Breaton, in Ultrafiltration Membranes and Applications, A.R. Cooper, Ed., Plenum Press, NY, 1979, pp. 373-404.

2. W. Eykamp, Industrial Ultrafiltration, paper no. 61, First International Conference on Separations Science and Technology, 191st National Meeting of the ACS, NY, NY, April 16, 1986.

3. L.E. Seargeant and R.A. Stinson, Journal of Chromatography, 173, 101 (1979).

4. N.K. Ghosh and W.H. Fishman, Biochemistry Journal, 108, 779 (1968).

5. J.F. Chao, V.P. Rollan, H.C. Hollein and C.R. Huang, Separation Science and Technology, 18 (11) 999 (1983).

6. C.S. Slater, T.G. Huggins, Jr., C.A. Brooks III and H.C. Hollein, Proceedings of the Third Annual Congress on the Economics, Scale-up and Automation of Biological Process Engineering, Baltimore, MD (1986).

7. T.J. O'Sullivan, A.C. Epstein, S.R. Korchin and N.C. Beaton, Chemical Engineering Progess, 80, 68 (1984).

8. P.G. Righetti and T. Caravaggio, Journal of Chromatography, 127, 1 (1976).

9. Worthington Enzyme Manual, Worthington Biochemical Corporation, Freehold, NJ (1977).

10. S.P. Colowick and N.O. Kaplan, Methods in Enzymology, Academic Press, NY, NY (1955).

11. C.S. Slater, T.G. Huggins, Jr., C.A. Brooks III and H.C. Hollein, Separation Science and Technology (in press).

12. W.F. Batt, in Membrane Separation Processes, P. Meares, Ed., Elsevier Scientific Publishing Co., NY, NY, 81-120 (1976).

13. Amicon Corporation, Publication #I-113G, Amicon Corp., Danvers, MA, (1985).

ISOLATION OF AGAROSE BY CHARGED FIBER TECHNIQUE

Chiau-Li Yeh
Shi-Fang Yang
Chi-Ching Chen
Department of Biology
National Taiwan University
Tainan, Taiwan 70101
Republic of China

Agarose in the form of aqueous gels has been widely used in electrophoresis, chromatography, culture media, immunological analysis, and cell cloning. It plays important roles in the biomedical and biotechnological fields. The demand for agarose has expanded rapidly in the past few years. A highly efficient mass production of agarose with good quality becomes extremely important. A fiber technique for the purification of agarose is being developed in this laboratory.

Bulk water treatment and heat transfer are two major problems in the large scale separation of agarose. Rapid and minimal processes of substrate transfer are required. A filtration media containing charged fiber was developed to achieve this purpose. In essence, the fiber was charged with anionic ion and the bulk hot substrate was introduced with pressure through a filtration mechanism. The end product is of high quality as compared to materials from other sources.

INTRODUCTION

It has been shown that agar consists of two components. One component is an ionic material called agaropectin; the other component is a neutral polymer called agarose (1). Agarose, because of its high gel strength at relatively low concentration, its transparency in gel form and neutral property, has been used extensively in many biomedical and biotechnological fields.

While all of these methodologies achieve their purpose relatively well, they are all time consuming and expensive. In fact, bulk water treatment and heat transfer are two major problems in large scale separation of agarose. In this paper, we report a fast separation scheme utilizing a charged fiber technique.

Agar or agarose gel has been used in immunological analysis of antigen and antibody by double diffusion (2), radial immuno diffusion (3), immuno electrophoresis (4), electro-immuno-assay (5), counter electrophoresis (6) and/or plaquing techniques (7).

The uses of agarose as electrophoresis medium cover most of the biological related applications. These include serum proteins (8,9), bacterial proteins (10), glycoproteins (11), nucleic acids (12-16), lipoproteins (17-21), and viruses (22).

Column chromatography using agarose gel particles has been applied in separation or analysis of nucleic acids (23,24), enzymes (25,26), antibodies (27), ribosomes (28), and viruses (29,30).

Affinity chromatography is a rapidly expanding use of agarose gel particles. It has been used in the separation of various biological components such as enzymes, proteins, nucleic acid, antigen, antibody, substrate, etc. (31-33).

In the field of microbiological application and cell cloning, agarose serves as a choice of culture media for clone isolation. Thus, extensive use of agarose in biomedical and biotechnological applications has expanded rapdily in recent years. As a result, an economical way of separating agarose is needed.

The isolation of agarose was first achieved in 1937 by acetylating agar, separating the acetylated agarose with chloroform extraction and deacetylating this fraction by saponification (34). Other procedures were developed thereafter. These include precipitation with cetylpyridinium chloride (35), differential solubilization in sodium iodide (36), ammonium sulfate (37), urea (38), and ionic matrix (39).

MATERIALS AND METHODS

Isolation of agarose from agar with the powder suspension method:

A. Preparation of crude material.

Generally agar solution was made to 2% concentration in 20 mM sodium citrate buffer, pH 5.2 at 80°C.

B. Removal of charged fraction in agar.

Agaropectin containing negative charge was precipitated by adding DEAE-cellulose powder suspension at 80°C.

C. Separation of agarose solution.

To obtain agarose solution, the precipitated agaropectin was removed by filtration with suction.

D. Gel formation of agarose was made at room temperature without stirring.

E. Washing of agarose gel was made possible after pressing the gel through mesh. Three volumes of ethanol were used in the washing cycle.

Isolation of agarose with charged fiber technique:

A. Preparation of charged fiber.

Sheets of cottoncloth were coupled with diethylaminoethyl derivatization for an exchange capacity of 0.7 meq/cm^2.

B. Preparation of crude material.

Agar solution was made to 2% concentraion in 20 mM sodium citrate buffer, pH 5.2 at 80°C.

C. Separation of agaropectin.

Removal of agaropectin was made by passing the hot agar solution through fiber sheets at 80°C.

D. Gel formation and wahing of gel were formed as described previously.

Determination of Gel Strength

A gel tester was used for determination of gel strength of various preparations at different gel concentrations from 0.5% to 1.0%. In all determinations, a gel tester of Nikkansui type manufactured by Kiyo Seisakusho Ltd., Japan was used.

Determination of Electroendoosmotic Flow

The electroendoosmotic flow was estimated from the migration distance of human serum albumin and the cathodal displacement of a 5% solution of dextran 70 after 1 hr electrophoresis in 50 mN barbital buffer, pH 8.6 on 1.5% agarose preparation. Localization of dextran zone was made possible by fixing with a solution of ethanol-acetic acid-water (70:5:25) (40). The electroendoosmotic flow was expressed as -Mr.

$$-Mr = -D/(A+D)$$

A : Migration distance of albumin

D : Migration distance of dextran

Estimation of Sulfate Content and Gelling Temperature

Sulfate content (41) and gelling temperature (42) of various agarose preparations were routinely determined as described for quality assurance.

RESULTS AND DISCUSSION

From previous experiments we have suggested the isolation of agarose by using the DEAE-cellulose ionic exchange method in acetate buffer (43). For large scale purification of agarose, there are several problems. Bulk water treatment is a major problem. Due to high viscosity of agar solution even at low workable concentrations, it requires a large water volume to begin with. Secondly, the purification can only be achieved in a solubilization form at a temperature higher than 75°C. Thirdly, its high viscosity makes the separation in filtration or precipitation more difficult. As a result, a sophisticated methodology with these problems in mind is essential. The conventional method using powder suspension of ionic material required tedious operations.

From our experimental results, we have achieved purification as judged by the criteria of sulfate content, electroendoosmotic flow, gelling temperature and gel strength (Table 1) with an acceptable quality comparable to those of commercial source.

Table 1. Properties of Agarose Preparations

Preparation	A	B	C	D
Preparation I	0.212	0.182	37.0	680
Preparation 105	0.241	0.194	38.0	752
SeaKem (ME)	0.235	0.172	36.0	975
Sigma I	0.255	0.261	38.0	764
Sigma II	0.232	0.205	38.5	875
Sigma III	0.250	0.128	36.6	821

Preparation I was made by the powder suspension method and preparation 105 was obtained by the charged fiber technique as described in Materials and Methods. A - sulfate content, % (w/v); B - electroendoosmotic flow, -Mr; C - gelling temperature, °C; D - gel strength, g/sq. cm.

SUMMARY

The major outcome of these experiments is that utilization of charged fiber techniques seems valuable in the breakthrough of those major problems in large scale purification of agarose.

ACKNOWLEDGEMENT

This work was supported by the National Science Council, Republic of China, NSC 74-0606-B006-08.

REFERENCES

1. Araki, C., Bull. Chem. Soc. Japan, 29, 543-4 (1956).
2. Ouchterlony, O., Acta Path. Microbiol. Scand., 25, 186 (1948).
3. Mancini, G., Vaerman, J. P., Carbonary, A. O., and Heremans, J. F., Protides Biol. Fluids, Proc. Symp., 11, 370 (1964).
4. Grabar, P., and Williams, C. A., Biochim. Biophys. Acta, 10, 193-94 (1953).
5. Laurell, C. B., Anal. Biochem., 15, 45-52 (1966).
6. Bussard, A., and Huer, J., Biochim. Biophys. Acta, 34, 258 (1959).
7. Jerne, N. K., Nordin, A. A., and Henry, C., In B. Amos and H. Koprowski (Eds.) Cell Bound Antibodies, Wistar Institute Press, Philadelphia (1963).
8. Laurell, C. B., Scand. J. Clin. Invest., 29, 71-82 (1972).
9. Laurell, C. B., Clin. Chem., 19, 99-103 (1973).
10. Goullet, P., Experientia, 25, 89-91 (1969).
11. Holden, K. G., Yim, N. C. F., Griggs, L. J., and Weisbach, J. A., Biochem., 10, 3105-9 and 3110-13 (1971).
12. Arlinghaus, R. B., Kaczmarczyk, W., and Polatnick, J., J. Virol., 4, 712-18 (1969).
13. Besson, J., Taulemesse, H., Berger, M., C. R. Soc. Biol., 167, 487-90 (1973).
14. Hayward, G. S., Virology, 49, 342-44 (1972).
15. Vasu, S., Rev. Roum. Biochim., 6, 75-86 (1969).
16. Zak, B., Weiner, L. M., and Baginski, E., J. Chromatog., 20, 157-60 (1965).
17. Rapp, W., and Kahlke, W., Clin. Chim. Acta, 19, 493-98 (1968).
18. Seidel, D., Wieland, H., and Ruppert, C., Clin., 19, 737-39 (1973).
19. Vogelberg, K. H., Utermann, G., and Gries, F. A., Z. Klin. Chem. Klin. Biochem., 11, 291-96 (1973).
20. Werner, M., and Jones, A. L., Clin. Chem., 18, 534-39 (1972).

21. Wieland, H., and Seide, D., Clin. Chem., 19, 1139-41 (1973).

22. Wolf, G., and Casper, R., J. Gen. Virol., 12, 325-29 (1971).

23. Besson, J., Radisson, J., Berger, M., C. R. Soc. Biol., 167, 1156-59 (1973).

24. Oberg, B., and Philipson, L., Arch. Biochem. Biophys., 119, 504-9 (1967).

25. King, A. M., and Nicholson, Biochem. J., 113, 17 (1969).

26. Lynn, K. R., J. Chromatog., 66, 375-76 (1972).

27. Bergin-Wolff, A., Hernandez, R., and Justin, M., Lancet, 11, 1278-80 (1971).

28. Stenesh, J., and Yang, K., J. Chromatog., 47, 108-110 (1970).

29. Fernelius, A. L., and Velicer, L. F., Arch. Gesamte Virusforsch., 25, 227-34 (1968).

30. Polley, J. R., and Webb, T. S., Can. J. Microbiol., 15, 1167-73 (1969).

31. Weetall, H. H., Separation and Purification Methods, 2, 199-229 (1973).

32. Guilford, H., Chem. Soc. Rev., 2, 249-71 (1973).

33. Turkova, J., J. Chromatog., 91, 267-91 (1974).

34. Araki, C., J. Chem. Soc. Japan, 58, 1338 (1973).

35. Hjerten, S., Biochim. Biophys. Acta, 62, 445-49 (1962).

36. Araki, C., Proc. Vth. Int. Seaweed Symp., 3-17 (1965).

37. Azhitskii, G. Y., and Kobozev, G. V., Lab Delo, 1967, 143-5 [CA66, 106120 (1967)].

38. Patil, N. E., and Kale, N. R., India J. Biochem. Biophys., 10, 160-63 (1973).

39. Duckworth, M., and Yaphe, W., Anal. Biochem., 44, 636-41 (1971).

40. Quast, R., J. Chromatog., 54, 405-412 (1971).

41. Young, K. S., Fisheries Research Board of Canada, Tech. Report No. 454 (1974).

42. Guiseley, K. B., Carbohyd. Res., 13, 247-56 (1970).

43. Chen, C. C., National Science Council Report (1985).

PHARMACEUTICAL APPLICATIONS

ULTRATRACE ANALYSIS AND SEPARATION SCIENCE TECHNOLOGY IN INDUSTRY

Satinder Ahuja
Pharmaceutical Division
Ciba-Geigy Corporation
Suffern, New York 10901 USA

H. Michael Widmer
Department of Analytical Research and Coordination
Ciba-Geigy Limited
CH-4002 Basel, Switzerland

Applications of separation techniques such as high performance liquid chromatography and capillary gas chromatography for quantitation and/or isolation of new drugs are described. Development of monitor systems, based on various separation technologies, which are useful for the surveillance of harmful pollutants are also presented.

INTRODUCTION

The discovery of new drugs mandates that meaningful and reliable analytical data be generated at various steps of new drug development (1). To assure the safety of a new pharmaceutical compound or drug, it is essential that the new drug meets the established purity standards as a chemical entity and/or when admixed with pharmaceutical excipients and animal feeds. This demands that the analytical methodology be sensitive enough to carry out measurements at low levels of the drug or its transformation products. Similarly, ensuring safety of drugs, cosmetics, environment, food and water requires low level analytical measurements in a variety of matrices (2). These low level analyses frequently have to be performed at parts per trillion (ppt) level. An example of the need of low level analyses is provided by 2,3,7,8-tetrachloro-dibenzo-p-dioxin (TCDD) which can cause abortion in monkeys at ppt levels (3). Allowing for a 100-fold margin of safety for human exposure, it can be calculated that the safe food level of TCDD would have to be less than 2 ppt. Another example highlights how low levels of polychlorinated biphenyls (PCB), 0.43 ppb, can weaken the backbone of trout by interfering in collagen synthesis (4). The analysis of backbones of these fish revealed excess calcium and deficienty in collagen and phosphorus. Since the fish were also deficient in Vitamin C, a cofactor in collagen synthesis, this led to the conclusion that the trout used Vitamin C for detoxification of PCB in lieu of skeletal development.

Analyses performed at parts per million (ppm) or μg level are frequently defined as trace analyses - an analytical landmark that was attained approximately 25 years ago. Ultra-trace analysis can be defined as analysis performed below ppm or submicrogram level. The majority of low level analyses today are performed at parts per billion (ppb) or ng level. A few analyses at parts per trillion (ppt) or pg level, are being performed by innovative analytical researchers; however, analyses down to femtogram (10^{-15}g) or attogram (10^{-18}g) levels have been reported in the scientific literature. Considering the current rate of progress, it will take us at least another 25 years to reach the 10^{-23}g level - perhaps the ultimate limit of detection.

Various detectabilities are reported in scientific literature (Table I). It is preferable to use gram or molar quantities. In either case molecular weight should be given to help evaluate detectability and quantifiability of the method. Furthermore, it is recommended that the following important analytical parameters should be reported for each method:

- Amount present in original sample/ml or g (APIOS)
- Minimum amount detected in g (MAD)
- Minimum amount quantitated in g (MAQ)

Table 1. Preferred and Commonly Used Analytical Units. From Reference (2) with permission.

Preferred Analytical Units		Commonly Used Units When Present/g	
g	Common Name	%	One Part per
1×10^{-6}	microgram (mg)	0.0001	million
1×10^{-9}	nanogram (ng)	0.0000001	billion
1×10^{-12}	picogram (pg)	0.0000000001	trillion
1×10^{-15}	femtogram	0.0000000000001	quadrillion

In addition, data such as accuracy, precision, linearity and specificity of the methodology should be provided.

The methods that are commonly used for ultratrace analyses are given in Table 2. Discussion here is limited to separation techniques such as HPLC and capillary GC. In addition, a monitor for hazardous materials, based on separation science technology, is described.

Table 2. Methods Frequently Used for Ultratrace Analyses. From Reference (2) with permission.

Method	Minimum Amount Detected, g
Mass Spectrometry	
Electron Impact	10^{-12}
Spark Source	10^{-13}
Ion Scattering	10^{-15}
Flame Emission Spectrometry	10^{-12}
High Pressure Liquid Chromatography (HPLC)	
Ultraviolet Detection	10^{-11}
Fluorescence Detection	10^{-12}
Gas Chromatography (GC)	
Flame Ionization	$10^{-12} - 10^{-14}$
Electron Capture	10^{-13}
Combination Techniques	
Liquid Chromatography/Mass Spectrometry	10^{-12}
Gas Chromatography/Mass Spectrometry	10^{-12}
Electron Capture - Negative Ionization Mass Spectrometry	10^{-15}

ANALYTICAL METHODOLOGIES

High Pressure Liquid Chromatography (HPLC)

HPLC is a very viable separation technique that is used in virtually every field of chemistry. This technique provides an excellent means of analysis and discovery of new compounds since compounds present even at ultratrace levels can be resolved from related compounds. One approach entails separation of potential compounds resulting as by-products, that cannot be resolved at normal inefficient crystallization techniques. Many of the by-products frequently have physicochemical properties and carbon skeleton similar to the parent compound with substituent(s) differing in position or functionality. Since it is not possible to theorize all by-products, some unusual compounds can be found and characterized with this approach. A more selective approach is based on changes brought about in a chemical entity to evaluate its stability with reactions such as hydrolysis, oxidation or photolysis. In this case, several theorized new and old compounds are produced. An innovative chromatographer can resolve and characterize both theorized and untheorized new compounds. Another interesting approach depends upon characterization of various degradation products produced in the matrixes used for pharmaceutical products. The compounds thus produced can be resolved from others by chromatography, and their structure determined by techniques such as elemental analysis, IR, NMR or mass spectrometry.

Two examples are discussed below relating to the analytical determinations of chemical compounds at ultratrace levels (5). In addition, the production of a new transformation product in a complex matrix is discussed which was discovered primarily because ultratrace analytical methodology was employed.

Purity Determinations. Analytical methods were needed to evaluate starting material for the synthesis of one of our drugs. This material has the following structure:

$C_6H_5CH_2O$ $COCH_3$

$C_6H_5CH_2O$

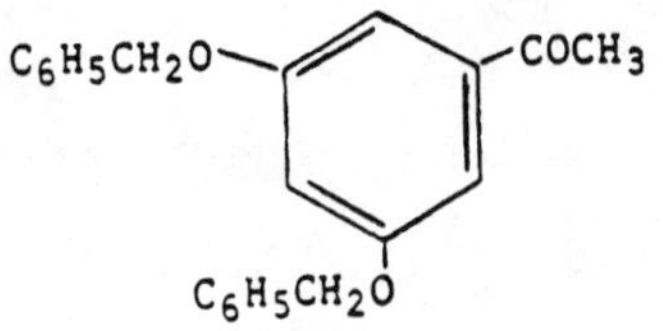

A batch (ARD 36792) of this material analyzed 96.3% by gas chromatography (6' x 4 mm 3% SE30 Column at 250°C) and 93.3% by titrimetry after oximation with hydroxylamine hydrochloride. Thin-layer chromatography performed with an optimum system (90 CCl_4/5 acetone/2 HOAc) showed two impurites totaling ~8%. This suggested titrimetry in combination with TLC would be adequate to control purity of this material. To check this hypothesis, a high performance liquid chromatographic method was developed (4 methanol/1 H_2O mobile phase was used with a C_{18} column and detector set at 225 nm) and the data on a sample (17-263) analyzed by DSC, TLC and HPLC are as follows:

DSC	-	99.7 mole % pure
TLC	-	No Impurity Detected
HPLC	-	3 Impurities: 0.056% (0.014, 0.016 and 0.026%)

No impurities were detected in this sample by TLC; however, HPLC showed three impurities totaling 0.06%. This compares with 99.7 mole % purity by DSC. Further comparisons on an alternate source sample revealed that the HPLC was the method of choice since it could help differentiate samples that appeared comparable by DSC: the reference sample was found to be less pure as it showed five impurities (0.12%) as opposed to only two impurities (0.02%) for the material from an alternate source.

Discovery of New Transformation Product in a Complex Matrix. It was necessary to conduct a long-term toxicity study to evaluate the safety of a potential drug with the following structure:

O_2 S X ONa CONHR

Molecular weight:	>400
Melting point:	224°C (decomp.)
Solubility:	Very slightly soluble in water Almost insoluble in acidic and basic solutions Soluble in methanol
pKa:	5.0

This compound was admixed with rat feed (Purina Laboratory Chow containing a minimum of 23% protein, 4.5% fat and maximum 6% fiber) with the following composition:

> Meat and bone meal, dried skimmed milk, wheat germ meal, fish meal, animal liver meal, dried beet pulp, ground extruded corn, ground oat groats, soybean meal, dehydrated alfalfa meal, cane molasses, animal fat preserved with BHA (butylated hydroxyanisole), vitamin B_{12} supplement, brewers' dried yeast, thiamin, niacin, vitamin A supplement, D activated plant sterol, vitamin E supplement, dicalcium phosphate, iodized salt, ferric ammonium citrate, zinc oxide, manganous oxide, cupric oxide, ferric oxide and cobalt carbonate.

A gas liquid chromatographic method was first investigated for analysis of this compound. Electron capture detector provided detectability down to 5 pg for the active component; however, investigations revealed degradation was occurring at the injection port. An HPLC method was then developed to circumvent this problem. The essential details of the method areas follow:

Column:	Bondapak C_{18}
Precolumn:	6 cm x 2.5 mm i.d. C_{18} Corasil
Solvent System:	70MeOH/33 H_2O/HOAC
Flow Rate:	0.8 ml/min

<u>Sample Preparation:</u>
Sample + 5 ml 1<u>N</u> HCl + 20 ml EtOAC
Shake. Centrifuge. Inject 25 μl of EtOAC layer

Analysis of 20 week old samples (Table 3) revealed that all samples had degraded significantly.

The HPLC chromatogram showed the presence of another peak. The compound representing that peak was isolated by preparative HPLC and characterized by field desorption mass spectrometry, NMR and other spectroscopic techniques to be the oxidation product ("4-Hydroxy compound" - Table 4).

The discovery of the "4-Hydroxy Compound" is primarily attributed to selective ultratrace methodology. It was possible to theorize that this compound could be produced by oxidation and amide hydrolysis on decarboxylation, the amine (RNH_2) and "ketosulfone". However, "4-Hydroxy Compound" was not available

Table 3. Analyses of 20-Week-Old Samples

Declared Content %	Found % of Label
0.06	69.3
0.02	58.1
0.004	49.4

Table 4. Selectivity of HPLC Method

	Retention Time (minutes)
"Ketosulfone"	5.0
RNH_2	6.9
RNCO	10.1
"4-Hydroxy Compound"	13.0
Parent Compound	18.0

to check this proposal. Reliability had to be placed in development of a sufficiently selective HPLC method that would resolve ultratrace levels of this compound. After this was accomplished, it could be shown that in this matrix oxidation is highly facile as compared to hydrolysis. Its eventual synthesis led to a compound with potential pharmacological activity.

Capillary Gas Chromatography

The results of a capillary chromatography study on a red wine extract show there is no problem to chromatographically resolving the main constituents. However, for separation of minor constituents in the sub ppm range, a one-dimensional chromatography approach provides insufficient resolution. The experience with red wine extract is by no means a special case: it demonstrates a general behavior of samples in natural and artificial matrices. Matrix effects are the most persistent problems in the daily work of the industrial analyst concerned with trace and ultratrace investigations (6). This is demonstrated in Figure 1. It represents chromatograms in which gasoline is analyzed by capillary GC. The main difference in the four chromatograms is in the change of sensitivity of the flame ionization detector setting. The top chromatogram is taken with an attenuation of 4096, the last with 1. The figure shows that there is one component present at a concentration higher than 10%, however, there are 10, 43 and 105 sample components at concentrations higher than 1%, 1000 ppm and 100 ppm, respectively (Figure 2). Similar relationships exist in cigarette smoke, polluted air, water and ocean pollution as well as industrial products.

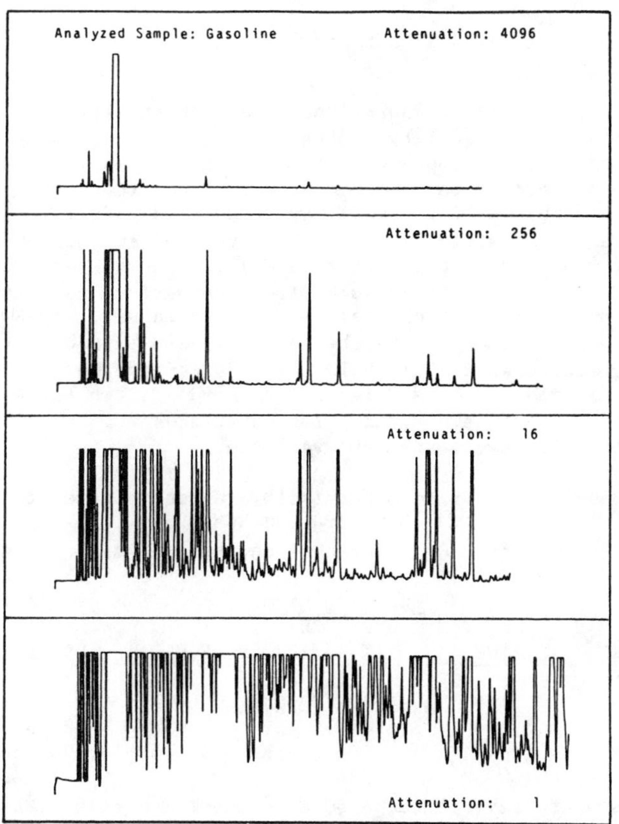

Figure 1. Gasoline analyzed by capillary GC (SE-54, 60 m).

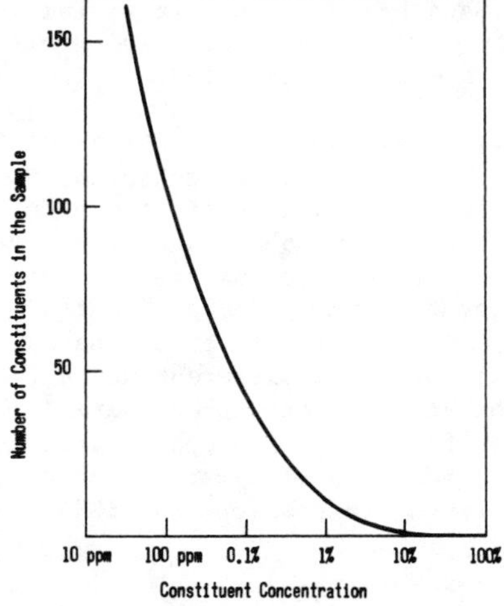

Figure 2. Plot of constituent number versus constituent concentration in gasoline.

HAZARDOUS CHEMICAL MONITOR

In the chemical industry there is a great awareness and interest about the quality of the products and in unwanted side effects of certain components, such as impurities with cargino-genic, mutagenic or teratogenic effects. Hazardous chemicals such as dimethyl sulfate (DMS) may be used as adducts in the synthesis of industrial goods, others such as epichlorohydrine (ECH) or bis(chloromethyl)ether (BCME) may be generated during a chemical process. Table 5 summarizes the maximum permissible working concentrations of these substances in air. In Switzer-land (CH) and Germany (GFR) they are called MAK-values; in the USA they are known as Threshold Limit Value-Time Weighted Average (TLV-TWA). Ciba-Geigy established its own Permissible Internal Exposure Level (PIEL) for substances such as BCME for which official levels are not available.

Table 5. Maximum Permissible Exposure Level for DMS, ECH and BCME.
From Reference 6 with permission.

	Exposure Level				Monitor
Species	USA TLV-TWA	GFR MAK	CH MAK	PIEL	Detection Limit
DMS	100 ppb	10 ppb	10 ppb	-	2 ppb
ECH	2 ppm	3 ppm	5 ppm	-	100 ppb
BCME	1 ppb	-	1 ppb	500 ppt	200 ppt

There are several techniques to check and control hazardous material in the air of working areas. In recent years the permanent surveillance of ambient air has become an important issue of the safety and health of people in industry, and the analyst has to conform with these needs and provide the appropriate methods and instrumentation.

In our company we developed a versatile and flexible monitoring system. This particular instrument is used for the surveillance of bis(chloromethyl)ether in ambient air. Our aim was to design and construct a mobile automated system to perform survey analyses at a rate of 1-5 analyses per hour on the hour. Figure 3 shows the operational principle of the instrument. It consists of compact module units which may be arranged for specific applicatons and serve different needs (7,8). The air sample is trapped on an adsorbent, then thermally desorbed and gas chromatographed, followed by an appropriate detection system, such as FID, FPD, ECD or a mass spectrometer. Detection limits down to ppt level are possible for BCME.

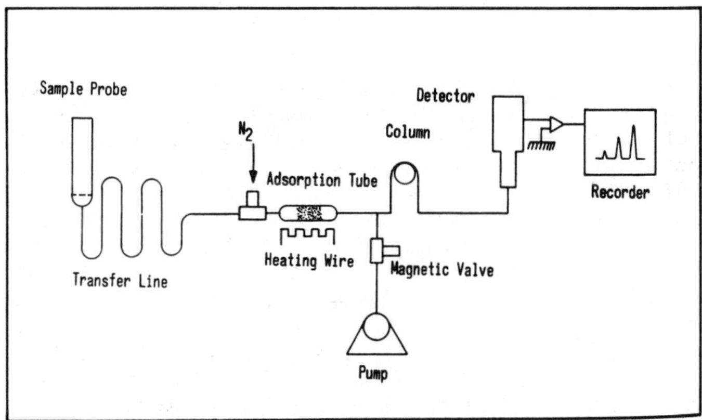

Figure 3. Schematic view of monitor (operational principle).

REFERENCES

1. S. Ahuja, "The Role of Ultratrace Analysis in R&D of Drugs," Indian Pharm. Congress, Varanasi, India, December 20-22, 1982.

2. S. Ahuja, Chemtech, 11, 702 (1980).

3. Chemical & Engineering News, August 7, 1978.

4. Ibid, September 25, 1978.

5. S. Ahuja, P. Liu and J. B. Smith, "Ultratrace Analysis of Pharmaceutical Compounds by High Performance Liquid Chromatography", FACSS Meeting, Philadelphia, PA, Sept. 19-24, 1982.

6. H. M. Widmer and K. Grolimund, Ultrahigh Resolution Chromatography, S. Ahuja, Ed., ACS Symposium Series 250, 1984, p. 199.

7. H. M. Widmer and G. Grass, Proc. 3rd Intern. Symp. Loss Prev. Safety Prom. Process Ind., 1980, Vol. 2, p. 3.

8. G. Grass and H. M. Widmer, Swiss Chem., 3, 117 (1981).

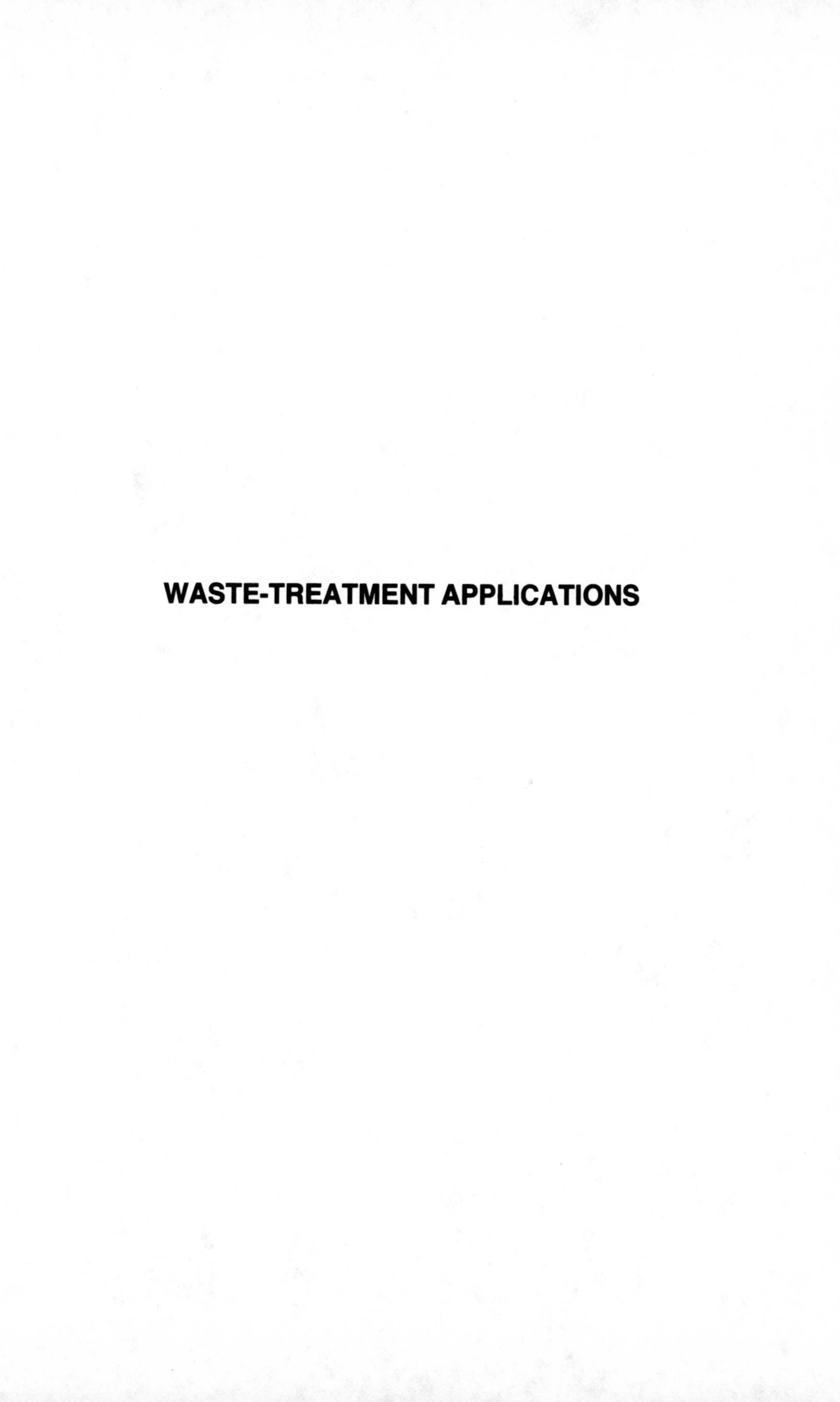

WASTE-TREATMENT APPLICATIONS

NEW DEVELOPMENTS IN WATER AND EFFLUENT TREATMENT

Christopher J.D. Fell

School of Chemical Engineering and Industrial Chemistry

University of New South Wales, Kensington, N S W, Australia

The current state of research activity in water purification and aqueous effluent treatment is reviewed. Despite the emergence of new separations technologies, conventional methods of water and effluent treatment are still receiving much research attention. This can be explained in terms of relative treatment costs.

It is shown how a multivariable optimisation approach can be used to predict the optimal design of reverse osmosis and ultra-filtration cartridges. Higher recoveries per pass, narrower channel dimensions and lower feed pressures are called for. With these modifications significantly lowered treatment costs should be possible.

Preliminary results obtained with a polyamide extended surface adsorbent fabricated on membrane principles appear promising. Adsorption of humic and proteinaceous materials can be readily reversed by a shift in pH, and adsorption capacity is high.

INTRODUCTION

Water and effluent treatment are fertile areas for the application of new separations technology. Within the last decade new water purification technologies to be adopted include Sirotherm (a thermally regenerable ion exchange resin) and ozonisation. New effluent treatment processes include the use of biological fluidised beds, enriched oxygen systems and anaerobic upflow sludge blanket reactors (1-3). The driving force behind these developments has been a demand for increased quantities of high quality process water in industry (especially industries such as medical, pharmaceutical and electronic) and tighter standards on effluent disposal.

However, water remains a cheap commodity and its processing costs must necessarily be low. For this reason, many of the more unusual separations technologies are immediately ruled out on grounds of cost, and, where effluent treatment is being considered, hard decisions must be made whether to go for on-site treatment, or to attempt to reach an accommodation with a municipal plant where substantial dilution with low strength domestic and commercial wastes will allow biological treatment to occur.

CURRENT AREAS OF RESEARCH ACTIVITY

Some measure of the current areas of research activity in both water purification and effluent treatment can be obtained by examining the frequency of subject entries in various categories in Chemical Abstracts over the period 1982-6.

The results of these surveys are shown in Figures 1 and 2. Entries over this period averaged some 2600 per annum for water purification and 5000 per annum for wastewater treatment.

Water Purification

Greatest research activity in water purification is in the low cost areas of coagulation and settling, and chlorination and deodorising. These are the types of treatment given to quality ground waters to prepare them for municipal use. Recent interest is in the use of new types of flocculants and the possible applications of magnetically retrievable flocs (Sirofloc) which can entrap suspended particulates. Ozonisation is seen as an alternative to chlorination, to prevent a residue of chlorinated organics in the treated water.

More complex waters must be treated chemically. Here, research activity has been directed to a better understanding of aqueous chemistry, particularly with regard to the precipitation of mixed species.

Other treatment methods to remove unwanted solutes are ion exchange, membrane technology and activated carbon adsorption.

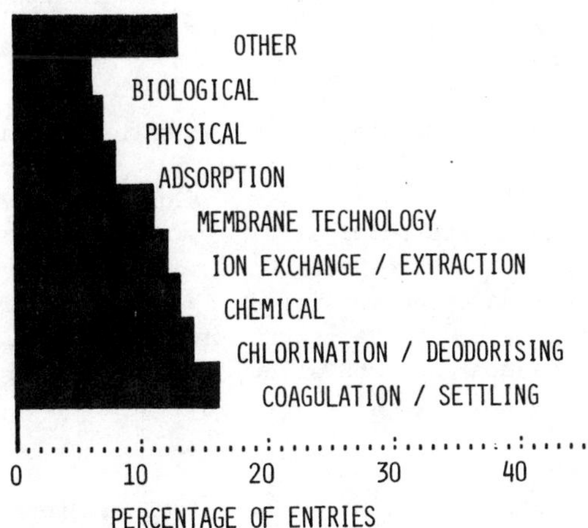

Figure 1. Subject entries on Water Purification in <u>Chemical Abstracts</u>. 1982-6

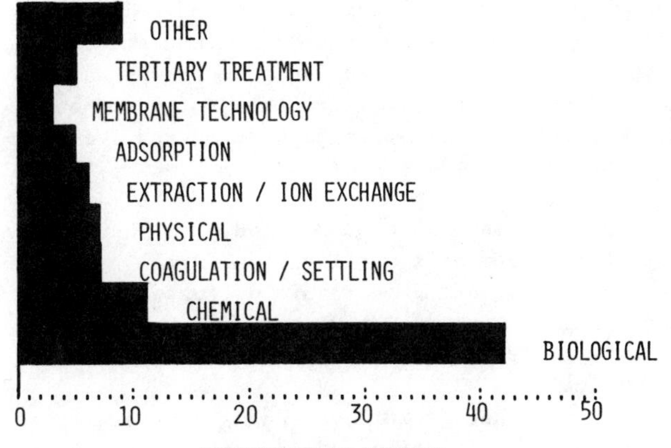

Figure 2. Subject entries on Water Treatment in Chemical Abstracts. 1982-6

The last technique is particularly effective where the water contains dissolved humic material. Despite the higher processing costs associated with their use, research into improved ion exchange resins and adsorbents continues. Membrane technology (cross flow filtration, ultrafiltration and reverse osmosis) seem set to become more widely used, with a growth rate in application of 15% per annum (4).

For ultrapure water, destined for use in the electronics industry, or for medical applications, multistage treatment is necessary. This usually involves a combination of membrane processes, ion exchange and chemical treatment. Sufficient experience now appears to have been obtained for preferred treatment plant designs to be specified (5). However, there is still a lack of highly sensitive instruments which will warn of non ionic solute breakthrough and allow remedial action to be taken.

Waste Water Treatment

Biological processes remain the most actively researched area of waste water treatment. Representing, as they do, a cheap and effective way of removing non-recalcitrant organic pollutants, these processes are in widespread use for the treatment of municipal waste water as well as on-site treatment of problem effluents. Recent work has been directed towards improving the density of viable cells in the treatment vessel, and, for aerobic systems, to improving the effectiveness of aeration. The conventional activated sludge process has been modified by incorporating particulates into the floc to give flocs which can be fluidised and readily settled. Such systems require much more intensive oxygenation and this is achieved by using enriched air or even pure oxygen. Interest in anaerobic systems stems from their production of methane. Ways to optimise the digester and to handle difficult organics are being explored. The nature of the microorganism population and its life cycle are now better understood (3).

Chemical treatment methods coupled with physical separation (settling, filtration or centrifugation) still remain the principal ways of treating inorganic pollutants, with greater attention being paid to the chemistry of the process to ensure minimum addition levels of precipitant.

Interest in other techniques such as extraction and ion exchange, membrane filtration and adsorption lies in their ability to remove dissolved material and, in some cases, to recover a by-product. Sales of the by-product and reuse of the purified water can help to defray the costs of these more expensive treatments. Particular developments of interest are the appearance of biphase liquid extractants, third generation ion exchange resins with high surface accessibility and capacity, and new types of adsorbents. Membrane technology has been making steady inroads into effluent treatment, particularly where the concentrate stream retained can be assigned a process value (6).

Research activity in tertiary treatment, by contrast, continues at a lower level than in the previous decade. This reflects the already extensive body of knowledge available on how to remove the nitrates and phosphates which are responsible for eutrophication.

COMPARATIVE COSTS

From data in the literature an approximate idea of the costs associated with each type of water treatment can be obtained. This information is summarised in Figure 3. The ranges reported are approximate only; they depend on the level of contaminant present and on the scale of the treatment plant and the nature of the water produced (potable or otherwise).

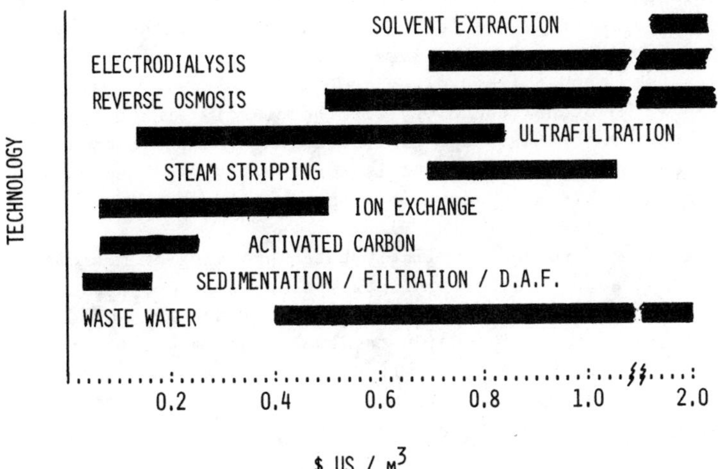

Figure 3. Costs for different water treatment methods.

It is seen that conventional treatment of groundwater is relatively inexpensive. By contrast, effluent treátment costs can be an order of magnitude more; sufficiently so for by-product sales to be important in assessing economic viability.

In making a comparison of the competing technologies of adsorption, ion exchange, aerobic biological processes and membrane technology, it should be recognised that the contact area employed to bring about the treatment process varies considerably, as shown in Table 1.

Table 1

Process	**Typical Contact Area m^2/m^3**
Activated carbon adsorption	5×10^8
Ion exchange	5×10^7
Conventional activated sludge	1000
Biological fluidised bed reactor	3000 - 4000
Membrane technology	500

In activated carbon adsorption and ion exchange the role of the receptor is passive. It must be regenerated when saturated. Biological processes have the advantage that the flocs serve as a continuously regenerating adsorbent, limited only by diffusion of substrate and oxygen to the cells constituting the flocs. By contrast, in membrane technology, all of the treated liquid must pass through the membrane itself. From the viewpoint of process synthesis (7) this is not ideal as it is based on the removal of the component present in the larger concentration (i.e. the water) rather than the dissolved contaminant. It is apparent that progress in membrane technology will be closely linked to the available area per cartridge. In developing new ion exchange resins and adsorbents, the capacity of the material and mass transfer to it become of dominant importance.

Figures quoted for the likely total spending on municipal, agricultural and industrial pollution control in the U.S.A. in the decade 1976-86 range to $400,000M (4). There is thus considerable economic incentive to develop new separations technologies applicable to water treatment. Similar figures undoubtedly also apply to the treatment of rain and infiltration waters for domestic and industrial use.

In the remainder of this review two areas will be explored in greater detail. Both are growth areas which are receiving considerable attention in research laboratories worldwide. They are the development of optimal ultrafiltration and reverse osmosis membrane cartridges, and the fabrication of tailored adsorbents to remove proteins and other organics from solution. Both show examples of how existing technology can be significantly adapted to provide more cost effective solutions to water and effluent treatment.

OPTIMISATION OF MEMBRANE CARTRIDGES

Reverse osmosis and ultrafiltration share many common features. In each, the process stream flows under pressure across a semipermeable membrane. Water passes through the membrane whilst the retained material builds up in the layer immediately adjacent to the membrane surface. This phenomenon is known as concentration polarisation.

The membrane itself is conventionally fabricated from a polymer either in an anisotropic integral form or as a layered (thin film) composite. There is still speculation on the way in which water passes through reverse osmosis membranes. One hypothesis suggests the existence of pores in the surface of the membrane. If such pores do indeed exist, they are too small to be viewed by the electron microscope. Another suggests a solution-diffusion mechanism. The polymers from which the active layer of reverse osmosis membranes are fabricated are hydrophilic and such a mechanism thus appears tenable. Ionic species present in the water are rejected, the extent of rejection depending on the properties of the active layer.

By contrast, in ultrafiltration, which is concerned with the removal of larger species (macromolecules, MW > 5000 Daltons), there appears clear evidence (8) that pores are indeed present at the membrane surface. They are in the size range 5 - 20 nm. The membranes in service are capable of rejecting molecules of substantially smaller dimensions and it is argued that this is a function of the rejection properties of the retained layer at the membrane surface.

The role of cross flow in both reverse osmosis and ultrafiltration is important. It serves to limit concentration polarisation by providing convective backmixing of retained species. In reverse osmosis, the buildup of high concentrations of retained material reduces the osmotic driving force that is experienced across the membrane and hence reduces water flux.

In ultrafiltration, the role of the retained species is less clear. Both osmotic and hydrodynamic mechanisms are invoked to explain the reduction in flux that occurs as concentration polarisation occurs. When protein solutions are being ultrafiltered, beyond a certain transmembrane pressure no further increase in flux is obtained with increasing transmembrane pressure. This is known as 'gel polarisation' and is said to occur when the surface concentration of retained material reaches some limiting value. Increased transmembrane pressure merely increases the thickness of the gel layer without increasing flux.

In the design of commercial reverse osmosis and ultrafiltration cartridges, the designer seeks to minimise the membrane area while at the same time keeping pumping costs for cross flow low. Four principal housing geometries are in present use. These are: hollow fibre, tubular, thin channel and spiral wound. They are illustrated in Figure 4.

The design of these various cartridges has previously been dictated by ease of fabrication, mechanical stability and in service performance. As yet, little attention has been paid to the optimal design of cartridges from a fundamental engineering viewpoint. There has, however, been a gradual transition from tubular designs through to hollow fibre and, more recently, spiral wound geometries. This trend is in keeping with the requirement for greater area of membrane surface per unit cartridge volume as suggested previously. However, costs for reverse osmosis and ultrafiltration equipment are still high, frequently being of the order of \$500 - 1000/m^2 of membrane surface for a fully configured system.

Such costs reflect both housing costs (the need to provide high pressure closures) and costs associated with extensive pumping equipment. They place an intolerable economic burden on the use of membrane technology in many water purification and effluent treatment applications.

Progress towards improved cartridge designs in which energy usage is optimal could significantly assist in improving the impact of membrane technology on water treatment.

Membrane Characteristics

For ultrafiltration membranes, where it is possible to assess pore open area, recent studies (8) have suggested that the open area is often low (0.1 - 2.0% of surface of membrane). This limits the intrinsic hydraulic permeability of the membrane and can affect the performance of the membrane in ultrafiltration service. It is desirable to choose a membrane with as high a pore open area as possible, consistent with its having acceptable rejection characteristics. A similar comment applies to reverse osmosis, although here the open area must be assessed by permeability and cut-off measurements alone.

In ultrafiltration in particular, there is emerging strong

a.

b.

c.

d.

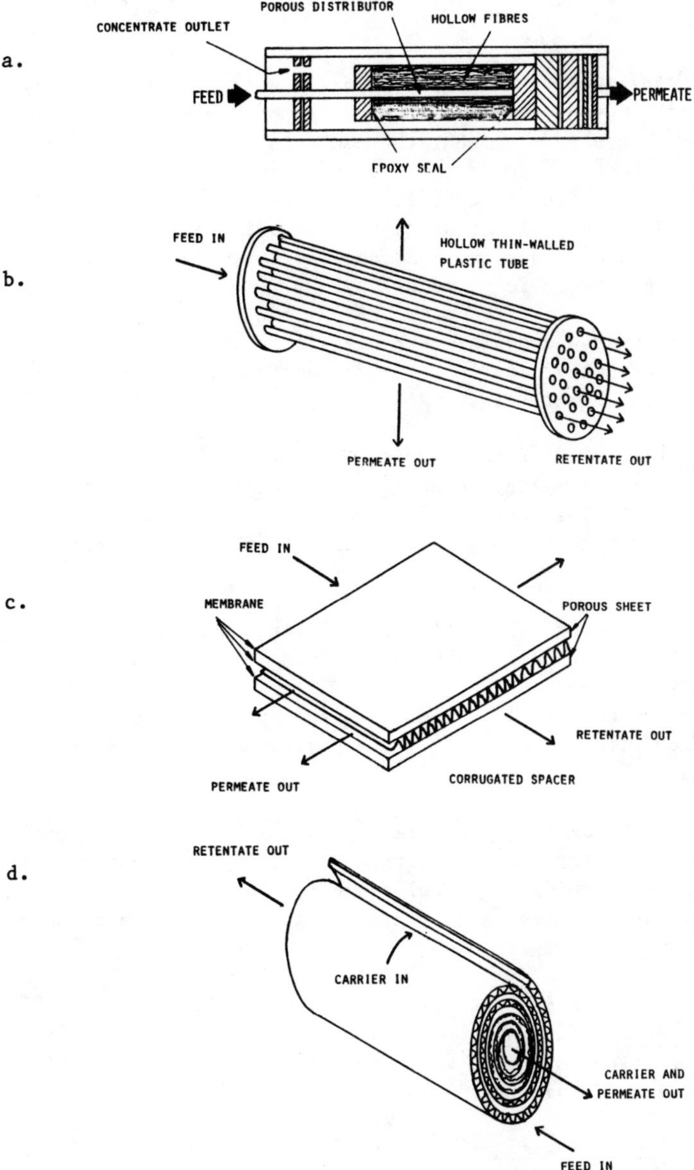

Figure 4. Principal housing geometries for membranes
a. hollow fibre b. tubular c. thin channel
d. spiral wound

evidence that substantial adsorption of solute takes placed onto the membrane (9). This gives rise to a non labile layer which may be up to 80 molecules thick. The layer determines membrane rejection.

It is likely that high open area membranes of appropriate rejection characteristics can be developed by modifying membrane surface chemistry such that the membrane material interacts with the solute being handled to give an adsorbed composite having smaller pores than the underlying membrane but retaining its high open area. This will necessitate a tailoring of the membrane for each process application, but can lead to in service membranes with a high hydraulic permeability.

For reverse osmosis membranes, it is obviously undesirable to allow extensive adsorption and proprietary descalants are available to limit this.

Optimisation of Cartridge Design

The aim in cartridge design is to engineer a balance between the applied transmembrane pressure, crossflow, and flow channel geometry such that the greatest possible flux of permeate (water) is obtained.

The mathematical modelling of reverse osmosis and ultra-filtration cartridges has been examined by Sherwood et al (10), Brian (11), Kimura and Sourirajan (12), Clifton (13) and the present authors (14,15). These treatments are in general concerned with predictions of the performance of existing designs.

The problem of optimal cartridge design can conveniently be cast into the form of a constrained multivariable optimisation problem in which flow channel geometry, feed pressure and cross flow are adjusted so as to minimise cartridge cost and the cost to provide the cross flow.

Such a treatment requires an objective function, a definition of state and design variables and a statement of transformation equations which relate the objective function to the state and design variables.

The objective function is defined as the total annual cost, and has components of the annual module cost and annual energy cost.

Annual module cost ($/year)
= membrane area (m^2) x membrane and housing cost ($/$m^2$)
x annual factor

The annual factor accounts for amortisation, interest charges and maintenance including membrane replacement.

Annual energy cost ($/year)
= feed flow x P x (8000/3600) x cost/pump efficiency
Here the cost is the $/KWH for a 8000 h/year operation.

State Variables are inlet flow, inlet feed concentration, membrane permeability and rejection. Alternatively a percentage recovery of feed can be set, where recovery is the volume of permeate/volume of feed.

Design Variables are: number of flow channels, channel length, channel height, channel width or channel diameter and supply pressure. By means of user specified variables it is possible to optimise the design of tubular modules; spiral wound modules and plate and frame modules.

Transformation Equations are:

(i) equations relating the concentration profile in the element to fluid flow parameters. Three cases are considered: entry region laminar, fully developed laminar and turbulent.

(ii) an equation for frictional pressure drop along the element.

(iii) equations for determining wall concentration and flux.

These are the transport equations for convective backmixing and membrane permeability.

Optimal Designs

A full description of the optimisation procedure for both reverse osmosis and ultrafiltration cartridges is given in Refs. 14 and 15. For the case of reverse osmosis, the buildup of retained salts at the membrane surface with increasing recovery per pass means that the cartridge design is chosen so that the energy provided in crossflow is most efficiently utilised. For ultrafiltration, the design is effectively constrained to operation at the onset of gel polarisation, for beyond this point energy is being used for overcoming needless hydraulic resistance. In both optimisation studies the Rosenbrock hill climb procedure was used and the flow channel length was sectioned to allow for piecewise solution of the relevant transformation equations.

For <u>reverse osmosis</u>, Figure 5 shows the variation of permeate cost with cross flow at a fixed membrane cost for a brackish water feed of 2500 ppm. It is found that to minimise permeate cost it is desirable to go to low transmembrane pressures, and flow channel geometries that are sufficiently small to allow laminar conditions to prevail. Thus, hollow fibre geometry with fibre diameters of << 1 mm are optimal. Similarly, with a spiral wound configuration, the gap between membranes should be as small as possible (< 0.1 mm after allowance is made for the thickness of the spacer) and the flow length should be kept short. As an example, using an optimum design for a brackish water cartridge, it should be possible to operate a spiral wound cartridge at an entry pressure of 10 - 15 bar compared with the manufacturer's specified pressure of 28 bar.

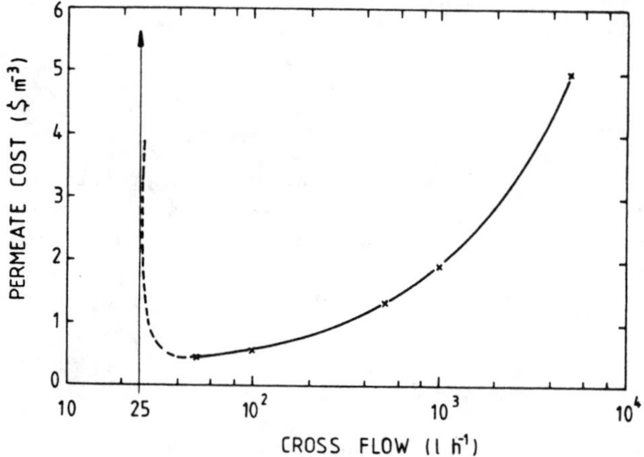

Figure 5. Variation of permeate cost with cross flow at a fixed membrane cost of $\$200/m^2$. [From Ref. 14 with permission]

For <u>ultrafiltration</u>, optimal designs are as shown in Figure 6. Here permeate cost is plotted against percentage of water recovered per pass. The values on the figure refer to different membrane costs. It is apparent that permeate costs will be minimised if high recoveries per pass are aimed at. This is further illustrated by Table 2. Typical optimal cartridge designs are reproduced in Figure 7. These represent idealised cases which may not be realisable in practice because of design constraints, but do again point to the need for designs in which effective use is made of convective mixing brought about by cross flow.

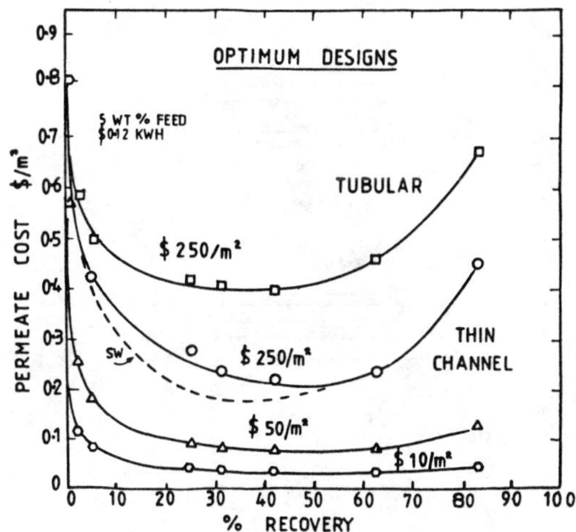

Figure 6. Predicted ultrafiltration permeate costs for different cartridge designs and percentage recoveries.

Table 2. Optimal tubular module design
Comparison of different % recoveries

% RECOVERY	N	D MM	L M	P ATM	PERMEATE COST $/M^3
0.5	105	4.7	0.82	0.6	1.28
5	201	0.42	1.14	3.1	0.50
25	1500	0.12	1.19	4.2	0.43

It is possible, in the case of ultrafiltration cartridge design to compare directly the permeate costs predicted for optimal cartridge designs with those for several commercially available ultrafilters. This is done in Figure 8. The permeate costs for these commercial units are generally much higher than could be achieved if the best possible design were to be adopted. Of the presently available ultrafilters, the Romicon and Nitto large diameter hollow fibre units appear the most attractive, although both are being operated at a recovery which must be considered low.

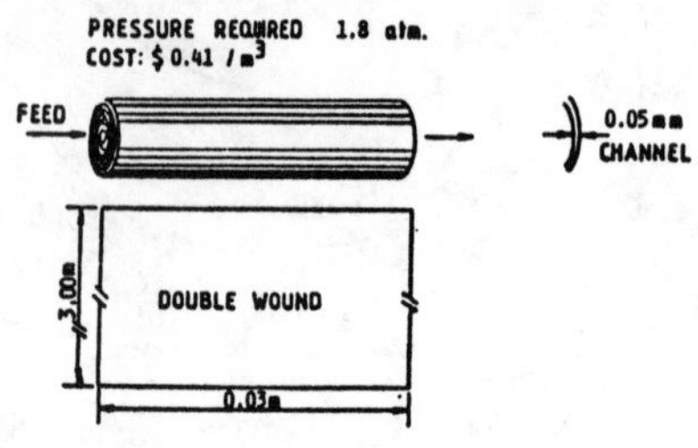

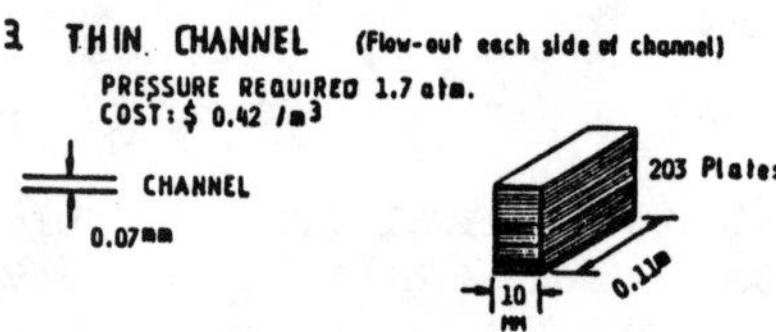

Figure 7. Optimal configurations for different module geometries.

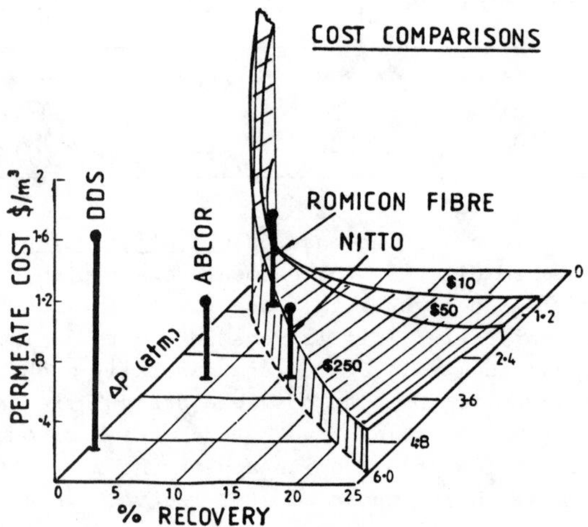

Figure 8. Permeate costs for various membrane costs and recoveries. The lines shown are (from top) for membrane costs of \$10, 50 and 250/m^2 respectively.

Opportunities for Development

The above analysis would suggest that it is possible to significantly improve the design of cartridges for both reverse osmosis and ultrafiltration. These new cartridges will operate with a higher recovery per pass and at a significantly lower entry pressure. From a mechanical viewpoint, the lower pressure will mean that simpler closures can be used, with low cost plastic housings being possible for ultrafiltration equipment. Similarly, pumping costs will be significantly reduced.

The present analysis does not take into account fabrication difficulties or problems of membrane fouling. The choice of a particular cartridge geometry can be influenced by assembly costs and questions of mechanical stability in operation, as well as by external considerations such as the effective surface are per unit module volume. It is believed that the problem of membrane fouling can be overcome by close attention to feed pretreatement and membrane-solute interactions. Notably, by using base membranes with large pores for ultrafiltration and relying on selective adsorption to give rejection capacity, one commercial design has been able to utilise the lower intrinsic bubble point of the membrane to incorporate air backflushing, with excellent results.

DEVELOPMENT OF TAILORED ADSORBENTS

The use of activated carbon adsorbents for the removal of humics, proteins and colorants from water is well established. However, some concern has been expressed over the existence of potential carcinogens in such carbons when they are used for potable water treatment.

A further drawback is that it is not possible for the carbon once used to be regenerated in situ, and the regeneration step, if practised at all, is usually carried out by the manufacturer of the carbon. For this reason the use of activated carbons is restricted to the removal of low concentrations of contaminants.

There is a need for readily regenerable, low cost adsorbents. Sporadic reports of polymer based adsorbents for specific applications, particularly in the downstream processing of biochemicals, have appeared in the technical literature in recent years. Attempts are currently being made in Australia to develop such adsorbents by making use of polymer coagulation technology first developed as part of the mechanisms occurring during the formation of anisotropic membranes for ultrafiltration. The present report is preliminary only, and represents work completed in 1986.

Structure of Polymer Coagulants

Studies of the mechanisms occurring when a polymer dissolved in an aprotic solvent (a dope, in membrane terms) is immersed in a non solvent (e.g. water) to form a membrane have been reported by Strathmann (17), Smolders (18) and others. The resulting asymmetric structure is as shown in Figure 9, with a thin dense upper or active layer, a transition layer and a grossly porous thick layer. Gittens et al (19) have categorised the nature of the structures in the grossly porous layer as O-type (where alveoli or macrovoids are freely connected) to P-type (where alveoli are separated by polymer material of the same thickness as the alveoli), to F-type (where finger like pores are present). Factors affecting the transitions between O,P and F-type structures have been studied by a number of authors including ourselves (20).

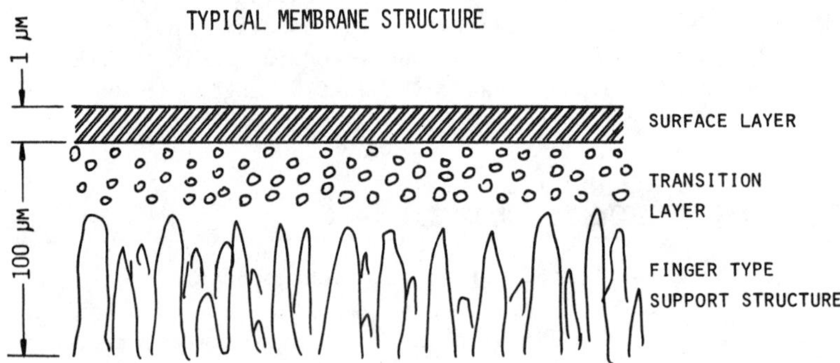

Figure 9. Structure of typical asymmetric membrane.

By adjusting the amount of polymer in the dope and the coagulation conditions, it is possible to coagulate an aliphatic polyamide dissolved in a mineral acid by immersion in water to give either of the structures shown in Figure 10. Both are examples of the O-type referred to by Gittens, one having a much more open configuration than the other. Such lace-like structures are often encountered in microporous membranes, but are then produced by a melt-spinning step rather than by water induced coagulation of a dope.

Adoption of O-structure membrane materials as possible adsorbents appears interesting provided that they can be produced in a usable form. It has been found possible to coat kieselguhr particles with dope and to coagulate this dope in water to give an O-type coating. The particles can then be used as an effective, low cost adsorbent.

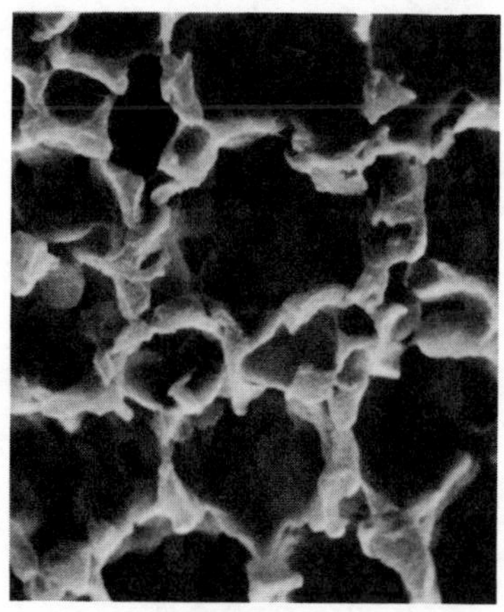

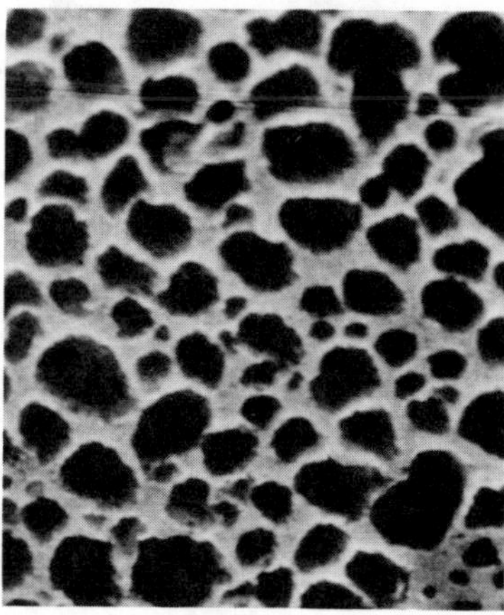

5μm

Figure 10. Structures of typical polyamide coagulants.

As an example of the use of this adsorbent, a column of adsorbent has been challenged with an aqueous solution containing 0.5% w/v of tannic acid. The resulting breakthrough curve is given in Figure 11. Substantial removal of tannic acid is possible under neutral pH conditions. The adsorption can be reversed by backwashing the column with alkali, and successful multiple cycling of the column has been possible. It is noted that the adsorbent is equally effective in removing dissolved protein. This result is not surprising as the chemical nature of polyamide is such that hydrogen-bonded adsorption will occur. Indeed, this property has been made use of in developing adsorbed composite membranes.

The results are, of course, preliminary, but sufficiently encouraging to suggest that the development of low cost adsorbents based on polyamide starting materials does have promise. Structural features of such adsorbents can be readily predicted by an extension of current studies on asymmetric membrane formation.

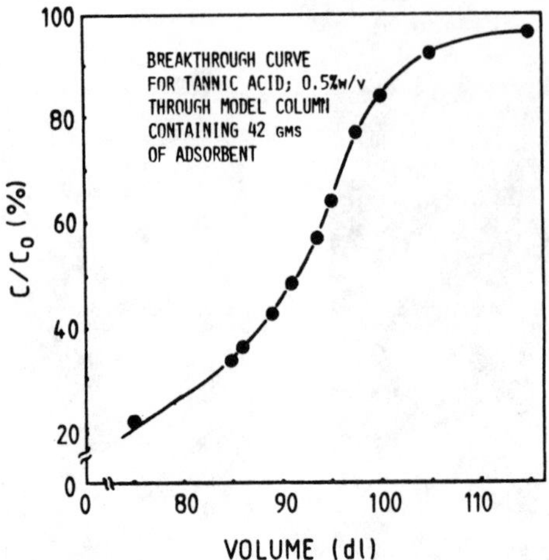

Figure 11. Breakthrough curve for tannic acid solution using packed bed of tailored polyamide adsorbent.

CONCLUSIONS

New developments in water and effluent treatment will be strongly influenced by cost factors. Where quality rain or infiltration water is available, minimal treatments will be required and the opportunity for application of advanced separations techniques will be poor. However, where available water is of poor quality or is brackish, there will be scope for treatment procedures ranging from membrane technology to adsorption. Treatment costs will dominate the decision to adopt any particular technology.

Effluent treatment does offer promise for the application of new separations technology as the cost of treatment by conventional means can be high. Although biological methods currently dominate aqueous effluent treatment, the use of chemical treatment, membrane technology or solvent extraction may be called for. The decision will rest on the toxicity of the contaminant and also on the possiblity of recovering a useful byproduct.

From the viewpoint of cartridge design for both reverse osmosis and ultrafiltration, there would seem to be considerable scope for design improvement. In particular, higher recovery, lower operating pressures and more permeable membranes are called for. All of these steps should significantly reduce treatment costs.

Although preliminary, the results on tailored adsorbents suggest that skills developed in membrane programs may be able to be used to develop extended surface materials suitable for use as adsorbents in the process industries.

REFERENCES

1. P. Coakley, in Sidwick, J.M. (Ed.) Topics in Wastewater Treatment, Soc. Chem. Ind., Blackwell, London (1985) p.1

2. P.F. Cooper, ibid., p.45

3. A.D. Wheatley, ibid., p.68

4. A.D. Little, Review for Swedish National Board for Technical Development, June, 1984

5. T. Mottershead, Ion Exch. Technol., 25 (1984)

6. T.V. Tran, Chem. Eng. Progress, March 1985, p.30

7. D.F. Rudd, Process Synthesis, Prentice Hall, N.J. (1973)

8. A.G. Fane, C.J.D. Fell and A.G. Waters, J. Memb. Sci., 9, 245 (1981)

9. A.G. Fane, C.J.D. Fell and A. Suki, J. Memb. Sci., 16, 1954 (1983)

10. T.K. Sherwood, P.L.T. Brian, R.E. Fisher and L. Dresner, Ind. Eng. Chem. Fund., 4, 113 (1965)

11. P.L.T. Brian, Ind. Eng. Chem. Fund., 4, 439 (1965)

12. S. Kimura and S. Sourirajan, Ind. Eng. Chem. Proc. Des. Dev., 7, 41 (1968)

13. M.J. Clifton, N. Abidine, P. Aptel and V. Sanchez, J. Memb. Sci., 21, 233 (1984)

14. D.E. Wiley, C.J.D. Fell and A.G. Fane, Desalination, 52, 249 (1985)

15. D.E. Wiley, C.J.D. Fell and A.G. Fane, Proc. 13th Aust. Conf. Chem. Eng., CHEMECA 85, Perth, Aust, (1985) p. 377

16. MEMTEC Ltd., Australia. Private communication (1986)

17. H. Strathmann, K. Kock and R.W. Baker, Desalination, 16, 179 (1975)

18. C.A. Smolders, in Ultrafiltration Membranes and Applications, R. Cooper (Ed.) Plenum, N.Y. (1980)

19. G.J. Gittens, P.A. Hitchcock and O.E. Wakley, Desalination, 12, 315 (1973)

20. M.S. Lefebvre, C.J.D. Fell and A.G. Fane, Ultrafiltration Membranes and Applications, R. Cooper, (Ed.) Plenum, N.Y. (1980)

INDUSTRIAL EFFLUENT APPLICATIONS OF MEMBRANE SEPARATION TECHNOLOGIES

Peter S. Cartwright, P.E.
C3 International, Inc.
2019 West County Road C
St. Paul, Minnesota 55113

European Office: Postbus 11252
2301 EG Leiden, The Netherlands

The membrane separation technologies of microfiltration, ultrafiltration and reverse osmosis have been used for a number of years to purify raw water for such applications as potable water production, medical/laboratory uses and industrial high quality rinsing requirements. Recently, these technologies have begun to find uses in effluent treatment applications where they offer such advantages as continuous operation, low energy utilization and simple, modular construction.

This paper describes the technologies in detail, and addresses the engineering design considerations of complete membrane processing systems. Also included are several case histories illustrating application of membrane separation technology to contaminated ground water clean up. Sample analysis, applications testing and development of total system design are traced in detail. In addition to technical considerations, the economic factors are presented.

INTRODUCTION

The membrane technologies of microfiltration, ultrafiltration and reverse osmosis are enjoying wide acceptance in water treatment applications ranging from potable water production to ultrapure water purification. The utilization of these technologies for effluent treatment is almost non-existent by comparison. I believe this can be primarily attributed to a lack of understanding of the capabilities of these technologies.

C3 International, Inc., is engaged in the investigation of membrane separation technologies with primary emphasis on waste treatment and food and chemical processing applications. This investigation usually takes the form of an analysis of the stream to be treated in the context of disposal restrictions and reuse or recovery options. After selection of the appropriate membrane polymer and device configuration, an applications test is performed to characterize the stream and obtain design data. If necessary, an on-site pilot test is run to gather long term operating data. From these tests, it is possible to design and provide all of the necessary engineering for construction or purchase of a total treatment system.

This paper summarizes the membrane technologies commercially available, outlines test procedures and cites five case histories illustrating the use of reverse osmosis in effluent treatment applications.

REVIEW OF TECHNOLOGIES

The membrane separations technologies addressed in this paper include:

- Microfiltration
- Ultrafiltration
- Reverse Osmosis

Although they all provide separation of liquid borne contaminants from the liquid, each utilizes a different separation mechanism and has specific advantages and disadvantages when compared to the others in a particular application.

Figure 1 depicts the mechanism of microfiltratrion. Generally, microfiltration involves the removal of particulate or suspended materials ranging in size from 0.1 to 10.0 microns (1000 to 100,000 angstroms).

Figure 2 depicts ultrafiltration, which is used to separate materials in the .001 to 0.1 micron range (10 to 1000 angstroms). Basically, ultrafiltration is used to remove dissolved materials while the suspended solids are removed by microfiltration.

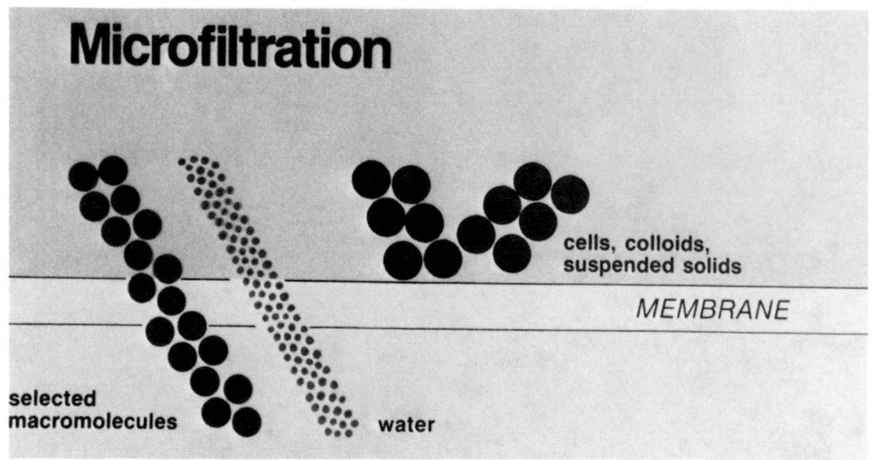

Figure 1. Microfiltration Mechanism

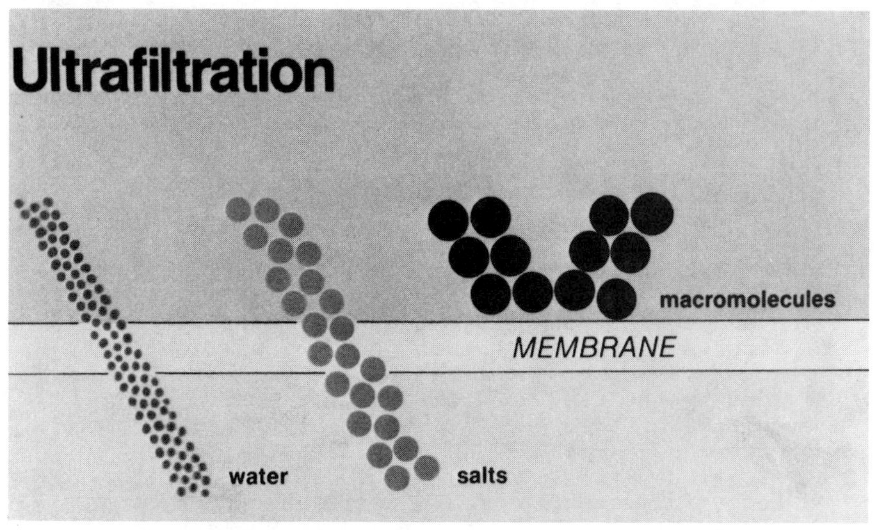

Figure 2. Ultrafiltration Mechanism

Figure 3 illustrates reverse osmosis which typically separates materials less than .001 micron (10 angstroms in size). Reverse osmosis offers the advantage of rejecting ionic materials which are normally small enough to pass through the pores of the membrane. As with ultrafiltration, reverse osmosis is used to remove dissolved materials.

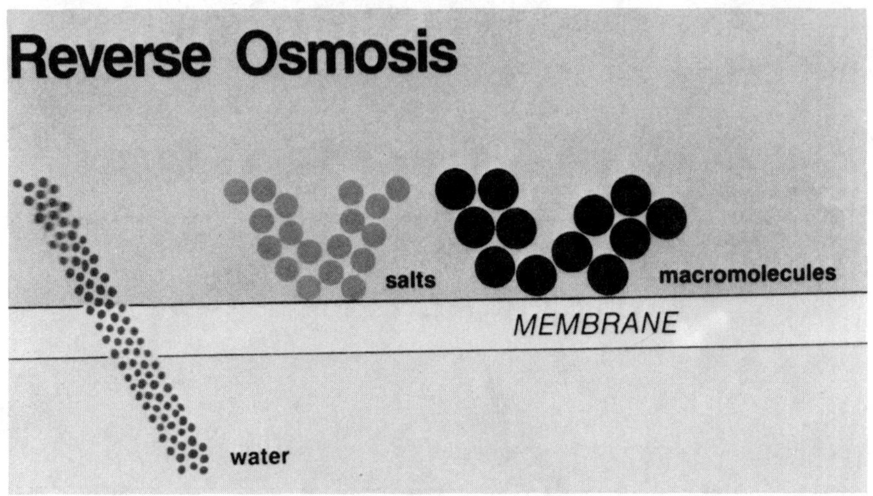

Figure 3. Reverse Osmosis Mechanism

Figure 4 depicts a general schematic for the membrane process of microfiltration, ultrafiltration and reverse osmosis. In these technologies, an important engineering design consideration is "recovery," which is defined as the permeate volume divided by the feed volume; in other words, the percentage of the feed flow which is pumped through the membrane. Typically, for effluent treatment applications, the recovery figure is at least 90%. As recovery is increased (to decrease concentrate volume), the concentration of solute and suspended solids in the concentrate stream increases.

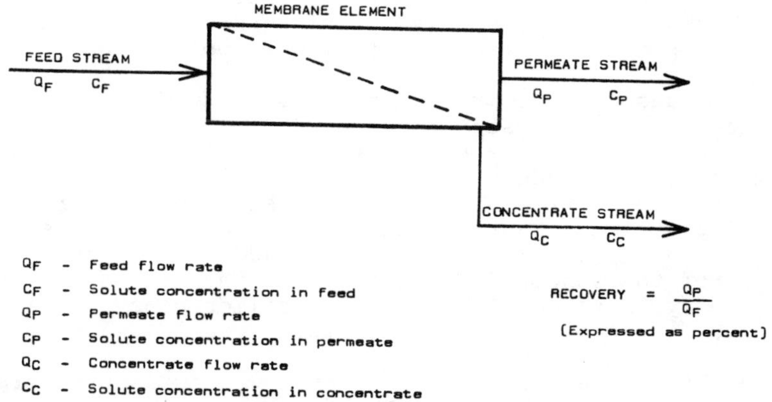

Figure 4. Schematic of Membrane Processes

For the processes of ultrafiltration and reverse osmosis which deal with dissolved materials, a property of the solution known as "osmotic pressure" $(\Delta\pi)$ becomes a limiting factor. Osmotic pressure is a characteristic of all solutions, and is loosely defined as the resistance of the solvent portion of the solution to passage through the membrane. Osmotic pressure is a function of both the particular solute as well as its concentration.

The recovery of the system can be controlled by restricting the quantity of flow in the concentrate stream, normally through the use of a concentrate valve. As recovery is increased, with the resulting decrease in concentrate flow, the concentration of solute in the concentrate stream increases which results in an increased osmotic pressure.

No membrane is perfect in that it rejects 100% of the solute on the feed side; this solute leakage is known as "passage." Expressed as "percent passage," the actual quantity of solute which passes through the membrane is a function of the concentration of solute on the feed side. Under high recovery conditions, the concentration of solute on the feed side is markedly increased, and therefore, the actual quantity of solute passing through the membrane also increases. Because most effluent applications demand that, in addition to a minimum concentrate volume, the permeate quality be high enough to allow reuse or to meet discharge regulations, the "Catch-22" predicament of permeate quality decreasing as recovery is increased can impose design limitations. Additionally, the increased osmotic pressure resulting as recovery is increased also imposes a design limit. Generally, pumping pressures in excess of 1000 psi are impractical for most applications.

Because of the propensity of suspended or precipitated materials to settle out on the membrane surface and plug the membrane pores, turbulent flow conditions must be maintained (Reynolds numbers in excess of 2000). For high recovery systems, this usually requires recycling a significant percentage of the concentrate back to the feed side of the pump. The addition of this portion of the concentrate stream into the feed solution obviously increases the dissolved solids concentration further increasing osmotic pressure.

All of these factors: recovery, osmotic pressure, permeate quality, recycle, etc., serve to underscore the value of testing the specific waste stream as throughly as possible. Because effluent streams often vary in analysis as a function of time, it is important that either a composite of a "worst case" sample be obtained for test purposes.

The applications test typically involves the evaluation of a 30-50 gallon sample of solution on a production-sized element. The element is mounted in a test machine with the engineering features of production systems. For a given element, the test can be completed within 1-2 hours. Figure 5 details the applications system design.

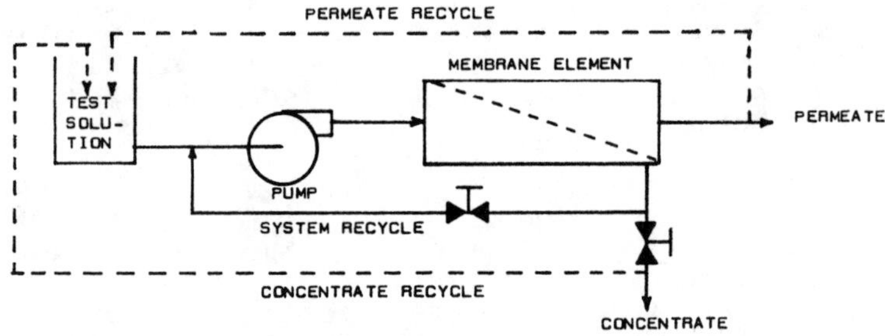

Figure 5. System Design

In general, waste treatment applications for reverse osmosis require membrane polymers with good temperature, pH and general chemical stability. Usually, the "mutally exclusive" requirements of producing a minimum concentrate volume (preferably zero) while producing a permeate stream acceptable for reuse or discharge to a POTW must be compromised. This requires a membrane polymer with the highest possible rejection characteristics. For the above reasons, thin film composite chemical compatibility studies must be made to ensure that the polymer is unaffected by the solutions being treated.

In addition to applications testing in order to select the optimum membrane polymer and element configuration and to accumulate such data as osmotic pressure effects, an on-site pilot test is often required to develop long term fouling data for pretreatment design considerations. Many waste streams contain chemicals which become insoluble as the solution is concentrated as recoveries are increased into the range of 90-95%. Often, commercially available dispersants will prove effective in minimizing the formation of insoluble compounds, but in general, the osmotic pressure effect and fouling resulting from concentration are both equally important factors in establishing the recovery limitations of a system design.

As a rule of thumb, fouling considerations preclude the use of hollow fiber membrane devices and promote consideration of spiral wound, tubular and plate and frame configurations.

CASE HISTORIES

The following provide examples of specific waste treatment applications and details of the testing required to design total treatment systems.

Printed Circuit Clarified Effluent Treatment

The client uses a subtractive printed circuit manufacturing process involving the etching of copper-clad laminate which produces significant quantities of dissolved copper. In addition, lead, nickel and chrome are found in the manufacturing effluent stream. The client is building a new manufacturing facility located in an area where municipal effluent treatment is not available. Although ambient temperature and humidity conditions favor atmospheric evaporation of the effluent to a lagoon, the type of soil available and concern with possible ground water contamination precludes this form of waste treatment. The client has decided to concentrate the effluent as much as possible and haul it away.

The waste treatment process selected involves the use of chemicals addition to produce insoluble heavy metal precipitates which are dewatered in a filter press prior to further evaporation or hauling. The chemical process utilized to effect precipitation is raising the pH with sodium hydroxide to 9-10, followed by addition of dithiocarbamate to provide a greater degree of precipitation. Following the clarification step (using a slant-tube clarifier), the effluent is filtered, run through activated carbon and treated with reverse osmosis to produce a permeate acceptable for reuse.

Figure 6 illustrates the total system.

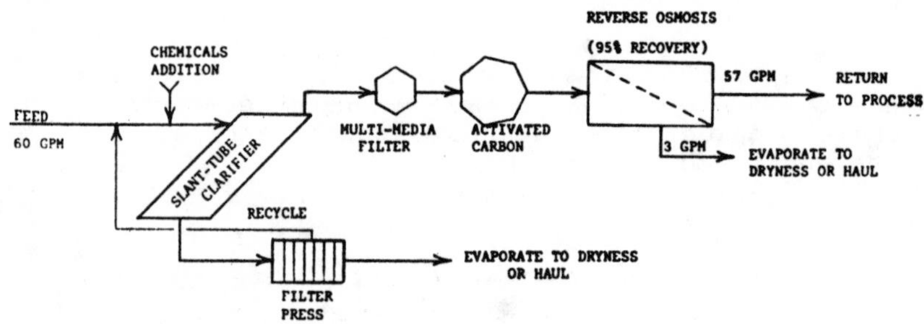

Figure 6. Printed circuit clarified effluent treatment.

Total feed flow to the treatment system is 60 gallons per minute and the reverse osmosis unit is designed to operate at 95% recovery. The 3 gpm of concentrate is to be combined with the sludge from the filter press for further evaporation or hauling.

Following is the analysis of the clarified effluent directed to the reverse osmosis unit:

FEED STREAM CHARACTERISTICS		
pH	=	12
TDS	=	2110 ppm
TOC	=	108 ppm
Cu	=	0.15 ppm
Ni	=	0.15 ppm
Pb	=	0.10 ppm
Cr	=	0.02 ppm

The following data summarize the performance of the reverse osmosis system:

R. O. SYSTEM PERFORMANCE

ANALYSIS	FEED	PERMEATE (95% RECOVERY)	CONCENTRATE (95% RECOVERY)
TDS (ppm)	2110	158	22,500
COPPER(Cu)(ppm)	0.15	0.05	--

The above data were developed as a result of the following scenario:

1) Engineering analysis to determine effluent stream characteristics and evaluate discharge and treatment options.

2) Testing of 50 gallon effluent sample to select specific treatment chemistries and best reverse osmosis membrane polymer and develop preliminary design.

3) 30-day pilot testing to optimize total system design.

4) Preparation of complete engineering specifications for total system procurement.

In summary, this system exemplifies the value of combining traditional treatment technologies such as precipitation/clarification with membrane processing to effect a total treatment system optimized for this particular set of conditions and restrictions.

Chemical Processing Effluent Treatment

A large manufacturer of chemicals associated with the food industry produces an effluent stream the concentrations of which exceed TOC discharge limits. Effluent volume is 80,000 gallons per day which was being removed by a hazardous waste hauler.

The raw effluent stream analysis is as follows:

TDS	-	340 ppm
TOC	-	4800 ppm
BOD	-	45,000 ppm
COD	-	19,700 ppm
pH	-	2.7
Toluene	-	1.4 ppm
Methanol	-	6,000 ppm
Acetic Acid	-	14,800 ppm

An engineering evaluation concluded the best approach would be air stripping to lower the concentration of volatile solvents (toluene and methanol), followed by reverse osmosis to dewater the effluent stream.

Application testing on a 50 gallon sample indicated that when run at the pH of the raw feed (2.7) the acetic acid was insufficiently ionized to allow sufficient removal by reverse osmosis. After pH adjusting the stream to a pH above 11, and utilizing the thin film composite membrane, 99% acetic acid removal was noted. The following is a summary of the data for this treatment:

CHEMICAL PROCESSING EFFLUENT TREATMENT
R. O TREATMENT

	FEED ANALYSIS	PERMEATE ANALYSIS BEFORE pH ADJUSTMENT	AFTER pH ADJUSTMENT
pH	2.7	2.7	11.1
Recovery	-	95%	98%
TDS (ppm)	340	204	496*
TOC (ppm)	4800	3000	870
BOD (ppm)	45,000	5800	2000
COD (ppm)	19,700	10,400	4600
Acetic Acid (ppm)	14,800	5030	160
% Acetic Acid Removal		66%	99%

*TDS higher than in feed stream because of the excess sodium ion added to raise pH.

Because of the high system recovery desired, the osmotic pressure of this solution was measured as a function of recovery. To effect the high removal of acetic acid, the feed stream must be adjusted to a pH of approximately 11. This results in an osmotic pressure of approximately 390 psi at 95% recovery. To allow for a minimum differential pressure of 200 psi (operating pressure minus osmotic pressure), the resulting reverse osmosis system was designed to operate at 600 psi.

In this application, the utilization of reverse osmosis preceded by the appropriate pretreatment technologies has resulted in a 95% reduction in hauling costs and recovery of process water which meets discharge regulations.

Battery Manufacturing Effluent

In the manufacturing of zinc-mercuric oxide batteries, an effluent is produced containing zinc and mercury at a high pH. This particular client generates a small volume of hazardous waste (60 gph), but the concentration of mercury in the effluent stream is extremely high. Applications testing of reverse osmosis indicated that a 2-stage system would be required to ensure sufficient mercury removal. The following summarizes these data.

	FIRST STAGE	SECOND STAGE
Operating Pressure (psi)	770	570
Recovery (%)	90	70
TDS Removal (% rej.)	93	93

An illustration of the total treatment system designed as a result of application testing is shown in Figure 7. Performance of the reverse osmosis system is summarized as follows:

	STATE DISCHARGE LIMITS	FEED STREAM	1st STAGE PERMEATE	2nd STAGE PERMEATE
TDS (ppm)	50	3170	250	22
Hg (mercury) (ppm)	0.128	42	1.08	0.007
pH	6-9	11.4	7.0	7.0

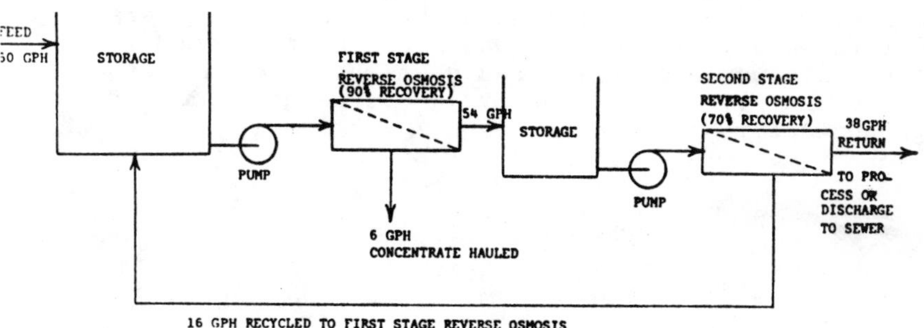

Figure 7. Battery manufacturing effluent treatment utilizing 2 stage reverse osmosis.

Total cost of the treatment system designed as a result of this testing is approximately $30,000.

Contaminated Groundwater Clean-up

A client must clean up groundwater contaminated with low molecular weight organic compounds (75-200 molecular weight) and sulfate and sulfite salts of sodium. The total flow rate is 25 gpm, and they plan to haul the concentrate away and discharge the permeate to a creek.

Application testing resulted in the following data for reverse osmosis treatment operating at 90%:

Parameter	Feed	Permeate (90% Recovery)	% Reduction
pH	9.5	6.2	-
TOC	361	41	89
BOD	430	51	88
COD	1,340	134	90
TDS	5,120	329	94
TSS	33	-	-
Oil/Grease	5.25	-	5
Sulfate	2,550	140	94
Sulfite	784	141	82

Cellulose acetate membrane polymer membrane was used to produce the above permeate quality, and on-site pilot test is currently underway to optimize total system design.

Contaminated Groundwater

A manufacturer of metal products has been discharging effluent to a lagoon for a number of years. Unfortunately, the lagoon was not properly lined and toxic waste has filtered down into the aquifer. Following is a typical analysis of the contaminated groundwater:

COMPOSITE SAMPLE

TDS	~21,000	mg/l
Antimony	2.0	mg/l
Arsenic	0.08	mg/l
Barium	80.	mg/l
Cadmium	0.04	mg/l
Chromium	390.	mg/l

Cobalt	9.2	mg/l
Copper	635.	mg/l
Lead	12.5	mg/l
Molybdenum	22.	mg/l
Nickel	1,600.	mg/l
Silver	0.09	mg/l
Thallium	0.8	mg/l
Vanadium	4.0	mg/l
Zinc	10.0	mg/l
Cyanide	3.	mg/l
Fluoride	8,600.	mg/l

Under court order, they are obliged to treat this ground water at a rate of 15 gpm. The purified water must meet discharge regulations and they evaporate the concentrate to dry solids and haul.

Testing indicated that 2-stage reverse osmosis was required and pilot testing was utilized to optimize total system design. Figure 8 illustrates this design.

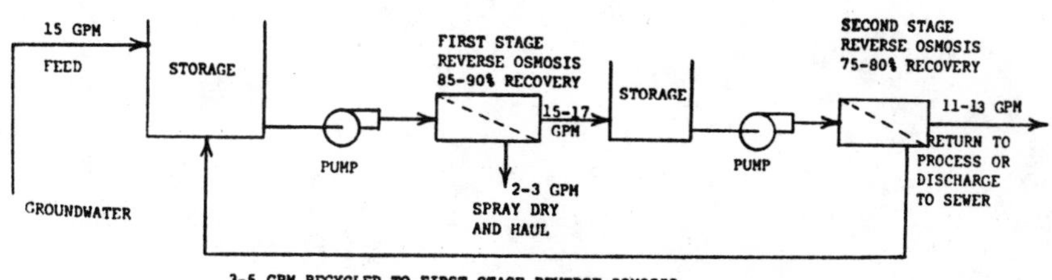

Figure 8. Contaminated groundwater treatment system.

SUMMARY

With regard to the future, certainly with increased emphasis on resource recovery and enforcement of RCRA and other similar legislation, the use of membrane technology will increase markedly as its contributions to effluent treatment become more universally understood and publicized. I feel no major breakthroughs in either reverse osmosis or ultrafiltration polymer development are required; however, there are a number of waste treatment and processing applications awaiting the development of a true UF membrane (no salts rejection) with a molecular weight cut-off in the range of 500-1000 daltons.

The opportunity to discuss case histories and exchange application information will do much in improving the confidence level in working with these new technologies. Basically, we are in a new industry and the challenges we face in the months and years ahead to provide and keep pure water are exciting indeed.

APPLICATION OF REVERSE OSMOSIS SPIRAL WOUND MEMBRANE MODULES IN THE TEXTILE INDUSTRY

S. N. Gaeta
De Martini S.p.A.
Via per Oropa 118
13051 Biella, Italy

Reverse osmosis spiral wound membrane modules were utilized to treat effluents from cotton dyeing industries. Economic and ecological advantages were analised. A "zero-discharge" system was designed and is in operation at a textile plant near Biella, Italy. The recovery of auxiliary chemicals and (hot) water from effluents from textile industries was studied. Maximum salts, surfactants, and COD rejections resulted 97%, 99,8%, and 98%, respectively. Effluents were concentrated up to 98%. Due to reuse of water and chemicals and to low operating costs needed by R.O. plants, membrane processes resulted more economic than traditional systems (biological, carbon adsorption, resins) (1).

Membrane materials were polyamides. Polyamides, membranes, and modules were produced by De Martini S.p.A., Biella, Italy. The modules operated in a pH range from 2 to 12 and at temperatures and pressures up to 50°C and 70 Atm., respectively. Sodium Chloride rejection ranged among intermediates values (60,80,90%). The excellent properties of these "loose" R.O. membranes allow the integration of R.O. in traditional textile processes quite simply.

INTRODUCTION

Textile plants use huge amounts of water (2,3), most of which is wasted. Traditionally, due to abundance and the low price of water, extra-water has been used to guarantee the best quality of textile products. Nowadays, while the requirement of fresh, good water for industrial uses has increased enormously, the supply of it remains unchanged. Industries are facing increasing shortages and the case of plants which have to shut down the production for lack of water is not rare (4).

Moreover, due to more strict government pollution control regulations, large volumes of waste water also mean higher volumes of effluents which have to be treated prior to disposal. In addition, textile waste water contains expensive auxiliary chemicals and energy which can be recovered making the membrane processes economically advantageous (5,6). Therefore, it seems necessary to manage textile waste water more efficiently by using new technologies.

Reverse osmosis membrane processes, rationally integrated in textile plants, seem to offer the possibility to reclaim water, auxiliary chemicals, and energy by allowing their recycle (7). In this way, the volume of waste water to be treated as well as the volume of fresh water to be pretreated, if necessary, is minimized. Therefore, by using R.O. processes the "money-wasting" water depuration processes can be converted into "money-making" processes by recovering water and species dissolved in it and by approaching zero-discharge systems.

In the past the extremely severe environments of textile effluents, i.e. wide pH range, highly aggressive chemicals and high temperature, imposed expensive pretreatments to textile effluents prior to feeding them to a R.O. plant (8). Nowadays, due to excellent chemical, mechanical, thermal, and transport properties of innovative membranes made by new polyamides, R.O. has a better chance to be advantageously applied to treat textile waste streams.

Water and Chemicals Used in a Typical Cotton Dyeing Plant

A typical cotton dyeing process comprises pretreatment, dyeing, and finishing of fabrics. Effluents are quite complex and heterogeneous solutions.

The preparation of goods for dyeing and printing is a far more important process than the production of white goods. Textile material to be dyed or printed must have the following properties: high and uniform dye uptake and absorptivity; completely free from husks; high degree of polymerization of the cellulose; adequate degree of whiteness to permit faultess dyeing of pale shades. The usual operations are: desizing, scouring, mercerization, and bleaching (9).

The dyeing bath is prepared to improve kinetics and equilibria of the dyeing process by adding appropriate chemicals.

A list of chemicals present in the effluents from pre-treatment and dyeing processes is shown in Table 1. The dyes used depend on the end product. They can have various chemical characteristics: ionic, polar organic, and non polar organic. Therefore, both electrostatics and non electrolytic interaction forces are expected to work between dye molecules and membrane polymer material.

Table 1. Important characteristics of waste waters produced during the pre-treatment and dyeing of cotton fabrics.

Process	Operations	Pollutants
Pre-treatment:	desizing, scouring mercerization, and bleaching	cotton impurities: proteins, pectins, oils, waxes, colors, suspended solids. chemicals: acids, alkaline detergents, bleaching chemicals, caustic soda.
Dyeing	dyeing	chemicals: dispersing agents, protective colloids, alkali, sodium sulphate, sodium nitrate, reducing agents, sequestering agents, levelling agents.

The finishing step improves the characteristics of the products by tailoring them to their end use, i.e., easy to wash, to iron, fire-retardant, etc.

Reverse Osmosis Processes

Reverse osmosis processes (10) use membrane modules capable of separating a solution in its components, soluble and insoluble. Reverse osmosis membranes reject low molecular weight salts, such as Sodium Chloride, and higher molecular weight species with a high efficiency. They are asymmetric membranes and are generally used in spiral wound, tubular,capillary, or hollow fibers modules.

Water and auxiliary chemicals can be recovered by using reverse osmosis processes. This is not possible when using other traditional treatments: biological, carbon adsorption, coagulation, ozonation. Moreover, reverse osmosis plants take less space than other systems.

The driving force in reverse osmosis systems is the pressure applied to reach the desired water flux through the membrane. By choosing adequate membrane characteristics (flux, rejection, stability) to the treatment which is required, it is possible, potentially, to obtain any kind of separation given any kind of waste-stream.

EXPERIMENTAL

Membrane and Modules Preparation

Asymmetric supported membranes were prepared by using a new class of polyamides discovered by Montendison, Italy (11) and developed and produced by De Martini S.p.A., Biella, Italy. The general formula can be summarized as:

$$\left[-\overset{\overset{\large O}{\|}}{C} - X - \overset{\overset{\large O}{\|}}{C} - N\underset{\underset{Rn}{|}}{\bigcirc}N - \right]_n$$

where X and Rn are different radicals depending on the desired polymer properties. They are characterized by molecular weight higher than 100,000 and generally narrow molecular weight distribution. They are completely amorphous and have Tg higher than 200°C. These polymers are suitable for preparing semipermeable membranes by using a phase inversion technique.

The asymmetric membrane is formed by casting a concentrated polymeric solution on a non-woven support and thereafter precipitating the film in a non-solvent coagulation bath. Liquid-liquid phase separation and gelation mass transfer phenomena control the process (12).

The membranes were assembled in 4" spiral wound membrane modules with an active surface membrane area of 5 m2. This configuration seems to be very efficient in the application studied (8). The modules used are commercialized by De Martini S.p.A. under the trade name Separem, Model 460 T and 480 T.

Pilot Plant

All the experiments reported were done in a pilot plant at De Martini S.p.A. where the modules were housed. It comprises a process tank, where was stored the feeding solution, and permeate and concentrate tanks, where the water permeated through the module and the concentrated stream were collected. A 10 micron filter was installed prior to the reverse osmosis unit. Sufficient instrumentation was included to permit the control of the unit continously. The membranes were cleaned by using products from the Ultrasil series. The axial flow rate was always 2000 lt/h and operating pressure and temperature were 30 Atm and 30°C, respectively.

RESULTS AND DISCUSSION

A SEM picture showing a typical structure of the asymmetric membrane prepared is reported in Fig. 1. The compact dense layer is about 5 micron thick, while the porous sublayer is about 95 micron thick with increasing porosity towards the lower surface. This structure is responsible for the characteristics of the "loose" membranes used in this study, i.e. NaCl rejection from 60 to 90% and high fluxes. Typical values of fluxes and rejections are reported in Table 2. Values of transport, thermal, mechanical, chemical and biological resistance of 4" spiral wound membrane modules used in this study are reported in Table 3.

The good transport, mechanical, and thermal properties are due to rigid and regular structure of the polymeric chains which allow a good interaction among molecules in the membrane itself, while maintaining highly porous structures. The excellent

Table 2. Flux and NaCl rejection for the polyamidic membrane used to manufacture Separem modules.
T = 25°C; P = 30 Atm ; C = 2000 ppm of NaCl;

Model	Flux, lt/m2d	NaCl Rejection, %
490 T	2000	90
470 T	2600	75
460 T	3000	60
400 T	3600	50
400 T	4100	40

Table 3. Characteristic properties of 4" spiral wound membrane modules manufactured with new polyamides by De Martini S.p.A. under the trade name Separem

Flux, gpd*	2.500
NaCl rejection, %*	from 40 to 90%
Maximum allowable temperature, °C	40
Maximum allowable pressure, Atm	60
pH allowable range	2-12
Chlorine resistance, ppm	50
Biological resistance	excellent after 2 years

* when tested at T = 25°C; P = 30 Atm ; C = 2000 ppm of NaCl.

chemical and Chlorine resistances can be attributed to the fact that no Hydrogen atoms are attached to Nitrogen or Ester groups (13). The resistance to biological attacks is also excellent. Membranes stored in tap water retain their properties unaltered after two years.

The results of a treatment done by a 480 T Separem module on an effluent from a cotton dyeing plant and the original composition of the effluent are reported in Table 4. The flux was quite constant over the period of time experimented. It was about 120 lt/h module after 140 hours. The permeate obtained resulted useful to be reused in the textile plant; it also met the standards set by government laws for discharging waste streams.

The textile effluent was composed by several streams coming from different textile processes and homogenized in an equalizing tank prior to feeding it to the reverse osmosis unit.

The initial waste water was concentrated up to 90%. The

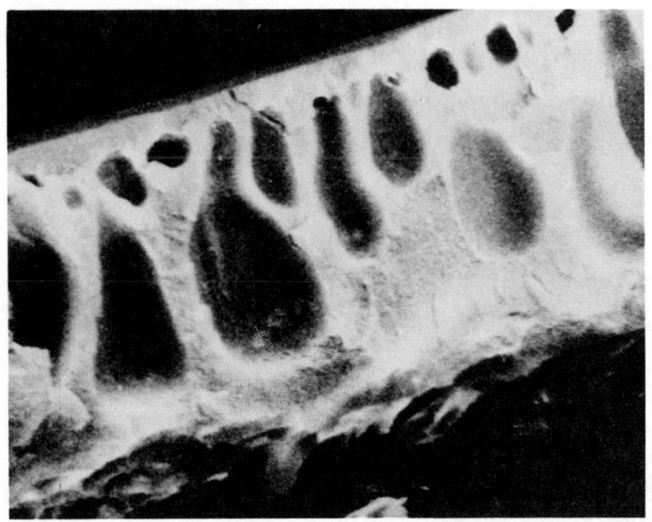

Figure 1. Structure of a Separem membrane.

concentrated stream out of the R.O. unit was biologically treated, thereafter discharged. In this way, the biological unit was very small and the possibility to recuperate the permeated water turned out quite useful due to lack of water in the area at the time of the experiments.

Table 4. Results of a treatment by R.O. of a textile effluent from a cotton dyeing plant using a 480 T Separem spiral wound module.

	IN	OUT	Rejection, %
COD	698	14	98
Non-ionic surfactants, ppm	12.43	0.01	99.9
Salts, ppm	5000	250	95
pH	8	6.7	
Color	visible	none	100

Data from a treatment by R.O. of an effluent which was

previously treated in a physico-chemical plant are shown in Table 5.

Table 5. Results of a treatment by R.O. of a textile effluent previously partially treated in a physico-chemical plant.

	IN	OUT	Rejection, %
COD	691	71	90
Non-ionic surfactants, ppm	537	0.8	95.8
Color	none	none	

In this case, due to an expansion of the textile plant, the existing physico-chemical treatment turned out to be inadequate to produce water which could meet standards set by government laws. By applying a R.O. unit after the existing process the water treated was good to be discharged or reused in the textile plant.

The flux obtained in this case was about 300 lt/h module and stable during the time of the experiment (140 hours). The high increase in flux respect to the experiment described earlier in Table 4 is presumably due to the fact that in this case was used a 460 T Separem module (characterized by higher flux respect to the 480 T Separem) and to the absence of fouling of the membrane caused in the first experiment by dye molecules. The dye present in the first experiment interacted physically with the membrane material generating a coating on it, therefore causing a flux reduction. Extraction experiments seem to confirm this hypothesis. In fact, color was extracted from a piece of the membrane taken from the module which was used in the first experiment. The membrane used in the second experiment here reported was color free.

In several cases it was found that water permeated through the membrane having a composition in COD and surfactants out of the limits set by government laws was still good to be reused in dyeing plants. This fact allowed the use of modules with higher fluxes and lower rejections making the potential use of R.O. plants economically more attractive. In Table 6 are reported some of these results. The membrane treatment resulted useful, in this case, to abate the color, so that the permeate

Table 6. Composition of water treated by R. O. using a 400 T Separem module.

	IN	OUT
Color	visible	0
COD	180	180
Surfactants		
anionic	0.8	0.08
non-ionic	6.7	3.2

could be reused in the dyeing bath.

In Table 7 are reported the results of a zero-discharge system put in operation at a textile plant near Biella, Italy. The process comprises an equalizing tank, a biological treatment, and a R.O. unit. The biological step abates color and COD, while the R.O. unit is used to recover water and salts used in the dyeing plant. The salts are recovered in the concentrate stream coming out of the R.O. unit. **Ninety percent of the salt is Sodium Sulphate. The recovery of this salt is economically attractive. In fact,** a typical cotton dyeing plant takes one Kg. of Sodium Sulphate for each Kg. of cotton produced. The salt to be reused had to be concentrated at 60 gr/lt. At this concentration its osmotic pressure is 21 Atm.

For this treatment was chosen a 470 T Separem module, whose Sodium Chloride rejection is 70%. In this way the Sodium Sulphate concentration in the permeated stream is about 2 gr/lt, which turned out to be adequate to allow the reuse in the textile plant of the water permeated. The flux obtained was about 200 lt/h module.

Table 7. Results of a R.O. treatment integrated with a biological treatment to obtain a zero-discharge system.

	after equalizer	after biologic	after R.O.
pH	9.6	5.2	5.2
COD	1050	40	0.8
Non-ionic surfactant, ppm	18	18	0.02
Sodium Sulphate, gr/lt	5	5	2
Color	present	none	none

The sludge from the biological step is converted in fertilizer and used in a field near the textile plant.

The recovery of Sodium Sulphate makes the R.O. unit economically convenient. In fact, by concentrating it up to 60 gr/lt saves 350 Italian Lire (Lit.) for each Kg. of treated textile material. The R.O. plant pay-off time, at these conditions of operation, is estimated to be 3 years. Details of this economic evaluation are reported in Table 8.

Table 8. Preliminary economic evaluation of the yearly saving due to the recovery of Sodium Sulphate from a textile effluent using R.O. processes. Base : 1000 m3 / d waste water treated.

cost:		
R.O. operation		Lit. 94.000.000*
plant and membranes		Lit 85.000.000**
	total	Lit. 179.000.000
recover:		
Sodium Sulphate***		Lit. 262.000.000
government tax for discharge****		Lit. 161.000.000
	total	Lit. 423.000.000
net gain per year		Lit. 244.000.000

* Hypothesis: energy consumption: 130 Kw; 230 days a year; Lit 130/Kw; 24 hr continuous operation.

** Yearly depreciation allowance .

*** 3000 Kg/d at Lit 380/Kg .

**** Tax to discharge waste in municipal facilities, **presumably Lit 700/m3.**

CONCLUSIONS

Shortage of water, increasing cost of waste water treatment, more strict regulations imposed on waste water disposal combined with the economical and ecological advantages of waste recycle (water and auxiliary chemicals) indicate R.O. membrane processes as an effective way to treat textile effluents. The quality of water from R.O. processes to be reused in textile plants can be tailored to specific needs by using membranes with adequate transport characteristics. Water to be reused in textile plants can contain impurity levels higher than fresh water and than the standards set by government regulations to dispose wastes.

A zero-discharge system was put in operation and it seems to give satisfactory results. High fluxes and rejections allow a favorable economic balance. The concentrate is transformed to fertilizer to be used in agriculture. This completely allows the zero-discharge operation. The only loss is the water evaporated from the sludge. Sodium Sulphate, used in dyeing operations, was completely recycled. The pay-off time for this R.O. plant is estimated to be 3 years.

The new membranes used, assembled in spiral wound modules, are stable to the aggressive environment of textile effluents making the integration of R.O. processes in textile plants quite easy.

ACKNOWLEDGMENTS

The work was partially funded by the Commission of the European Communities, BRITE program, BRITE project P-1170-7-85. The typing assistance of Miss. A. Cerutti is gratefully acknowledged. The author whishes to thank Dr. L. Giro for his stimulating conversation during the preparation of this paper.

REFERENCES

1. E. Drioli, R. Napoli, in Processi di osmosi inversa nel trattamento delle acque di scarico dell'industria tintoria, IRSA Ed., Proc. II Convegno sul "Trattamento delle acque di scarico industriale", 45, 1978, p. 197.

2. "The Use of Water by the Textile Industry", Textile Res. Conf., Wira, Leeds, United Kingdom (1973).

3. BASF, Manual Cellutosic Fibres, Tapp & Toothill Ldt, London, U.K. (1985).

4. W.C. Tworeck, W.R. Ross, N.J. van Rensburg, Textile Industries Dyegest S.A., 11, 2 (1984).

5. T.T. Fulmer, America's Textiles, 14, 89 (1985).

6. C.A. Brandon, Textile Industries Dyegest S.A., 2, 2 (1985).

7. G.R. Groves, C.A. Buckley, R.H. Turnbull, Journal WPCF, 51, 27 (1979).

8. E. Drioli, R.M.A. Napoli, U. Sabatelli, F. Bellucci, University of Naples, Proc. of the "Primer Symposium Int. Sobre Residuos Industriales y Ambiente", Caracas, 1976.

9. S.M. Doshi, S.A. Tarabaakar, and U.K. Misra, Textile Engineering, 4, 15 (1978).

10. H.K. Lonsdale, in Membrane phenomena and processes, T.Z. Winnicki and A.M. Mika Eds., Wroclaw Technical University Press, 1986, p. 223.

11. P. Parrini, Desalination, 48, 67 (1983).

12. R.E. Kesting, in Materials Science of Synthetic Membranes, D.R. Lloyd Ed., ACS Symposium Series, 1985, p. 131.

13. T. Kawaguchi and H. Tamura, Journal of Applied Polymer Science, 29, 3359 (1984).

THE USE OF SULFURIC/PHOSPHORIC ACID TREATED PEAT FOR RADIOACTIVE WASTEWATER TREATMENT

Richard V. Bynum and James D. Navratil
Rockwell International
Rocky Flats Plant
P.O. Box 464
Golden, CO 80402

Patrick MacCarthy
Department of Chemistry and Geochemistry
Colorado School of Mines
Golden, CO 80401

Peat is a relatively inexpensive material which possesses a native cation exchange capacity. Efforts to utilize peat have been hampered by its low permeability to water and its tendency to severely leach in water at pH 6. These disadvantages have been significantly minimized by treating the peat with a combination of concentrated sulfuric and phosphoric acids, resulting in a particulate material which is permeable to water and resistant to leaching. The acid treatment also increases the cation exchange capacity of the peat. This paper describes preliminary results of both column and batch studies of the modified peat for use as an actinide adsorbent.

INTRODUCTION

The removal of actinides from aqueous solutions is a topic of considerable environmental concern, and one which is likely to attract greater attention in the future. In order to prevent the release of radionuclides into the environment, it is necessary to remove these species from contaminated water prior to its discharge. In dealing with process wastewaters from nuclear facilities, one must seek a process which is both technically efficient in removing the contaminants and also economically attractive for large scale application. Two of the processes currently used for removing actinides from contaminated waters are ion exchange on zeolites and adsorption on activated carbon.

The use of sulfuric acid and sulfuric/phosphoric acid treated peat as a cation exchanger has been described in a number of earlier applications (1). The advantages of the chemically modified peat compared to the raw peat are that it is particulate in nature and is permeable to water and suitable for column use. In addition, the treated peat possesses an enhanced cation exchange capacity, is resistant to leaching up to a pH of 9, and experiences minimal swelling or contraction as the pH and ionic strength of the solution are changed. The acid treated peat has previously been shown to remove heavy metal cations from solution under a wide range of conditions.

The purpose of this work is to investigate the potential of acid treated peat for removing plutonium and americium from process wastewaters at the Rocky Flats Plant. The sulfuric/phosphoric acid peat is compared to sulfuric acid treated peat, zeolite, and coconut carbon in these tests. Both batch and column experiments were performed as previously described (1). The primary advantage of ion exchangers prepared from peat is that they can be manufactured by a simple process from inexpensive starting materials.

RESULTS OF BATCH EXPERIMENTS

The uptake of plutonium from 10 ml. spiked buffer solutions (0.05-0.1 mg/l Pu) by 0.25g of either sulfuric acid or sulfuric/phosphoric acid treated peat is shown in Figure 1. Over the pH range of 7-9, the H_2SO_4/H_3PO_4 peat removed 97% Pu.

The uptake of plutonium decreases at pH 7. This is in contrast to results obtained with sulfuric acid only treated peat where the plutonium uptake did not decrease until pH 3. Additionally the plutonium uptake by the sulfuric/phosphoric treated peat was not found to decrease until pH 9 whereas the sulfuric acid treated peat's ability to bind plutonium was found to decrease at pH 7. This is not surprising in view of the relative basicity of the sulfate and phosphate anions.

The results of sulfuric acid treated peat, coconut charcoal and zeolite are shown in Figure 2. The results show that under optimal pH conditions the modified peat was more effective than the other two adsorbents in removing the actinides in batch experiments.

PRELIMINARY RESULTS OF COLUMN EXPERIMENTS

In previous column experiments (1), sulfuric acid treated peat outperformed zeolite and carbon as far as plutonium breakthrough capacity and elution behavior were concerned. The results of preliminary column studies with sulfuric/phosphoric acid treated peat indicate a much better capacity for plutonium than the sulfuric acid peat. These studies are continuing.

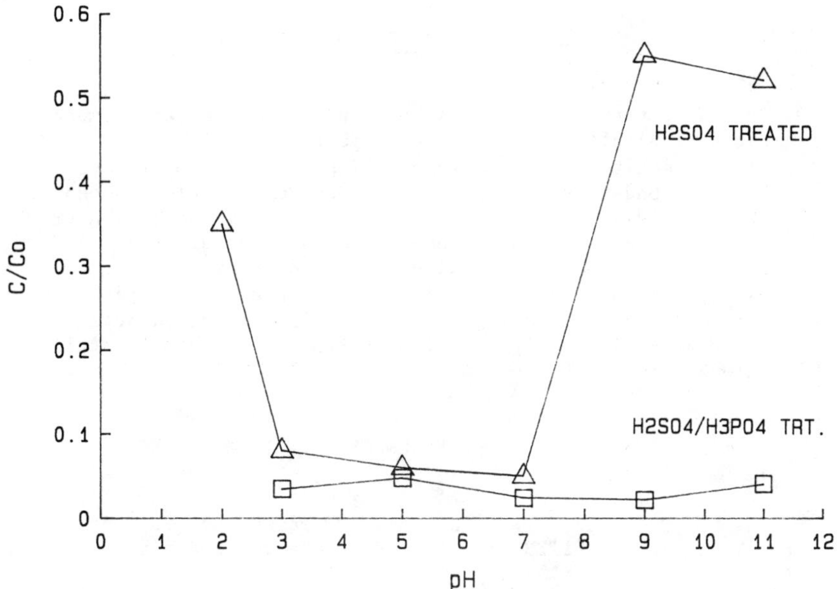

Figure 1. Sorption of plutonium onto peat as a function of plutonium solution pH. C/Co is the ratio of the plutonium concentration after contact with peat to the initial plutonium concentration.

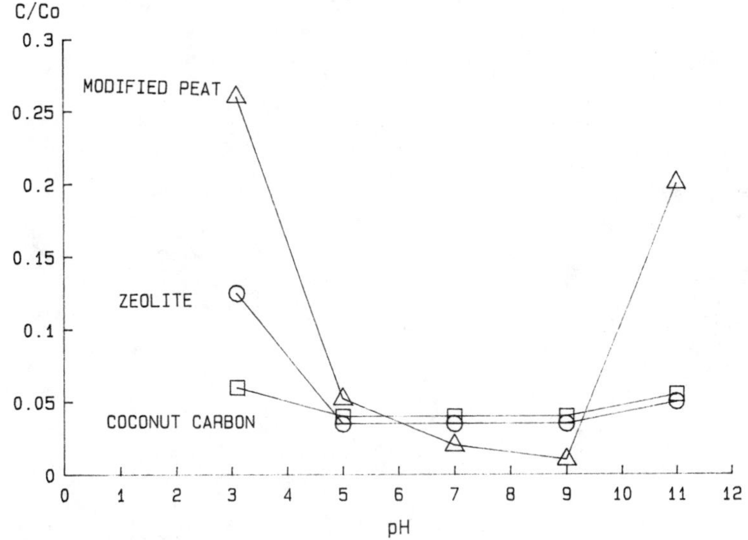

Figure 2. Sorption of plutonium onto peat, zeolite and coconut carbon as a function of plutonium solution pH. C/Co is the ratio of the plutonium concentration after contact with the sorbent to the initial plutonium concentration.

SUMMARY

Overall, the sulfuric/phosphoric acid treated peat compares favorably with sulfuric acid treated peat and with coconut carbon and zeolite, the results of which have been presented previously. Under optimal pH conditions the sulfuric/phosphoric acid treated peat was more effective in removing the plutonium in batch experiments than the peat treated with sulfuric acid alone and was also more effective than coconut carbon and zeolite. Additionally, no significant leaching of the peat was noticed at pH's of up to 11, which is a significant advantage over untreated peat and peat which had been treated by sulfuric acid alone.

REFERENCES

1. C. M. Smith, J. D. Navratil, P. MacCarthy, Solvent Extr. Ion Exch., 2, 1123 (1984), and references therein.

TREATMENT OF LOW-LEVEL LIQUID RADIOACTIVE WASTES BY ELECTRODIALYSIS

J. A. Del Debbio and R. I. Donovan, Westinghouse Idaho Nuclear Company, Inc., P. O. Box 4000, Idaho Falls, Idaho 83403, J. E. Lundstrom, Ionics Inc., Watertown, MA 02172

This paper presents the results of pilot plant studies on the use of electrodialysis (ED) for the removal of radioactive and chemical contaminants from acidic low-level radioactive wastes resulting from nuclear fuel reprocessing operations. Decontamination efficiencies are reported for strontium-90, cesium-137, iodine-129, ruthenium-106 and mercury. Data for contaminant adsorption on ED membranes and liquid waste volumes generated are also presented.

INTRODUCTION

The treatment of low-level liquid radioactive wastes for removal of residual radioactivity and hazardous chemicals prior to discharge is a primary concern in the nuclear industry. Large volumes of these wastes are generated in nuclear reactor operations, by the use of radioisotopes and during decontamination operations. At the Idaho Chemical Processing Plant, located at the Idaho National Engineering Laboratory, aqueous wastes from spent nuclear fuel reprocessing operations are collected and evaporated producing an acidic condensate containing trace radioactivity at the rate of about 30 M^3/day (8000 GPD). Evaporator bottoms are disposed of as high-level waste. The condensates after assuring that release limits for radionuclides are met, along with other service wastes are released to a percolation pond. The condensates are the main contributors to contaminated discharges. Although current releases are in compliance with Department of Energy regulations, future increases in processing rates and more stringent guidelines may require additional decontamination prior to release.

Condensates are essentially dilute nitric acid solutions containing a variety of trace amounts of radionuclides and chemical contaminants the most important of which are listed in Table I. Ion-exchange methods have proven capable of removing strontium, ruthenium, and cesium isotopes, but because of the need for frequent regeneration, are practical only at nitric acid concentrations below 0.01 Normal. The feasibility of using electrodialysis to treat up to 39 M^3/day (10,000 GPD) of condensates is currently being evaluated with a one-hundredth scale pilot plant built by Ionics, Inc. of Watertown, MA. Test results for the removal of the contaminants in Table I are discussed in this paper.

Table 1. Contaminant concentrations in evaporator condensates.

Contaminant	Half Life (yr)	Concentration mCi/L[a]	mg/L[a](ppm)
Strontium-90	30	4.4E-6	3.3E-8
Cesium-137	30	5.0E-6	5.7E-8
Iodine-129	1.6E7	8.4E-6	4.8E-2
Ruthenium-106	1	1.8E-6	5.2E-10
Mercury	Stable	--	1.4
Nitric Acid, Normality	Stable	0.001-0.3 0.1 Avg.	

[a] These values represent those for the condensate feeds reported here. Over a period of a year, these concentrations may vary by a factor ±100.

Electrodialysis is a process by which aqueous solutions are deionized by the electromigration of cations and anions through semipermeable membranes (1, 2). Figure 1 illustrates the basic concept and indicates the relationship between the current passed and the number of equivalents transferred. Cations and anions migrate with respect to electrical polarity through selectively permeable membranes, thus producing an ion- depleted product. The ion transfer rate (equivalents/sec) is directly proportional to the current passed times the current efficiency. The current efficiency is the ratio of the theoretical current needed to transfer a given number of equivalents to the actual current passed.

Deionization on a large scale is accomplished by an ED membrane stack schematically represented in Figure 2. The ED stack consists of alternating cation-permeable and anion-permeable membranes separated by spacers (not shown) and bounded by cathode and anode compartments. Cations migrate through cation-permeable membranes toward the negative pole while anions migrate through anion-permeable membranes toward the positive pole. The result is a deionized product and a concentrated brine. Electrode compartment rinse streams (catholyte and anolyte) serve to remove the gaseous products of electrode reactions. One cell pair, including spacers, and an electrode compartment are shown in Figure 3. Spacers maintain membrane shape, control flow distribution and provide mixing which helps reduce boundary layers through which ion migration is slow.

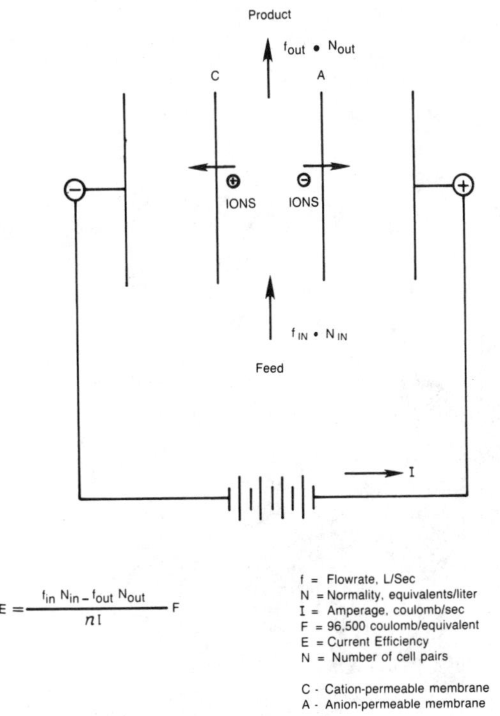

Figure 1. The relationship between current and ion transfer through a membrane.

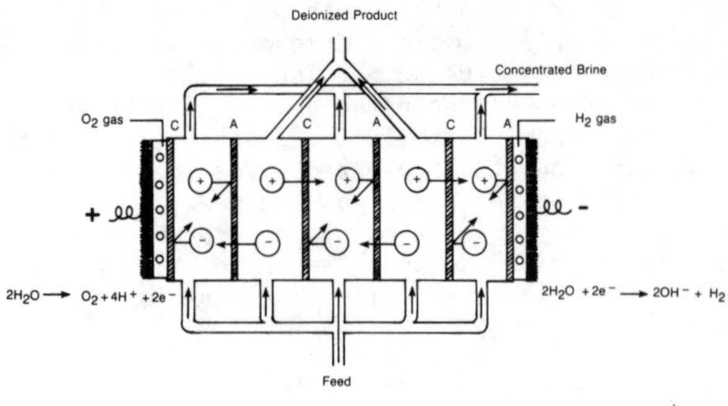

Figure 2. Schematic of an electrodialysis process.

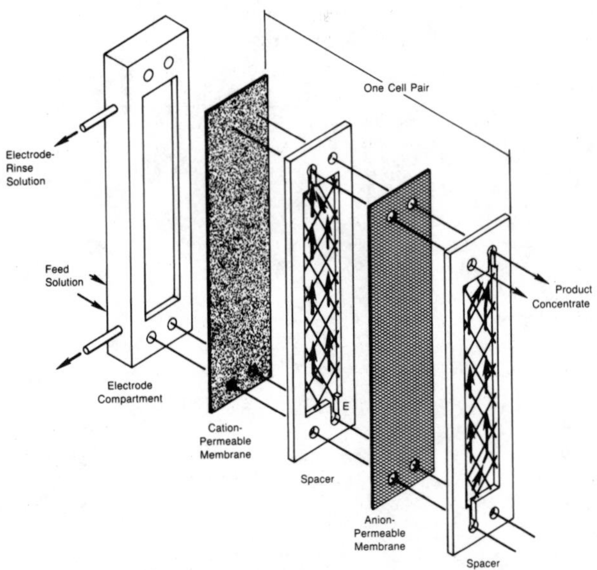

Figure 3. Main components of an electrodialysis membrane stack showing flow distribution by spacers.

The membranes used in ED systems are ion-exchange resins in sheet form. They are cross-linked polymers having anionic groups, such as sulfonate ($-SO_3^-$) making them cation-permeable, or quartenary groups ($-NR^{3+}$) making them anion-permeable. Important properties of membranes include mechanical and chemical stability, low electrical resistance, selective permeability, and water exclusion.

By far the widest application of ED is desalination of brackish water. Other applications include metals recovery from electroplating operations, desalting proteins and cheese whey and deacidification of fruit juices. Electrodialysis can generally process feeds in the pH range of 1 to 12. However, during the course of the tests reported here, feeds having pH values as low as 0.53 have been processed with no apparent detrimental affects.

EXPERIMENTAL

Process Description and Test Parameters

Waste treatment by the ED pilot plant is a two-step process involving decontamination and contaminant concentration as conceptually illustrated in Figure 4 for an acidic cesium nitrate feed solution. Feed enters the pretreatment membrane stack where partial acid neutralization by hydroxyl ions generated by the cathode reaction takes place. Also, nitrate ions migrate through an anion-permeable membrane into the recirculating catholyte of the concentration cell where they are balanced by hydrogen ions. Thus, catholyte acidity is maintained and precipitation of contaminant hydroxides is prevented. The feed then enters alternating compartments of the decontamination stack where contaminant cations and anions migrate through semipermeable membranes into a brine resulting in a decontaminated product. The brine is circulated through a concentration cell where cations and anions are separated by migration into a catholyte and anolyte. Anolyte acidity is maintained at about 1 Normal by water addition, thus causing overflow from the anolyte tank. The catholyte, which reaches a steady state acidity of about 0.1 Normal, overflows due to electroosmotic water transfer.

Shown in Figure 5 is a schematic of the ED Pilot Plant. There are two decontamination stacks in series, S-1 and S-2 consisting of 126 and 24 cell pairs, respectively, and two electrical stages each. The pretreatment stack, S-3[1], is coupled to the concentration cell, S-3. Both S-3 and S-3[1] contain neutral cell compartments bounded by anion-permeable membranes through which water from the water make-up tank is circulated. These compartments serve

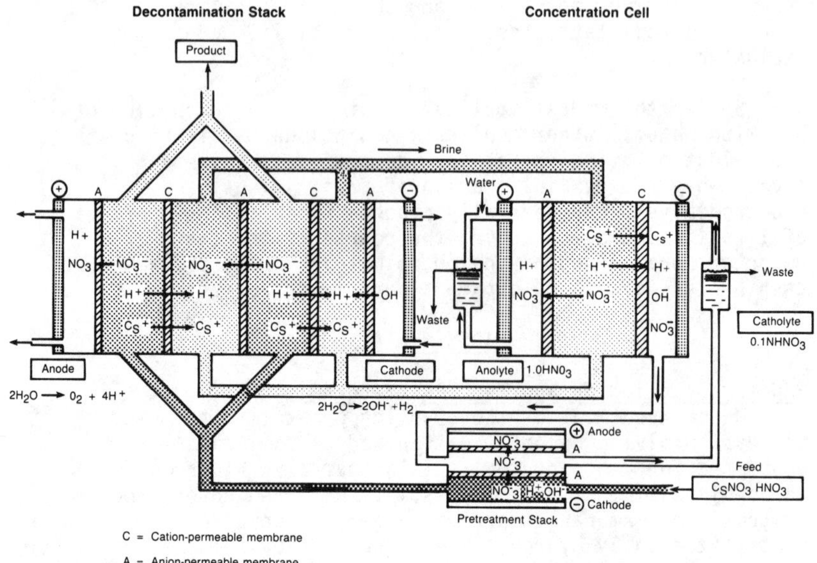

Figure 4. Waste decontamination and contaminant concentration in the electrodialysis pilot plant.

to limit back-migration of hydrogen ions from the anolyte, which can reduce current efficiency. The solution in this loop is called the neutral cell solution. Pictures of the equipment appear in the Appendix.

The primary feed to the pilot plant is reduced in acidity by entering it in the recirculating S-1 feed loop which consists of the pretreatment stack (S-3[1]) and the S-1 feed tank. The solution (secondary feed) from the S-1 feed tank, which is now reduced in acidity, joins the brine exiting the first electrical stage (E1) of S-2 and feeds stack S-1. The product from S-1 is further decontaminated in S-2. By appropriate stream splitting, the feeds to S-1 and S-2 also serve as electrode rinse streams and are recycled to the S-1 feed tank.

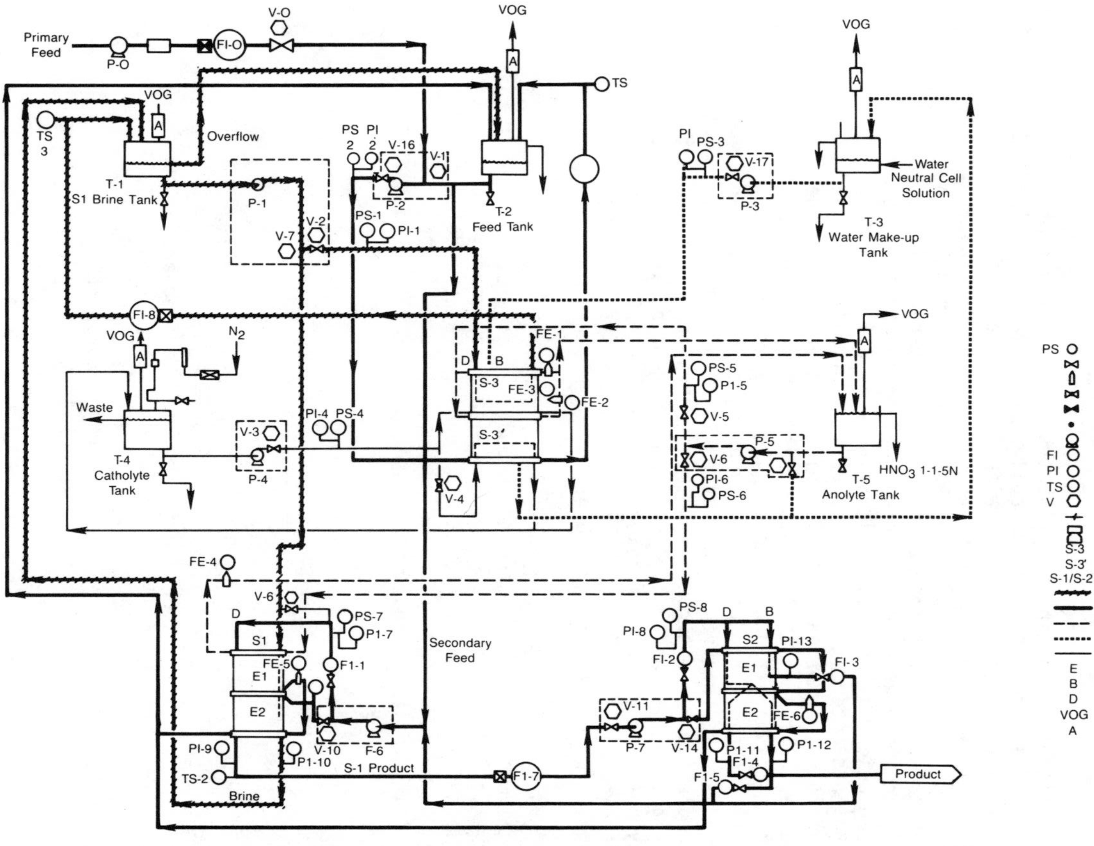

PS	Pressure Switch
	Pressure-Reducing Regulator Self Contained
	Office Plate in Quick Change Fitting
	Needle Valve, Regulating
	Drain Valve (Normally Closed)
•	Material Class Break (See Note)
	Pump
FI	Flow Indicator
PI	Pressure Indicator
TS	Temperature Switch
V	Valve Designation Numbers
	Poly Tubing
	Filter
S-3	Concentration Cell
S-3′	Pretreatment Stack
S-1/S-2	Decontamination Stacks
	S1 Brine
	S1/S2 Feed, Electrode Rinse
	Anolyte
	S3/S3 Brine (Neutral Cell Solution)
	Catholyte
E	Electrical Stage
B	Brine
D	Diluate
VOG	Ventilation Off-Gas
A	Solid Adsorbent

Figure 5. Schematic flowsheet for electrodialysis pilot plant.

Primary feed acidities ranged from 0.0065 to 0.27 Normal nitric acid with feed rates from 18 liters/hour for feeds up to 0.2 Normal nitric acid, and 12 liters/hour for those of higher acidity. Membrane stack pressures ranged from 69 to 310 K Pa (10-45 psi). Pressure switches automatically shut down all pumps in the event of pressure upsets. All stacks were operated at constant voltages, while currents varied with nitric acid concentration. Voltages and currents were as follows: S-1, 50v, 0.2-2.5 a; S-2, 20 v, 0.01-0.7 a; S-3/S-3^{1} 5 v, 40-70 a.

Test Procedures and Analytical Methods

The ED pilot plant was operated batchwise, processing 400 liters of condensate per batch. Acid concentrations of the brine, anolyte, neutral cell solution, and S-1 feed were continuously monitored with in-line conductivity instrumentation calibrated for nitric acid. Catholyte acidity was monitored by pH measurement and final acid determination performed by titration. Current efficiencies of each membrane stack were determined on the basis of how much acid was removed. Acid removal in S-1, S-2, and S-3^{1} was determined by pH measurement of samples taken before and after each stack, while acid removal by the concentration cell (S-3) was determined by titration of influent and effluent samples. Decontamination efficiencies were determined by analysis of the primary feed and of the product collected in a 400 liter tank. Cumulative percent contaminant distribution, among the brine, anolyte and catholyte from run to run was estimated by analysis of samples taken from the respective tanks (shown in Figure 5) and by measuring the total volume, including overflows, at the end of each run.

Strontium-90 was determined by chemical separation followed by beta counting techniques. Iodine-129 was chemically separated and counted with a low energy photon spectrometer. Cesium-137 and Ruthenium-106 were determined by gamma ray spectrometry using a germanium detector. Mercury analyses were done by atomic absorption spectrometry.

RESULTS AND DISCUSSION

Ten runs were conducted, during which approximately 3460 liters of condensates were processed in 245 hours. Average contaminant concentrations in the feed and product and decontamination efficiencies are listed in Table 2. Decontamination efficiencies were high except for mercury, which may be present in the form of mercuric chloride complexes. Chloride concentrations in the feed varied from

19 to 93 mg/L. Under these conditions, the predominant mercuric chloride complex may be $HgCl_2$, which, being neutral, does not undergo electromigration, but may be partially removed through adsorption. Also, in some cases $HgCl_2$ may have further chlorinated to $HgCl_3^-$ by means of a sufficient chloride concentration in the membranes. If this were the case, $HgCl_3^-$ would pass into the brine in the S-1 stack and then on to the anolyte in the concentration cell. Analyses have indicated that of the total amount of mercury processed, 25% entered the anolyte and 38% was adsorbed on the membranes (see Table 3).

The total amount of each contaminant processed in ten runs and its percent distribution are listed in Table 3. According to the data, a large percentage of all contaminants became adsorbed on the membranes and an appreciable percentage of Cs-137 and Hg migrated into the catholyte and anolyte, respectively. The data suggest that the primary removal mechanism for Sr-90, I-129, and Ru-106 may be adsorption or a combination of electromigration and adsorption with very slow movement through the membranes. Contaminant adsorption, however, can be confirmed only by membrane analysis after sufficient radionuclide deposition. Analysis of solid adsorbents placed in small tank ventilation lines (see Figure 5) indicated no significant volatilization of Ru-106 or I-129, although some volatilization from membrane stacks may have occurred. Filters placed in the primary feed line showed no significant radionuclide adsorption.

Figure 6 indicates the adsorption behavior of the contaminants. With the exception of Cs-137, an apparent steady-state adsorption was reached after Run 7. Additional tests are required to determine long-term contaminant adsorption and migration characteristics.

Table 2. Average contaminant concentrations and decontamination efficiencies.

Contaminant	Feed (mCi/L)	Deionized Product (mCi/L)	Decontamination Efficiency (%)
Strontium-90	4.4E-6	1.3E-7	>97
Cesium-137	5.0E-6	1.9E-7	96
Iodine-129	8.4E-6	9.9E-7	88
Ruthenium-106	1.8E-6	2.7E-7	85
Hg, mg/L	1.4	0.48	42-91 Avg. 65
Nitric Acid, Normality	0.15	2.1E-3	99

Table 3. Contaminant distribution in the ED pilot plant.

Contaminant	Distribution (%) Catholyte	Anolyte	Brine	Product	Adsorbed[a]
Sr-90	9.6	5.5	1.1	2.9	81
Cs-137	30.3	8.4	6.9	5.2	49
I-129	0.29	2.9	0.27	11	85
Ru-106	2.1	5.6	1.9	15	74
Hg,mg	0.86	25.3	0.49	35	38

a Adsorbed on the membranes.

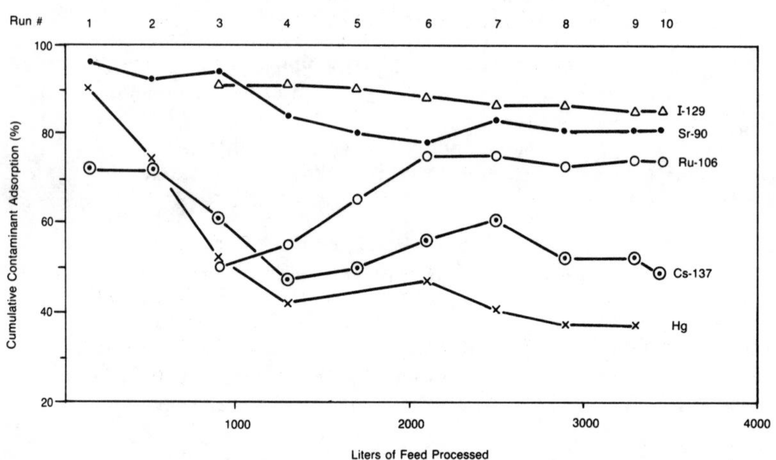

Figure 6. Contaminant adsorption on membranes vs volume of feed processed.

Total product radioactivity for the strontium, cesium, iodine, and ruthenium isotopes averaged 1.6 E-6 mCi/L with an average decontamination of 92%. Deacidification was 99%. Product pH ranged from 2.79 to 3.6 with an average of 3.3, thus enabling the efficient use of ion-exchange methods should further decontamination be required. Average current efficiencies of the membrane stacks for nitric acid removal were estimated to be as follows: S-1, 10%; S-2, 85%; S-3, 55%; S-3[1], 50%.

The volume of anolyte waste generated, expressed as a percentage of the total feed volume processed, ranged from 0.9% for the 0.0065 Normal nitric acid feed to 24% for the 0.27 Normal nitric acid feed, with an average of 12%. The average for catholyte waste was 1.7%. Radionuclide concentrations in these wastes ranged from 2.4 E-4 to 2.8 E-7 mCi/L low enough to be treated as low-level waste.

CONCLUSIONS

The Electrodialysis Pilot Plant has proven capable of removing greater than 90% of the Sr-90, Cs-137, and Ru-106 and nearly 90% of the I-129 from dilute acid wastes. High acid and nitrate removal will enable the use of ion-exchange methods if additional waste stream decontamination is required. Mercury removal was low (approximately 65%) and is probably dependent on the chemical form present. A study of the chemical nature of mercury in the condensate and its behavior in the ED system may shed light on approaches that can be taken to improve removal efficiencies.

Although contaminant adsorption on the membranes was considerable in the early runs, the adsorption appeared to reach steady-state levels. Thus, an ED plant processing 30 M^3/day of low-level wastes could operate for extended periods before significant contamination build-up occurred. For example, less than one curie of Ru-106, the principal gamma-emitter, would be deposited in one year, assuming average feed concentrations of 1 E-4 mCi/L and 70% adsorption. Techniques for removing contaminants from the membranes are being evaluated.

The membranes have been exposed to dilute nitric acid solutions for a period of about one year without any apparent detrimental effects. Membrane life is expected to be three years for membrane stacks and one year for the concentrator cell. The combined catholyte and anolyte waste volumes generated were low (approximately 14% of feed volume) and could be treated as low-level wastes.

ACKNOWLEDGEMENTS

The authors wish to thank S. H. Hinckley for conducting the many long pilot plant runs and P. W. Walden for competently performing membrane stack maintenance involving disassembly and assembly operations. Both individuals contributed many helpful suggestions. We also would like to thank the radiochemical analysis group, B. R. Hunter, supervisor, and the spectrochemical analysis group, D. A. Pavlica, supervisor, for performing sample analyses while coping with a heavy sample load.

REFERENCES

1. L. E. Applegate, "Membrane Separation Processes", Chemical Engineering, June 11, 1984, p 64.

2. H. Strathmann, "Electrodialysis and Its Application in the Chemical Process Industry", Separation and Purification Methods, Vol. 14, No. 1, pp 41-66 (1985).

CATECHOLAMIDE CHELATORS FOR ACTINIDE ENVIRONMENTAL AND HUMAN DECONTAMINATION

R. J. Bergeron
Department of Medicinal Chemistry
University of Florida
Gainesville, FL 32610

J. D. Navratil
Rockwell International
Rocky Flats Plant
P.O. Box 464
Golden, CO 80401

A brief overview is presented concerning a joint program between the University of Florida and Rocky Flats which focuses on the design, development and testing of catecholamide chelators for human and environmental actinide decontamination. The recent efforts of a study concerning the ligand's ability to sequester plutonium from blood serum is highlighted.

INTRODUCTION

A series of structurally novel compounds, referred to as H-octadentate catecholamides, has been synthesized (Figure 1). Each of the points of the H has a catechol moiety fixed to it. It is a system which is not difficult and costly to synthesize. Furthermore, the synthesis of the catecholamide is designed with the idea of eventually being able to fix the ligand to a matrix as well as to tailor the ligand's geometry to the metal. Clearly, a ligand which binds metal ions indiscriminately would be quickly rendered useless as other ions present would soon saturate the metal binding sites. In keeping with this problem we have designed the synthesis such that the dimensions of the chelator backbone can be varied until the actinide binding properties have been optimized.

To evaluate these ligands for environmental and human decontamination, we have studied plutonium removal from buffered aqueous solutions and blood plasma via the ligand sorbed onto Amberlite XAD-4. The results of these studies are described.

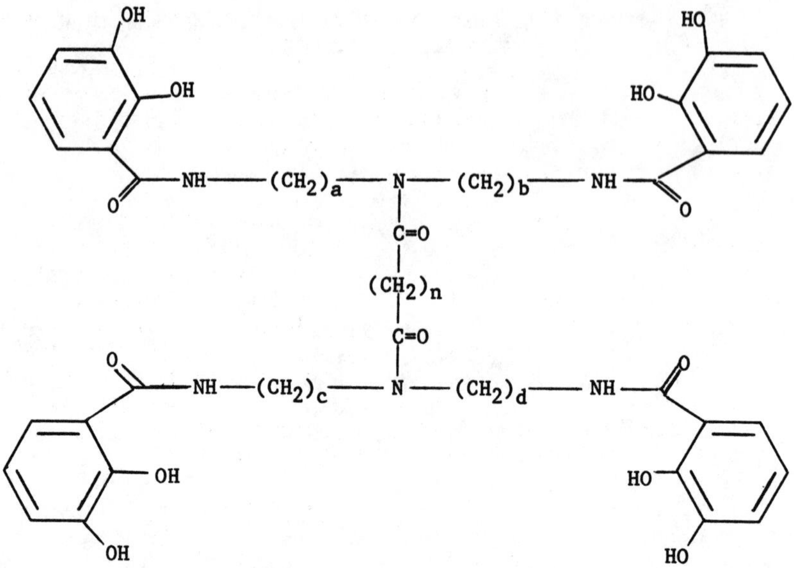

cpd #	n	a	b	c	d
1	2	3	3	3	3
2	2	4	4	4	4
3	2	3	4	3	4
4	3	3	3	3	3
5	3	4	4	4	4
6	3	3	4	3	4
7	2	3	3	3	4
8	2	3	3	4	4
9	2	3	4	4	4
10	3	3	3	3	4
11	3	3	3	4	4
12	3	3	4	4	4

Figure 1. Structures of Catecholamides.

PLUTONIUM REMOVAL FROM BUFFERED AQUEOUS SOLUTIONS

The ligands were dissolved in methanol and mixed with Rohm and Haas Co. Amberlite XAD-4, a nonionic macroreticular polystyrene-divinyl benzene material (1). The methanol was allowed to slowly evaporate, thus producing a resin sorbed with ligand. The plutonium solution was prepared by diluting a purified stock solution with pH 7.3 phosphate buffer and filtering through Whatman 42 paper.

Weighed amounts of XAD-4 containing the octacoordinate catecholamide were added to 10 ml of the buffered plutonium solution under nitrogen and equilibrated overnight on a rotary mixer. The solution was then separated from the resin by filtering through Whatman 42 paper followed by centrifugation at 15,000 rpm for 30 minutes to remove any fines. The plutonium concentration in the buffer was then determined using a Nuclear Measurements Corporation proportional alpha counter with an argon/methane atmosphere. The final concentrations were corrected for plutonium polymer absorption by the XAD-4.

Table 1 shows the Kd's of ligands at loadings of 10^{-6} and 10^{-7} moles ligand per 0.25g XAD-4. It is very difficult to arrive at a correlation between Kd's and ligand structure, but compounds 4,6,7, and 10 yield the higher Kd's at both ligand loadings.

Table 1. Results of plutonium removal from buffered aqueous solutions.

	Kd's x 10^6	
Compound Number	A	B
1	2.69	28.8
2	0.75	--
3	0.96	7.50
4	5.30	115
5	2.25	27.4
6	21.2	45.0
7	7.76	81.5
8	3.75	1.85
9	20.9	62.6
10	6.60	5.31
11	0.37	3.95
12	4.40	38.5

A - 1 X 10^{-6} Moles ligand per 0.25g XAD-4
A - 1 X 10^{-7} Moles ligand per 0.25g XAD-4
Kd = Concentration of plutonium on the resin per mole of ligand divided by
Concentration of plutonium remaining in solution per ml of solution
These numbers represent an average of at least 3 independent experiments.

PLUTONIUM REMOVAL FROM BLOOD PLASMA

The plasma experiments were carried out as follows. The ligands were dissolved in methanol and added to the washed XAD-4 resin. Blood plasma was spiked with a purified stock plutonium solution to a concentration of 4.2×10^{-7} moles per liter. To each of three vials was added 0.25 g of the loaded resin and 10 ml of the plutonium spiked plasma. Controls were each prepared with 0.25 g of unloaded XAD-4 and 10 ml of plutonium spiked plasma. The vials were placed on a tube rotator and allowed to equilibrate overnight. The solutions were then pipetted away from the resins and placed in Nalgene Oak Ridge Type centrifuge tubes. The samples were centrifuged at 15000 rpm for 30 minutes to remove any fines. Aliquots(2.00 ml) from each sample were affixed to each of two stainless steel planchets and analyzed with a Nuclear Measurements Corporation proportional counter.

The results are presented in Table 2. It is interesting to note that compounds 3,7, and 9 give the best results at a ligand loading of 10^{-6} moles ligand/0.25 g XAD-4 whereas at 10^{-3} moles ligand/0.25 g XAD-4, compounds 9 and 6 remove the most plutonium. These results are very encourging since almost 40% of the plutonium was removed with only one contact between the plutonium and ligand.

Table 2. Results of plutonium removal from blood plasma.

Compound Number	Plutonium Removal (%) A	Plutonium Removal (%) B
1	8	3
2	0	0
3	21	26
4	17	0
5	28	8
6	36	9
7	-	26
8	-	3
9	38	20
10	30	6
11	22	0
12	21	0

A - 1×10^{-3} Moles ligand per 0.25 g XAD-4
B - 1×10^{-6} Moles ligand per 0.25 g XAD-4
These numbers represent an average of at least 3 independent experiments.

SUMMARY

The catecholamide "H" shaped ligands show excellent promise as actinide chelators for environmental and human decontamination. Greater than 99% of the plutonium in an aqueous buffer solution was removed in a single pass. Additionally, 38% of the plutonium was removed from plutonium spiked blood plasma in a single contact. Future studies will evaluate the performance of the ligands for actinide removal from waste water and plutonium clearence from animals.

REFERENCES

1. R. J. Bergeron, S. J. Kline, J. D. Navratil, and C. M. Smith, Radiochim. Acta, 35, 47 (1984).

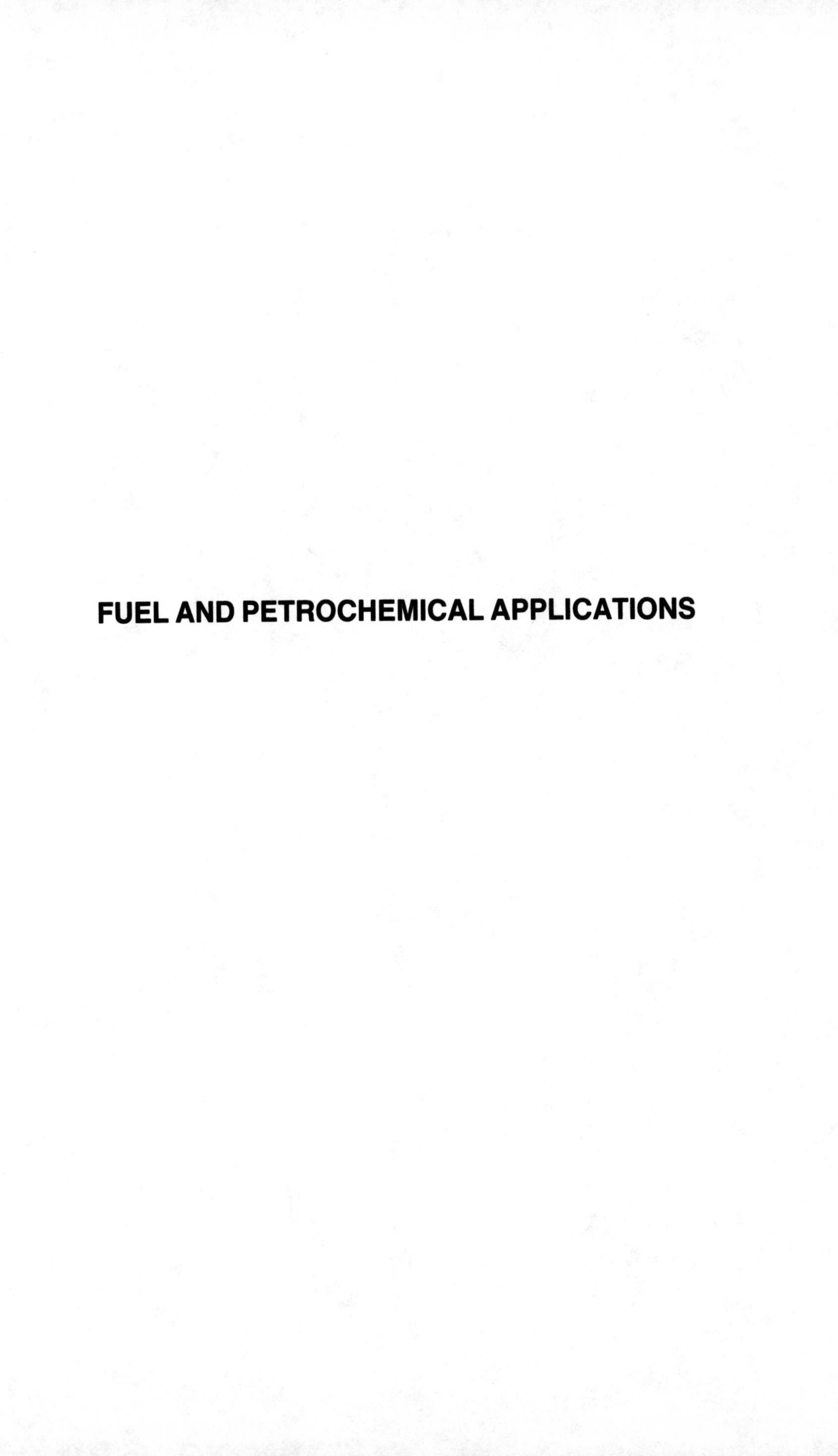

FUEL AND PETROCHEMICAL APPLICATIONS

APPLICATIONS OF SEPARATIONS TECHNOLOGY FOR THE CONTROL OF POLLUTANTS IN ADVANCED COAL COMBUSTION SYSTEMS

Howard Feibus
Peter J. Muchunas
Department of Energy
Germantown, MD. 20545

To help ensure that the use of coal, the Nation's most abundant energy resource, can continue and expand while minimizing the associated environmental impact, the U.S. Department of Energy conducts an extensive program to develop advanced coal combustion systems. The scope of the program includes pre-combustion cleanup, combustion processes, and post-combustion cleanup. In each of these areas, separation technology is used to remove pollutants from the desired product stream.

The pre-combustion cleanup program aims to develop techniques to separate 95 percent of the mineral matter and 90 percent of the sulfur from the parent coal. The successful development of these techniques will enable coal to be used in equipment which was designed to burn oil and natural gas. In addition, these techniques will be applicable to coals used in existing power plants, in order to reduce the emissions of sulfur.

The pre-combustion cleanup program also involves the development of techniques to separate particulate matter and sulfur compounds from the products of gasification. The applications are for combined cycle systems, and separation must be accomplished at high temperatures and pressures so that the gas will be sufficiently clean for processing in either a combustion turbine or a fuel cell, and can meet environmental emissions requirements.

A number of advanced combustion processes are under development, including fluidized bed combustors in which sulfur is controlled in situ. Both slagging and non-slagging combustors are also under development. These devices are being developed to be retrofitted to existing boilers so that coal can be burned to displace oil and natural gas. They are being designed to separate particulate matter and sulfur from the product gas stream before it enters the furnace.

The post-combustion cleanup program aims to develop devices to simultaneously separate compounds of sulfur and nitrogen, and devices for the separation of sulfur compounds. The first objective is directed at devices for new units and aims to achieve 90 percent separation. The primary application for the development of advanced sulfur dioxide separation devices is for retrofit to existing units for acid rain precursor control. The objective is to attain a minimum of 50 percent separation at a cost of less than $500 per ton of SO_2 removed.

The post-combustion cleanup program also involves the development of techniques to separate particulate matter and sulfur and alkali compounds from the products of combustion. The applications are for pressurized fluidized bed combustion and coal-fired turbines.

These applications of separation technology will be discussed, and the devices under development will be described.

INTRODUCTION

Coal is this Nation's most abundant energy resouce. It comprises more than 80 percent of our recoverable fossil energy resources, yet supplies less than 25 percent of the Country's primary energy; while our least abundant resource--oil--continues to be the Nation's dominant fuel source. Clearly, it is in the national interest to increase the use of coal to meet future energy needs. Unfortunately, the utilization of coal presents a number of complications. A primary concern is the control of sulfur dioxide (SO_2), nitrogen oxides (NO_x), and particulates; products generated when coal is combusted. Improved systems to control these emissions will require the development of advanced technologies that are capable of more efficent performance at less cost than the systems currently available. Accordingly, to help ensure that the use of coal can continue and expand, the U.S. Department of Energy (DOE) conducts a comprehensive research and development (R & D) program to develop coal cleaning, flue gas cleanup, and combustion systems. The scope of the program is very broad, covering every stage of the coal utilization chain: pre-combustion, combustion, and post-combustion. In each of these areas, programs have been initiated whereby separation technologies are used to remove pollutants from the desired product stream. The overall goal is to develop the technology through proof-of-concept by providing a sufficent data base that will permit the private sector to make a commercialization decision.

The purpose of this paper is to discuss the applications of the separation technologies for each program, and to describe the major devices under development. Programs associated with each area of the coal utilization chain that are underway are:

- o Pre-combustion
  - Coal Preparation
  - Gas Stream Cleanup for Integrated Gasifier Combined Cycle and Molten Carbonate Fuel Cell systems

- o Combustion - Fluidized Bed Combustion (Atmospheric and Pressurized)
  - Advanced Combustors

- o Post-combustion
  - Flue Gas Cleanup
  - Gas Stream Cleanup for Pressurized Fluidized Bed Combined Cycle and Direct Fired Turbine systems

COAL PREPARATION

The essence of coal preparation, or coal cleaning, is the separation of impurities from coal. The commercial practice of coal cleaning is currently limited to separation of ash and pyritic sulfur (inorganic), based on differences in the specific gravity of coal constituents and the differences in surface properties of the coal and mineral matter. Approximatly 35% of todays steam coal is physically cleaned with reductions in pyritic sulfur ranging from 20 to 50%.

The DOE research program is investigating cleanup processes that are classified as either physical or chemical. The objective of the physical cleaning research is the development and testing of improved physical cleaning systems that could increase pyritic sulfur and ash removal effectiveness from the present 50% of the feed coal, up to 95%, at a cost of $12 per ton, and BTU recoveries of greater than 80%. DOE is also pursuing the development of chemical cleaning technologies for the removal of organic sulfur. No chemical cleaning technology is now in commercial use. This program seeks to remove a total of 90+% sulfur and 95% ash.

The goal of the coal preparation program is to develop the technologies that could reduce the ash and sulfur contents of domestic coals to the point where the final product would be able to compete with oil and/or natural gas in both new or retrofit applications. The final fuel could be finely pulverized coal, coal slurry fuels. The end-use targets are not only utility and large industrial boilers designed for oil, but also small industrial and commercial boilers, residential furnaces, and even heat engines such as turbines and diesels. Additionally, the program has direct application toward the control of SO_2 from existing sources. In the near term (nearer, that is, than for chemical cleaning), advanced physical coal cleaning technologies may be able, economically, to reduce the emissions of suspected acid deposition precursors from existing coal-fired boilers in both the utility and industrial sectors, while retaining the same source of coal.

As a major part of its physical coal cleaning activity, DOE entered into a Project Agreement with the Electric Power Research Institute (EPRI) in September 1984 to develop and test advanced physical fine coal cleaning devices at a proof-of-concept scale of nominally one ton of coal per hour. The devices to be evaluated are to be selected on the basis of their ability to significantly advance the state of the art of sulfur and ash removal with favorable economics. Tests are to be conducted at the Coal Cleaning Test Facility (CCTF) operated by EPRI at Homer City, Pennsylvania.

In accordance with this agreement, DOE issued a Request For Proposal (RFP) in February 1985. The purpose was to solicit proposals for test and evaluation of contractor-developed advanced physical fine coal cleaning processes capable of significantly reducing the pyritic sulfur content of any given bituminous coal (28 mesh x 0 or finer), as compared to reduction achievable by state-of-the-art coal cleaning processes. The resultant coal product is to contain no greater than 2% ash and less than 30% moisture, and to operate at Btu recoveries of at least 80%. Two contract awards were made under this RFP. One was to Bechtel National, Inc., for the testing of a microbubble process, and the other to Advanced Energy Dynamics (AED) to test an electrostatic coal cleaning technology. They will be tested in succession at the CCTF, the first being the AED technology scheduled for installation and test in August 1986. A brief description of these two systems follows:

Electrostatic Separation

The AED Ultra Fine Coal (UFC) Cleaning System operates on the principle of fracto-charging. It is based on the phenomenon that the fresh surfaces created when any solid material is broken emit electric charge. When a mixture of two types of particles is introduced into a system in which at least some of the particles are broken, a differential charge is created. One component is charged positively, and the other negatively. Under the right conditions, the electric charge can be extremely high. The result is an entirely new method for separating the components of the mixture. In coal, these components are the organics, which are combustible (the coal itself), and the inorganics, which are noncombustible (iron pyrite containing sulfur, and ash-forming minerals). The organics are charged positively, and the inorganics are charged negatively. The system operation is rather simple. Pulverized coal is fed into a charger where the particles are accelerated to a high velocity and projected against or along surfaces that cause breakage of the particles. In the process, pyrite and mineral matter collect negative charges, while coal collects a positive charge. The charged particles then exit into a collector, where an electric field is maintained between slowly rotating electrode discs. Alternate discs are maintained at positive and negative potential. The coal collects on the negatively charged discs, while the pyrite and mineral matter collect on the positive discs. The discs are then separately scraped to extract products and rejects. The resulting product has significantly lower sulfur and ash than the raw coal feed.

The UFC cleaning system was originally designed to clean minus 400-mesh coal; however, it is also capable of cleaning coal up to 140-mesh in size. The system takes advantage of the fact that pulverizing coal to a fine size liberates significantly more pyrite and ash, making these materials more accessible for removal. Tests at bench scale on a variety of coals have demonstrated that the UFC system has the ability to achieve total sulfur reductions of from 50-60%; ash reductions of more than 80%; and BTU recoveries in excess of 90%.

Microbubble Flotation

The device to be built and tested by Bechtel is a microbubble flotation cell for ultrafine coal recently developed by Bergbau Forschung of Germany, along with proprietary additives developed by Energy International, Inc. Microbubble flotation is a further development of froth flotation. In conventional flotation, very fine particles are often not recovered because of what has been termed "low collision efficency." It has been proposed that the low efficency with small particles is due to the inability of the small particles to cross the hydrodynamic barriers surrounding the rising large bubbles. Research has demontrated that in addition to coal mineral characteristics and the reagents used, air bubble size plays a vital role in flotation. To be effective then, ultrafine coal flotation requires both extremely small bubbles and the means to impart adequate forces at the bubble-solid interface to permit coal/bubble attachment.

The Bergbau Forschung concept greatly increases the separation efficency of fine particles. Its cell consists of a conical vessel into which a conditioned slurry is introduced through several inlets. The chemicals used to condition the slurry are proprietary reagents used to create hydrophilic pyrite surfaces. The slurry is then distributed to multiple aerators where it is mixed with air and water. The aerator, placed outside the separation vessel, is used to generate microbubbles and provide the solid/bubble contact. The bubble loading takes place in milliseconds, and with adequate force to ensure maximum bubble loading with even the finest particles. With the introduction of the aerated slurry into the separator, the air bubbles rise to the surface without being disturbed by turbulence within the vessel.

In addition to these two advanced systems, DOE is also supporting research on other physical cleaning technologies that are capable of cleaning fine (28 mesh to 325 mesh) and ultrafine sizes (minus 325 mesh). The status of these efforts is summarized below:

True Heavy Liquid Cyclone

The use of "true heavy liquids" such as Freon (DuPont's trade name for the fluorocarbon refrigerant) as a separating medium has been shown to yield close to the theoretical limit of separations of coal from ash and mineral constituents at very fine sizes in the laboratory. The true heavy liquid is used to serve as the fluid in which the particles are separated by gravity. Currently, the parameters that affect the performance of the cyclone are being studied. This technology has been tested in the ton-per-hour size range, and conceivably could be introduced into the commercial marketplace within the next few years.

Froth Flotation

Froth flotation utilizes differences in surface characteristics to separate coal from its impurities. The froth flotation technique treats finely-ground coal with an oil-based substance that adheres to the coal. The mixture is fed into a "froth flotation cell" where air bubbles, created by a mechanical agitation device, attach themselves to the coal and rise to the surface. The impurities remain in the tank. Froth flotation is the only technology used routinely today to clean coal 28 mesh (595 microns) and finer: Still, the separation efficiency is usually inferior to that of float/sink testing, and pyritic sulfur removal tends to be poor.

Selective Coalescence

Given the presence of an appropriate medium, and aided by various additives, small coal particles can be caused to selectively coalesce, i.e., agglomerate, into larger particles which can then be separated from the undesirable impurities that do not coalesce. DOE is researching both the basic physical mechanisms that are involved in this phenomenon, and novel non-aqueous media, such as liquid carbon dioxide, that have the potential of yielding exceptionally high energy recoveries (greater than 95%) with better than 90% ash and pyritic sulfur removal. None of these DOE-sponsored efforts have progressed beyond the laboratory scale to date.

As stated above, chemical coal cleaning is required to remove the organic form of sulfur from the coal. One chemical process being pursued by DOE has demonstrated the ability to remove over 90% of the total sulfur and 95% of the ash from selected coals. The TRW "Gravimelt" process exposes the coal to molten sodium hydroxide (the alkali), or to a mixture of sodium hydroxide and potassium hydroxide for durations on the order of 3 hours at temperatures in excess of 700°F. The coal is then rinsed with water, followed by a weak acid solution. A recently-completed economic assessment estimates that coal could be cleaned to these levels at a cost range of about $40-$50 per ton.

Another alkali displacement process, developed by General Electric and now pursued for DOE by TRW, uses microwave energy. The process begins with a wet mixture of fine coal and sodium hydroxide that is passed through a microwave heating chamber. Normally, the exposure is less than one minute, although one or two additional passes may be necessary for some coals. The coal is then rinsed in water followed by a weak acid solution, just as in the previous process. Laboratory tests have achieved ash removal of over 90%, but recent results on sulfur removal have been less than 90%. These laboratory-scale concepts must be refined further in order to provide the engineering data needed for scaleup, and to assess the required technological adjuncts such as alkali regeneration.

The Gravimelt technology is now being tested in a modular-unit circuit at a scale of 20 pounds of coal per hour. Additionally, caustic regeneration is considered as a critical problem and is being pursued separately. Similarly, we are proceeding toward a 20-pound-per-hour modular circuit to test the microwave process; this program should be completed by the end of the 1986 fiscal year.

DOE is also supporting research in the microbial degradation of organic sulfur in coal. In this process, microorganisms are grown in fermenters and then fed into a coal reactor containing pulverized coal in a slurry at ambient temperatures. After a given retention time, the product is dewatered producing a clean coal. In the latter part of 1985, DOE awarded two contracts to Atlantic Research Corporation (ARC) for microbial desulfurization of coal. One of the contracts was to demonstrate 90% organic sulfur removal using its unique microbe CB1™; the other was to optimize ARC's Microbially Augmented Ash and Pyrite Physical Separation Process (MAAPPS) to provide clean coal of less than 1% ash and less than 0.5% pyritic sulfur. The MAAPPS process uses a microbe, Thiobacillus ferrooxidans, to alter surface properties of fine ash and pyrite to make them more hydrophilic. The altered surface characteristics make it easier to depress the pyrite and ash in a column froth flotation system, which permits good Btu recovery with high removal of pyrite and ash. In addition, a contract was awarded to IGT to isolate and test microbes capable of removing organic sulfur from coal.

GAS STREAM CLEANUP

The Gas Stream Cleanup program is a cross-cutting activity for the development of technologies to control and/or remove detrimental fossil fuel gas stream contaminants at high temperatures and pressures, to protect downstream power conversion equipment, and to meet or exceed environmental emission standards. Research and development is being carried out in two stages of the fuel chain: pre-combustion and post-combustion. Under precombustion, processes are being developed to remove contaminants from fuel gas streams produced by gasifier systems prior to use in gas turbines and molten carbonate fuel cells; under post-combustion, in direct coal-fired turbines and from gas streams from pressurized fluidized bed combustors prior to use in gas turbines.

The contaminants of concern include particulates, sulfur, halogens, alkali compounds, trace metals, and hydrocarbons.

Turbine Cleanup - PFBC

The major application of a PFBC (a description of which follows in this paper) is in a combined cycle system for electrical generation.

When turbines are operated in PFB combustor gas streams, they are subject to corrosion, erosion, and deposition. Corrosion in PFB turbines is primarily caused by the presence of alkali sulfate compounds in the liquid state. Erosion results from the impact of high velocity particulate material. Deposition is caused by a buildup of fine particles on the turbine surfaces. Tests have demonstrated that the simultaneous occurrence of erosion and corrosion exhibits a synergistic effect; that is, the combined rate is greater than the sum of the two processes occurring independently. Since the PFB gas turbine operates on sensible heat from the gas stream, it is essential that cleanup processes be accomplished at or near bed temperature (1550°F to 1750°F).

To eliminate the effects of erosion/corrosion, cleanup research is directed toward control of particulates and alkali compounds. Particulates are the major contaminant of environmental concern for PFBC power generating systems, since the PFBC has demonstrated the capability to meet New Source Performance Standards (NSPS) (Table 1) with respect to NO_x and SO_x emissions. Some of the particulate removal systems (positive type filters, etc.) being developed are anticipated to simultaneously meet both the turbine cleanliness requirements and the NSPS.

Table 1. Environmental emissions new source performance standards (lb/MBtu)

Pollutant	Existing Utility NSPS(1)	Existing Industrial NSPS(1)	Proposed Industrial NSPS(2)
NO_x	.6	.7	.7
SO_2	Reduction of 70% 90% as mined with max. of 1.2	1.2	Proposal Pending
Particulate	0.3	.1	.05 (w/o FGD)

(1): 250 MBtu/hr and greater
(2): 100 MBtu/hr and greater

Gas streams from a PFBC, if untreated, would greatly exceed the contaminant levels considered acceptable for each design, in terms of both alkali and particulates. Therefore, the turbine cleanliness goals have been established at .01 gr/SCF for particulates greater than 5 microns, .03 lbs/million Btu of heat input for particulates, and .02 PPM of alkali.

There are a number of cleanup devices that have been investigated, and the three most promising are now being tested at the subpilot scale. These are:

- o Granular Bed Filter - High temperature capable granules (sintered bauxite for example) in a moving bed system are used to collect particles from a dirty gas stream. The dirty granules are transported to a separate cleaning area where particulates are collected and the clean granules are returned for further collection of particulates. Both inertial impact and surface adhesion of particles make the process work.
- o Ceramic Crossflow Filter - Ceramic membranes with controlled porosity are assembled into a compact design package. Multiple units are installed in a pressure vessel. Particulates are collected on the membrane surfaces and are periodically removed from the surface and collected for disposal.
- o Electrostatic Precipitators - Impressed high voltage electrical potentials are developed on multiple electrodes within a vessel where particulate loaded gases are channeled. The electric potential drives the entrained particles to collector electrodes where they adhere. Periodically, the collectors are shaken and the collected particles are removed for disposal. Removal efficiencies of greater than 99% have been demonstrated on PFBC entrained ash.

With respect to alkali, a number of sorbents, such as activated bauxite and emathelite, have been identified at bench scale as having the sorbent capacity to be feasible for alkali removal applications.

Turbine - Gasification

Gasification conversion is accomplished by introducing air or oxygen and steam into a reactor vessel containing coal under controlled conditions of temperature, pressure, and flow. Several generic gasifier reactor types are available to produce coal gas. These include fixed bed, fluidized-bed, amd entrained flow gasifiers. The name refers to the method of movement of the coal within the gasifier reactor vessel. An important application for gasification is in an integrated gasifier combined cycle for electric power production.

It is anticipated that the technology concepts for the development of alkali and particulate removal from a high temperature pressurized gasifier gas stream are to a large extent similar to that of the PFBC cleanup effort. However, the gas stream contains two contaminants not present in the PFBC gas stream - ammonia and hydrogen sulfide (H_2S). Specific goals for particulates and alkalies after dilution with combustion air are the same as for the PFBC. Ammonia and sulfur removal must be sufficent to meet NO_x and SO_2 environmental standards.

At present, there are no known high temperature (above 1000^oF) control methods for ammonia. Process control methods are being studied to determine their potential for controlling ammonia production.

For hydrogen sulfide, a zinc ferrite sorbent has performed quite satisfactorily, to date, and appears promising for attaining the H_2S removal goals. In a zinc ferrite device, a pressure vessel is filled with extrudates or balls of zinc ferrite material. The sulfur laden gas stream is percolated through the beds of zinc ferrite where H_2S is both absorbed and chemically bound to the granules. The zinc ferrite granules can be regenerated by reverse flow of an oxident and steam.

Direct-Coal-Fired Turbine

Firing gas turbines with coal in the form of coal water slurry or micronized coal has the potential to displace significant amounts of oil and gas in the utility and industrial sectors. One of the ways being investigated is to use an off machine type combustor. This will, however, subject the turbine to particulates, sulfur oxides, nitrogen oxides, and alkali trace metal compounds which, unless controlled, will erode/corrode the turbine and pollute the environment when emitted downstream.

Specific cleanup goals are to meet NSPS for sulfur, particulates and NO_x at up to 2300^{0}F and 30 atmospheres. Levels for particulates and alkalies to protect the turbine are yet to be determined; tentative levels are based on fuels derived from petroleum and are .02 PPM for alkali and .01 gr/SCF for particulates greater than 5 microns.

Control of particulates may be achieved through use of advanced combustors such as slagging combustors (addressed later in this paper). These show the potential to control the evolution of particles from the combustor by forming agglomerates of large ash particles which can be removed from the combustor. The primary emphasis on R & D efforts is on sulfur control prior to the turbine. Sulfur dioxide control currently concentrates on the use of sorbents, such as finely divided hydrated lime, that is likely to be within or closely coupled to the combustor. The effectiveness of this approach is limited to below 2000^{0}F. While nitrogen oxides also present a potential problem, it is believed that low NO_x combustors can be used, and thus NO_x will not have to be controlled beyond those achieved in the combustion process.

Control of alkali metal compounds is also to be addressed. Due to the high combustion temperature, the potential exists for elevated levels of vapor phase alkali metal compounds in the post-combustion gases.

The most significant characteristic of the combustion gases in the direct coal-fired turbine system is the temperature at which hot gas cleanup is to be accomplished. Temperatures on the order of 1,850^{o}F to 2,300^{o}F will be encountered just prior to the turbine. This represents a considerable increase above the feed gas temperatures encountered in PFBC/turbine and gasification/ turbine systems. Contaminant control techniques for these other turbine systems are thought to be applicable to the direct coal-fired turbine system, but numerous technical uncertainties must be addressed for the higher temperature situation. Examples of uncertainties include durability of materials and material selection, performance of control devices, and specific operating regime.

Gasification for Molten Carbonate Fuel Cells

The molten carbonate fuel cell is an electrochemical energy conversion device defined as electro-chemical cells that can continuously transform the chemical energy of a fuel and oxidant to electrical energy by isothermal process involving an essentially invariant electrode-electrolyte system. The fuel supply being considered for Fossil Energy's coal-based molten carbonate fuel cell program is that derived from the low-Btu gasifier. The operational performance of the molten carbonate fuel cell is greatly diminished when operating on contaminant laden low-Btu coal derived fuel gas, and the problems differ from turbine systems. The sulfur and chlorine tolerances are extemely low and are projected to be less than 1 PPM. Sulfur contributes to the major fuel cell operational problems, electrolyte creepage, and corrosion, which results in failure of the wet seal and leakage of reactants from the cell. Tests with small-scale cells on bottled gas mixtures have shown the need to reduce fuel gas H_2S content to 1 PPM or below.

Commercial low temperature (wet) methods could possibly remove particles, sulfur, chlorine, and trace metals to acceptable levels for MCFC operation; however, the conventional low temperature systems are somewhat more complex and more expensive than the improved high temperature, high pressure systems under development. In addition, these systems, to date, have not been utilized extensively on commercial coal gasifiers. Studies conducted, to date, have provided some insight as to the fuel cell tolerance levels for the various contaminants. These contaminant levels are being used as goals from which to base the removal specifications for the various control concepts. For H_2S, the zinc ferrite sorbent, based on lab experimental results, has been selected for conducting on-line gasifier H_2S removal tests.

FLUIDIZED BED COMBUSTION

A fluidized bed combustor is a combustion chamber for converting the chemical energy of fuel, primarily coal or waste products, into thermal energy for process heat, steam, or electricity.

In fluidized-bed combustion, crushed coal is mixed into a bed of inert ash and limestone or dolomite, and the individual particles are suspended by an upward flow of air, so that the entire mass assumes the characteristics of a fluid while burning takes place. Depending on the pressure regime selected, a system may be an atmospheric fluidized-bed (AFB) or a pressurized fluidized bed (PFB). For the purposes of this paper, only the PFBC will be discussed. A big advantage of the fluidized bed combustion process is its ability to provide in-situ emissions control. The operating temperatures of 1500-1750° F are well below the thermal NO_x formation point. By including a suitable sorbent as part of the bed material, such as limestone or dolomite, the SO_2 released during combustion can be adsorbed, eliminating a downstream scrubber except for particulate collection. Two reactions take place to accomplish SO_2 removal: limestone is calcined to produce calcium oxide and then combined with SO_2 to form calcium sulfate. Cyclones, electrostatic precipitators, or other cleanup devices previously discussed under Gas Stream Cleanup can be used to control particulates from a PFBC.

PFB

The goal of the PFB program is to develop a technology data base by 1990 from which industries and utilities could develop and operate coal-fired PFB systems capable of economically generating heat, steam, and power from coals of all ranks and sulfur content in an environmentally acceptable manner. The major application of PFB is in a combined cycle configuration for the generation of electric power. The potential utility market consists of replacement of retiring coal-fired steam units with new plants, and repowering of oil and gas-fired steam units. During the past 10 years, R&D has progressed to the point where sufficent data is available to design and construct a PFB demonstration plant capable of achieving overall efficiencies of 40%. As a result, although DOE is conducting definitive R&D activities formulated to assure that the technology continues to advance to commercial readiness, program focus has shifted to the development of a second generation advanced cycle. The objectives of these advanced concepts are to increase efficency from 40 to 45% and to reduce the cost of electricity 20%. A contract is expected to be awarded this fiscal year, under a multi-phase cost-shared arrangement, for an advanced system that offers substantial improvements in performance reliabilty and operability.

ADVANCED COMBUSTORS

A coal combustor is a device mounted on a boiler or heater in which coal is burned to produce usable heat by converting the chemical energy of the coal to thermal energy.

Combustors in varying sizes and configurations have been used by the industrial and utility sectors for years. Environmental constraints imposed by the NSPS, however, have prohibited the full realization of their performance potential. The high operating temperatures necessary for substantial improvements in thermal efficiency have invariably resulted in unacceptable levels of NO_x, while their use with high sulfur coal has produced unacceptable levels of SO_x.

An advanced combustor is a device intended to extend or expand the state of the art in such a way as to encourage and promote the conversion of oil and gas-fired facilities to coal use through improvements in the economics of manufacturing, installing, operating, and maintaining these devices and through environmental improvements.

DOE's Advanced Combustor Program, as currently structured, is directed toward the development of a combustor that will control or remove objectionable sulfur and particulate matter from coal-derived fuel, will maintain low levels of NO_x before it is injected into retrofitted oil or gas designed boilers or heaters, and will be applicable and useful in the control of acid rain.

The goal of the program is the establishment of a technology base that will permit private industry to commercialize coal burning equipment in such a way as to replace the use of oil and natural gas in all segments of use, from large utility boilers down to private residential users.

The separate parts of the program are made up of the projects briefly described below:

- Near Term Retrofit Concepts
  - TRW, Inc. - This low emission system would operate in two stages. In the first stage, coal would be partially burned, and molten ash and sulfur removed. The existing firebox of the boiler would then serve as the second stage where combustion would be completed.
  - AVCO Everett Research Laboratories, Inc. - This concept involves a compact slagging combustor mounted on the outside of a boiler designed for either oil or gas firing, and would remove coal ash before the combustion gases enter the boiler where heat exchange takes place. The design is intended to be more compact than the TRW concept, and more efficient in ash removal than devices now available.

- Conceptual Evaluation
  - Science Applications International, Inc. - A wet oxidation process that would eliminate the release of nitrogen and sulfur oxides and dry particulate material to the environment.
  - Energy and Environmental Research Corporation - A Vortex Containment Combustor which would selectively remove ash particles in a combustor much smaller than conventional designs.
  - Solar Turbines, Inc. - An impact separator which would remove particulates from the gas stream before it enters the boiler.
  - Management and Technical Consultants, Inc. - A Pulsed Combustor concept that would burn Coal Water Mixtures in boilers designed for oil firing at higher than current efficiencies.
  - Foster Wheeler Development Corporation - Explosive Communition of coal to enhance combustion. Coal mixed with water is raised to high pressure. This pressure is then dropped sharply, allowing the water in the coal to expand, shattering the coal particles and improving combustion.
  - Avco Everett Research Laboratories, Inc - A novel approach to pulsed combustion of coal will be studied.

- New Concepts

  The program is currently being expanded by a competitive solicitation, which will result in additional contracts directed toward the development of combustors for light industrial, commercial, and residential applications.

FLUE GAS CLEANUP

Commercially available flue gas cleanup systems (lime/limestone scrubbers) typically operate by passing the flue gases through the mist or slurry of lime and/or limestone. The sulfur dioxide in the gases reacts with these alkalies to produce sulfite or sulfate products. The nitrogen oxides in the gases are only negligibly reduced.

The DOE flue gas cleanup program can be roughly characterized into two major subdivisions: simultaneous SO_x/NO_x control, and SO_2 control. The first area aims at development of processes for application to new construction.

Objectives are to attain simultaneous capture of both NO_x and SO_2 at the 90% level, using processes characterized by production of either a saleable by-product or a dry, non-hazardous waste. The cost goal is a minimum reduction of 20% as compared to a limestone scrubber for SO_2 control coupled with a selective catalytic reduction process using ammonia for NO_x control. Avoidance of sludge producing processes is considered to be important since the use of sludge ponds in typical limestone scrubber processes requires considerable space (one acre, to a depth of one foot per megawatt per year) and the potential exists for ground water contamination as a result of pond leachate. In effect, the end result in such a case may prove to be the conversion of an air pollution problem into a water pollution problem. The second area is aimed at providing a low cost technological option for moderate reductions of SO_2 emissions from existing powerplants such as may be required under proposed acid rain control legislation currently under consideration by the Congress. Objectives here are to attain 50-70% SO_2 capture levels at a maximum cost (levelized basis) of $500 per ton of SO_2 removed (1980 dollars), with waste/by-product characteristics similar to those described for the simultaneous NO_x/SO_x control project element.

Simultaneous SO_x/NO_x Control

The primary market for this technology will be for new and replacement plants in the utility and industrial sectors. The ongoing simultaneous SO_x/NO_x control program element includes the following project areas:

- o Electron beam processes
- o Dry regenerable processes
- o Spray dryer processes
- o Advanced separation technologies
- o Applied research

Electron Beam Processes

Irradiation of flue gas with electron beams produces, under the proper temperature and humidity conditions, an abundant supply of free radicals, ions, monoatomic species and secondary electrons which promote the oxidation of SO_2 and NO to sulfuric and nitric acids, respectively. Two separate processes based on this phenomenon are being developed. The first uses a lime spray dryer upstream of the electron beam chamber. The second uses ammonia injection to react with the acids and form ammonium salts.

Electron Beam Spray Dryer.

In this process, the flue gas is contacted in a spray dryer with a lime slurry. This step provides the desired temperature and moisture conditioning of the gas, along with initial SO_2 removal as would be anticipated in a conventional spray dryer process. Moreover, a surplus of lime is provided to react with the acids. Unreacted SO_2 and NO are oxidized to their respective acid states in the electron beam reaction chamber. The acids then react with the unreacted hydrated lime to form a mixture of calcium sulfate and calcium nitrate. These reactions take place both in the electron beam reaction chamber, as well as on the filter cake in the fabric filter baghouse. A proof-of-concept test and evaluation of this process was recently completed. A 5000 ACFM slipstream from TVA's Shawnee steam plant was utilized as the gas source. Radiation dosage had only a slight effect on SO_2 removal, but strongly affected NO_x capture. Both SO_2 and NO_x removal were highly dependent on reagent stoichiometry.

Electron-Beam/Ammonia Process

In this process, flue gas is passed through a cyclone used as an initial particulate removal stage; the gas is then cooled and humidified prior to being irradiated in the presence of ammonia. The resultant product is anticipated to be a mixture of ammonium sulfate and nitrate, which is then collected and may have use as a fertilizer product. This process was successfully tested in a Japanese pilot plant heating iron ore sintering exhaust gas. Construction of a 15000 SCFM proof-of-concept test and evaluation unit was completed in 1985 at the Indianapolis Power and Light Company's E. W. Stout plant near Indianapolis, Indiana. This unit operates on a boiler slipstream. Process operation will be evaluated with incoming flue gas SO_2 concentrations up to 4000 ppm.

Dry Regenerable Processes

Processes currently under development in this category include those based on the use of alumina supported copper oxide and sodium based sorbents. Two versions of copper oxide based processes are being evaluated. One utilizes a fluidized bed absorber for solids-gas contacting, while the second uses a novel moving bed contactor. Although the latter is at an earlier stage of development, it potentially offers the opportunity to combine the functions of SO_2, NO_x, and particulate control in one process vessel, thereby simplifying the overall cleanup system and reducing capital cost requirements.

Copyright Oxide Processes

Process chemistry has been well characterized, both in the current work and in a previous project using fixed bed reactors, and is summarized below.

For SO_2 removal:

$$CuO + SO_2 + 1/2\ O_2 \longrightarrow CuSO_4$$

For NO removal:

$$4\ NO + 4\ NH_3 + O_2 \xrightarrow{CuSO_4} 4\ N_2 + 6\ H_2O$$

This latter reaction is essentially the same as current selective catalytic reduction processes being marketed for NO_x control. Regeneration can be achieved with a variety of reducing gases. In terms of cost, safety, and availability, natural gas (methane) would most likely be the reductant of choice. The key regeneration reaction would then be:

$$CuSO_4 + 1/2\ CH_4 \longrightarrow Cu + SO_2 + 1/2\ CO_2 + H_2O$$

The copper formed in the regeneration process is oxidized upon exposure to the flue gas which contains excess oxygen and is then ready for reuse. The concentrated SO_2 stream can then be further processed using commercially available technology to produce a sulfuric acid or elemental sulfur byproduct.

Extensive testing of this process was conducted in a 40-inch by 48-inch rectangular cross section fluidized bed absorber 12 feet high. Gas was supplied from a 500 lb/hr combustor utilizing a 3% sulfur coal. Current plans are to run a minimum of 1000 absorption/regeneration cycles per sorbent sample in a small scale life cycle test rig. This effort will then be followed by a scale up to the proof-of-concept level (3 to 5 MW) where an actual utility boiler gas slipstream would be treated.

The Moving Bed Copper Oxide process in terms of chemistry is similar to the fluidized bed process. Although relatively less developed than the fluidized bed process, it offers the cost savings potential of integrating simultaneous SO_x/NO_x control with particulate capture via the use of a novel contactor/ filter. High temperature fabric serves as a filter cloth to trap particulate matter. The copper oxide/alumina sorbent granules are not only used for SO_2 and NO_x control, but also serve to continually clean the face of the filter cloth removing the particulate and maintaining reasonable pressure drops. The flyash can then be separated from the sorbent using conventional pneumatic classification techniques. To date, work on this process has been performed only at small scale, and using a simulated flue gas.

NOXSO Process.

This is a dry, regenerable process that is being developed for simultaneous NO_x/SO_x control at the 90% level. The sorbent used is Na_2CO_3 deposited on an alumina substrate. A fluidized bed absorber is used for gas-solids contacting, which is carried out at a temperature of approximately 250°F. An extensive small scale research effort utilizing both simulated and actual flue gas was recently completed to elucidate the process chemistry, which proved to be much more complex than previously thought. SO_2 is captured in the form of Na_2SO_4, which is then regenerated at high temperature with H_2S, resulting in an elemental sulfur by-product. NO_x is captured in the form of $Na_2N_2O_2$, $NaNO_2$, and $NaNO_3$. These compounds decompose to NO during regeneration and the NO can be recycled to the boiler. The sorbent has been tested over 20 absorption/regeneration cycles, with satisfactory results. Future plans are to scale up this process to operate using the combustor, absorber, and regenerator previously used in testing of the fluidized bed copper oxide process.

Spray Dryer Processes

A small scale test facility producing 40 SCFM of simulated flue gas and incorporating a 31-inch diameter spray dryer is being used to evaluate the potential for simultaneous SO_x/NO_x control. Various additives will be evaluated to determine their potential. The most promising will be scaled up and tested in an existing 22 MW industrial spray dryer. Initial testing was accomplished using NaOH as an additive to hydrated lime slurry. This approach was selected based on the positive results reported for this system by other researchers.

Advanced Separation Technologies

The purpose of this project is to review the literature and other available information on advanced separation techniques to determine their applicability in developing a less expensive, reliable flue gas cleanup process for postcombustion cleanup of SO_2 and NO_x. The advanced separation techniques considered are listed below.

- Solvent (liquid-liquid) extraction
- Gas-separation membranes
- Electrodialysis
- Inverse thermal phase separation
- Solid electrolyte decomposition cells
- Oxidation and concentration cells with fluid electrolyte
- Ion exchange
- Resin and zeolite adsorption
- Supercritical and near-critical extraction
- Electrophoresis
- Chromatography
- Liquid-membrane emulsions

Included are several applications of advanced separation techniques where the flue gas is contacted directly with the separation device, but in most cases, the advanced separation technology is used in the regeneration of a wet scrubbing liquor. The advanced separation processes listed above have received little or no attention for flue gas cleanup applications. The primary reasons for this are the enormous volume of flue gas to be processed in typical utility applications, the dilute concentrations of SO_2 and NO_x in the flue gas, the presence of contaminants such as fly ash and trace metals, and the high temperatures involved.

Of the techniques considered, two (solid electrolyte decomposition cells and fluid electrolyte oxidation and concentration cells) are currently in the early stages of research in the ongoing DOE program. Research on zeolite adsorption for SO_2/NO_x control has been completed with essentially negative results.

Of the remaining ten approaches, the literature survey indicated that the following warrant further investigation:

- o Solvent (liquid-liquid) extraction
- o Gas separation membranes
- o Electrodialysis
- o Inverse thermal phase separation

The outlook for solvent extraction is promising. A serious problem in many wet simultaneous flue gas cleanup processes is caused by nitrogen and sulfur species reacting to form aminosulfonates and imidodisulfonates, greatly complicating the chemistry and subsequent regeneration of the scrubbing liquor. This can probably be minimized or even reversed by using solvent extraction on either the nitrogen or the sulfur species to separate one from the other. To evaluate the use of this advanced separation technology, solubilities and distribution coefficients for the nitrogen and sulfur species of interest in the solvent are required. Since this data is not generally available, further research would be required before a more detailed evaluation of the potential for this application could be performed.

The outlook for removing both NO_x and SO_2 from flue gas with membranes would be promising if a membrane could be developed that would be permeable to SO_2 and NO_x but not to CO_2, O_2, and N_2. Existing NO-removal membranes have never been tested in the presence of oxygen, and probably would not work well because the ferrous ions employed in these membranes would be oxidized to ferric ions, thereby reducing the NO flux. Ferric ions might be regenerated to ferrous ions, but it would be easier to perform this regeneration if the liquid membranes were not immobilized, as they tend to be in a practical installation. Further research into the removal of NO_x by gas separation membranes under actual conditions expected to occur in the flue gas from a coal-fired boiler is needed. The outlook for removing SO_2, but not NO, from flue gas is promising. It would probably be necessary to design a recovery system that was not affected by water permeating with the SO_2. Also, there is about fifty times as much CO_2 as SO_2 in flue gas, so it will probably be necessary to tolerate some CO_2 in the permeate or use two stages of membrane filters. It may be possible to recover and sell CO_2 with this arrangement. Promising membrane materials for use if CO_2 is not desired in the permeate are polyethylene glycol in polyacrylate, a copolymer of poly (oxyethylene) glycol carbonate and polycarbonate on silicone rubber, and polyvinylidene fluoride with sulfolene.

The possibility for new flue gas cleanup processes, using electrodialysis, appears intriguing. For example, flue gas could be scrubbed with an absorbent that absorbed both SO_2 and NO_x. This absorbent could then be wholly or partially regenerated by electrodialysis. Such a process has not been attempted. An experimental investigation would be required to determine the technical feasibility of this process. Allied Chemical is currently developing the soxal electrodialysis process for control of SO_2 only.

Inverse thermal phase separation is a promising technique for separating spent liquid absorbent into two components, either to effect chemical separation or to conserve steam in steam-stripping regeneration. However, research is needed in such areas as composition of the phases, solubilities, thermodynamic properties, chemical stability, etc.

A very preliminary economic analysis was conducted comparing membrane systems with the Allied Chemical electrodialysis process and a limestone scrubber coupled with a selective catalytic reduction process for NO_x control. All of the systems are configured for controlling SO_2 and NO_x at the 90% level with the exception of the Allied process which controls SO_2 only. Freon 113 is used as the sweep vapor in the swept membrane system. First year annual revenue requirements are shown in Figure 1.

At present, in-house research is being conducted on electrodialysis for use in a simultaneous NO_x/SO_2 control process. In addition, a Program Research and Development Announcement was issued this Spring to solicit new research approaches on the potential application of advanced separation techniques to flue gas cleanup processes.

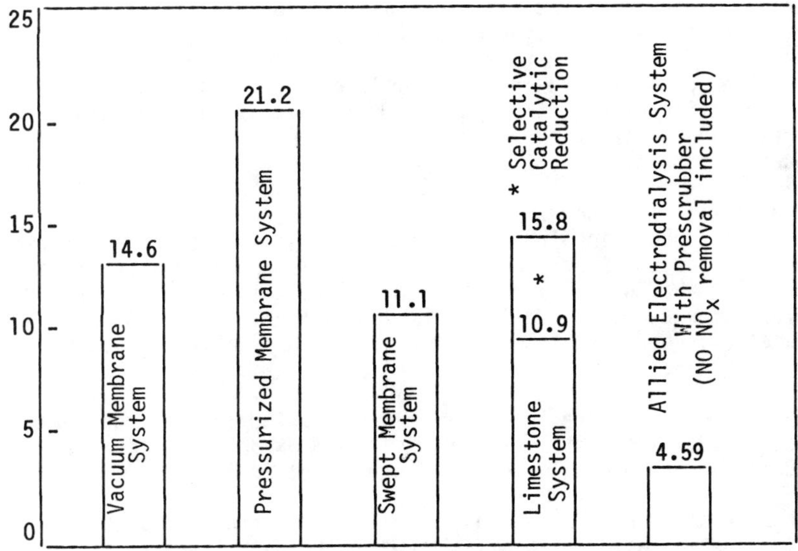

Figure 1. First-year annual revenue requirements for three gas separation membrane FGC systems, a limestone system, and an electrodialysis system (1984 mills per kWh). From reference (1) with permission.

Applied Research

In addition to the simultaneous NO_x/SO_x removal processes described in this section, an active research program is also underway. Areas of effort include:

- Determination of thermodynamic and thermochemical properties of relevant compounds.
- Research on the control of NO via the use of chelating agents.
- Application of microwave energy to enhance flue gas cleanup processes.
- Electrochemical membrane cell process research.
- Zirconia cell based solid state decomposition process research.

Sulfur Dioxide Control

The primary objective is to develop new technological options that would be available to retrofit existing boilers under an acid rain control scenario. A second objective is for the development of Furnace Sorbent Injection for application to new boiler construction utilizing low sulfur coals (primarily western subbituminous and lignites) where SO_2 requirements are 70% as opposed to 90% for higher sulfur coal.

Acid Rain Control Technology

This project area was instituted in late 1985 in an effort to identify alternative technological options for the reduction of SO_2 emissions from existing coal-fired boilers.

Three projects are currently underway. One is to evaluate duct injection of dry lime hydrate into a humidified flue gas stream. The other two involve induct spray drying types of processes with hydrated lime slurries.

Dry Hydrated Lime Duct Injection

This process involves the pneumatic injection of dry hydrated lime into the duct upstream of the particulate control system to react with the SO_2. The duct is humidified with water or steam and/or temperature controlled. Gas temperature is a critical parameter. Poor results were obtained at temperatures of 250°F and above.

Current plans are to scale this process up to the proof-of-concept level where it will operate on a 5 MW boiler slipstream at Ohio Edison Company's Toronto Power Plant in Toronto, Ohio.

This slipstream will be drawn from a boiler firing a 3.6% sulfur coal. Testing will be done with both baghouse filters and electrostatic precipitators in order to determine performance with, and effects on, both types of particulate collection devices.

In-Duct Spray Drying

The dry scrubbing process using a lime slurry for SO_2 removal from flue gas is a mature, commercialized technology. It does, however, require the use of a large reaction vessel providing sufficient volume to produce a nominal 10-second residence time. This spray drying technology is a direct adaptation of spray drying in the food industry where for a number of reasons conservative design criteria are required. Recent research, however, suggests that for moderate levels of SO_2 reduction and with improved techniques for producing smaller droplets, the residence time can be reduced to perhaps one second. Such an approach could allow for elimination of the spray dryer vessel by injecting the slurry directly into the duct work. Two concepts are under development. One will utilize an electrostatically enhanced rotary atomizer to inject a lime slurry. This process will undergo proof-of-concept test and evaluation at the Ohio Power Company's Muskingam River Plant at Beverly, Ohio. A slipstream will be utilized from a boiler burning a 4.3% coal. The second process will employ a dual fluid nozzle to inject a pressure hydrated dolomotic lime slurry. Testing will be performed on a slipstream from a boiler burning a 2.2% sulfur coal (washed from a 3.2% sulfur coal).

Furnace Sorbent Injection

Dry injection of calcium based sorbents into coal-fired boilers for SO_2 control is conceptually a very simple, cost effective approach. The process involves the calcination of compounds such as $CaCO_3$ or $Ca(OH)_2$ to CaO which reacts with SO_2 to form $CaSO_4$. Prior work with low sulfur coal-fired boiler injection using $CaCO_3$ proved disappointing. Emphasis then shifted to evaluation of $Ca(OH)_2$ sorbents, especially those produced by pressure hydration as opposed to the normally used atmospheric hydration process.

It was found that the former process produces a more reactive material which is attributed to higher specific surface areas resulting from smaller particle sizes. Results of small scale testing of this sorbent proved to be very promising and the process was scaled up and tested at the Ottertail Power Company's Hoot Lake Station using a 50 MW tangentially fired, lignite fueled boiler. This unit was previously evaluated using a limestone sorbent so the results allow for direct comparison of the two sorbents under similar operating conditions. Concurrently, the research on sorbents is continuing, focusing primarily on the incorporation of additives or promoters to increase sorbent activity.

Tests at Hoot Lake were conducted with both pressure hydrated calcitic and dolomitic sorbents. Short-term tests (up to 4 hours) were initially conducted to optimize the number and location of the sorbent injection points, as well as to evaluate the effects of sorbent type and feed rate. This was followed by a long-term test (30 hours) of the calcitic hydrate to evaluate effects on boiler and precipitator performance and operability.

CONCLUSION

If coal is to increase its role in meeting our energy needs, improved combustion systems must be developed that will control air emissions. The DOE's extensive R&D program is designed to control these emissions through application of separations technology. This technology is being applied at each stage of the fuel chain: precombustion, combustion, and post-combustion. An overview of selected projects under each of these areas that have the potential to control the emissions in advanced coal combustion systems was presented.

REFERENCES

1. R. J. Walker, C. J. Drummond, and J. M. Ekmann, Evaluation of Advanced Separation Techniques For Application To Flue Gas Cleanup Processes For The Simultaneous Removal of Sulfur Dioxide and Nitrogen Oxides, Report DOE/PETC/TR-85/7, Pittsburgh Energy Technology Center, Pittsburgh, Pennsylvania, 1985.

KRW HIGH TEMPERATURE COAL GAS DESULFURIZATION

Soung S. Kim and David A. Lewandowski
KRW Energy Systems Inc.
P.O. Box 334
Madison, PA 15663

The KRW high temperature coal gas desulfurization program consisting of gasifier in-bed desulfurization and external bed desulfurization is described. Gasifier in-bed desulfurization is a sulfur removal process using dolomite and limestone as sorbents in the gasifier. Test results from over 500 hours of in-bed testing in a 15 ton/day fluidized-bed coal gasification plant indicated that 92 percent sulfur removal can be achieved. This high percentage of sulfur removal was attained while using a Pittsburgh coal containing 4.5 percent sulfur at calcium to sulfur molar ratios of 1.5 to 2.0.

The external bed desulfurization system using zinc ferrite sorbent was designed based on the experimental results obtained from a bench-scale unit at the Morgantown Energy Technology Center. The zinc ferrite sorbent is capable of removing sulfur compounds in the hot coal gas stream at a level of 10 parts-per-million. A scaled-up external bed desulfurization system is being installed at the KRW Process Develoment Unit (PDU). Upon mechanical completion of the system in 1986, PDU testing will be implemented to demonstrate the feasibility of this zinc ferrite cleanup concept and to develop a two-sorbent system, integrating external bed with in-bed desulfurization.

INTRODUCTION

The KRW high temperature coal gas desulfurization program is jointly funded by the United States Department of Energy (DOE) and the KRW Energy Systems Inc. (formerly Westinghouse Synthetic Fuels Division) for removal of sulfur and particulates in the product gas stream of coal conversion processes. The goal is to provide an environmentally clean and cost effective process of coal utilization for electric power generation.

KRW has been developing a fluidized-bed coal gasification process for various applications of coal gasification, including electric power generation through combined cycle plants and fuel cells. Since 1974, the KRW Process Development Unit (PDU) located in Madison, PA has processed a variety of coals (including Eastern, Western and South African coals) in over 10,000 hours of operation. Currently, KRW is conducting a gasifier mechanistic study and downstream unit process development program under a 32-month contract with the United States Department of Energy.[1] Removal of sulfur and particulates in the product gas is a key part of this program.

The KRW high-temperature coal gas desulfurization program is progressing in two phases to develop a process using two sorbents: 1) dolomite or limestone in a fluidized gasifier bed, and 2) zinc ferrite in an external fixed bed. Desulfurization tests injecting dolomite and limestone into the gasifier with the coal feed (in-bed desulfurization) provided promising results. Over 500 hours of PDU testing since December 1984 indicated that 92 percent sulfur removal can be achieved through the in-bed concept. The second phase of the program, using an external fixed bed system with zinc ferrite sorbent, was initiated in 1984. Installation of this system on the PDU is nearly complete and initial system testing is scheduled to begin in June 1986; five PDU tests are planned for 1986.

A unique aspect of the KRW high-temperature desulfurization program is its two sorbent system. The external zinc ferrite system will be integrated with the in-bed dolomite/limestone system later in 1986. In this integrated system, sulfur dioxide from the zinc ferrite regeneration will be processed in the bottom of the in-bed desulfurizer/gasifier in order to convert the sulfur dioxide to calcium sulfate, and then the calcium sulfate will be removed with the ash agglomerates formed during the coal gasification process. Unreacted sulfur dioxide will be reduced to hydrogen sulfide in the gasifier and will be absorbed by the in-bed sorbent, dolomite or limestone.

The two sorbent system provides such benefits as:

- High sulfur removal efficiency by zinc ferrite sorbent.
- Bulk sulfur removal by inexpensive dolomite or limestone.
- Economic conversion of sulfur dioxide to disposable solids in the gasifier bottom thus eliminating the need for a sulfur recovery process.

The following sections provide the PDU test results of the in-bed desulfurization tests and the design features of the external bed desulfurization system.

KRW HOT GAS CLEAN-UP PROCESS

A schematic flow diagram of the KRW hot gas clean-up process is shown in Figure 1. A fluidized bed gasifier with 15 tons-per-day of coal feed capacity can be operated with or without dolomite sorbent feed.

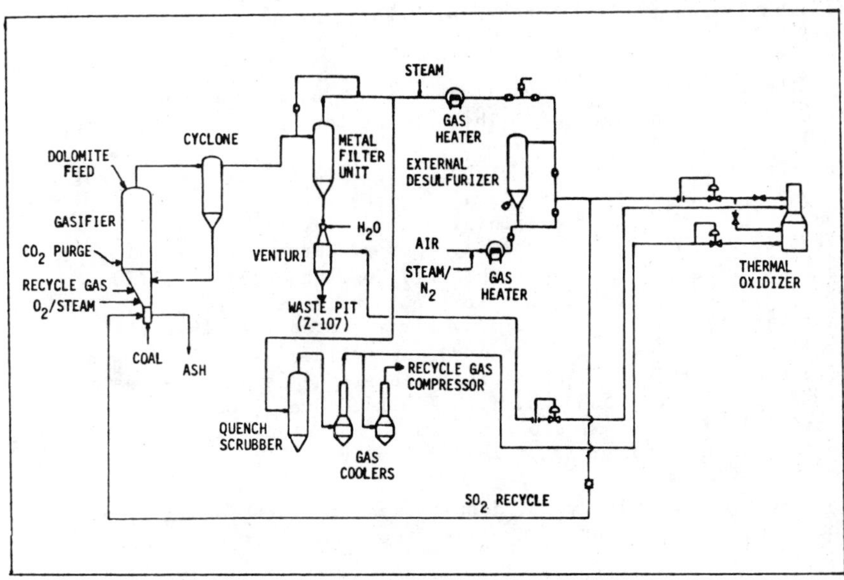

Figure 1. KRW Hot Gas Clean-up Process

Raw product gas produced in the gasifier is passed through a cyclone for removal of bulk solids entrained in the gas stream. The gas is then passed through an absolute filter, which contains sintered metal filter elements, for the removal of remaining particulates before it is introduced into the external desulfurizer. The particle-free product gas is directed downward in the desulfurizer and the desulfurized gas is then flared in a thermal oxidizer. Gas carrying solids from the metal filter unit are scrubbed in a venturi scrubber and are also flared in the thermal oxidizer. The solids are removed into a pit from the venturi scrubber as a slurry.

During regeneration of the desulfurizer sorbent the regeneration gas, an air and steam mixture, will be directed upward in the desulfurizer and be sent to the thermal oxidizer. At this time, the gasifier product gas will bypass the desulfurizer to be cooled and flared in the thermal oxidizer. A part of the cooled clean gas is recompressed and recycled for the gasifier operation. During integrated operation with the in-bed desulfurization, the regeneration off-gas from the desulfurizer which contains SO_2, steam, and nitrogen will be recycled to the gasifier.

IN-BED CONCEPT AND TEST RESULTS

KRW has applied the in-bed or in-situ desulfurization concept to its pressurized, ash agglomerating, fluidized-bed coal gasification system, and has achieved high-temperature desulfurization with both limestone ($CaCO_3$) and dolomite ($CaCO_3 \cdot MgCO_3$) sorbents. These sorbents were selected for testing in the PDU because of natural abundance and low cost.

The gasifier in-bed desulfurization system is shown schematically in Figure 2. The sorbent (dolomite or limestone) is loaded into a lockhopper and pressurized, and then it is injected into the gasifier either by pneumatic transport using recycle gas or by gravity feed. The sorbent mixes with the coal char in the gasifier fluidized bed. Hydrogen sulfide, produced during the coal gasification process, reacts with the calcium oxide (or calcium carbonate) forming calcium sulfide. This calcium sulfide falls into the lower section of the gasifier where it is oxidized by air injected into this region to form calcium sulfate. The calcium sulfate is then removed with the ash through a mechanical feeder and is disposed as a landfill. Sorbent and char fines carried over from the fluidized bed gasifier are captured in a cyclone and returned to the gasifier via a non-mechanical valve.

Reaction chemistry for desulfurization is shown in Figure 3. The major sulfur emitting species from a coal gasifier is hydrogen sulfide.

Desulfurization with calcium carbonate in the sorbent can occur by two different reaction paths. Calcium carbonate either reacts directly with hydrogen sulfide (direct reaction path) or it first calcines, forming calcium oxide and carbon dioxide, before reacting (sequential reaction path). The reaction path followed in the gasifier is a function of the temperature and the carbon dioxide partial pressure in the reaction zone as shown in the phase diagram of Figure 4 which gives the stable species of dolomite. Higher temperatures and lower carbon dioxide partial pressures favor formation of calcium oxide. It should be noted that dolomite contains magnesium carbonate which calcines under typical gasifier operating conditions to magnesium oxide; however it does not react with hydrogen sulfide to a significant extent.

The reaction path followed is important since it determines the final equilibrium concentration of hydrogen sulfide. Examples of hydrogen sulfide equilibrium concentrations for both reaction paths are shown in Figure 5. Equilibrium concentrations are pressure dependent for the direct reaction path, but independent of pressure for the sequential reaction.

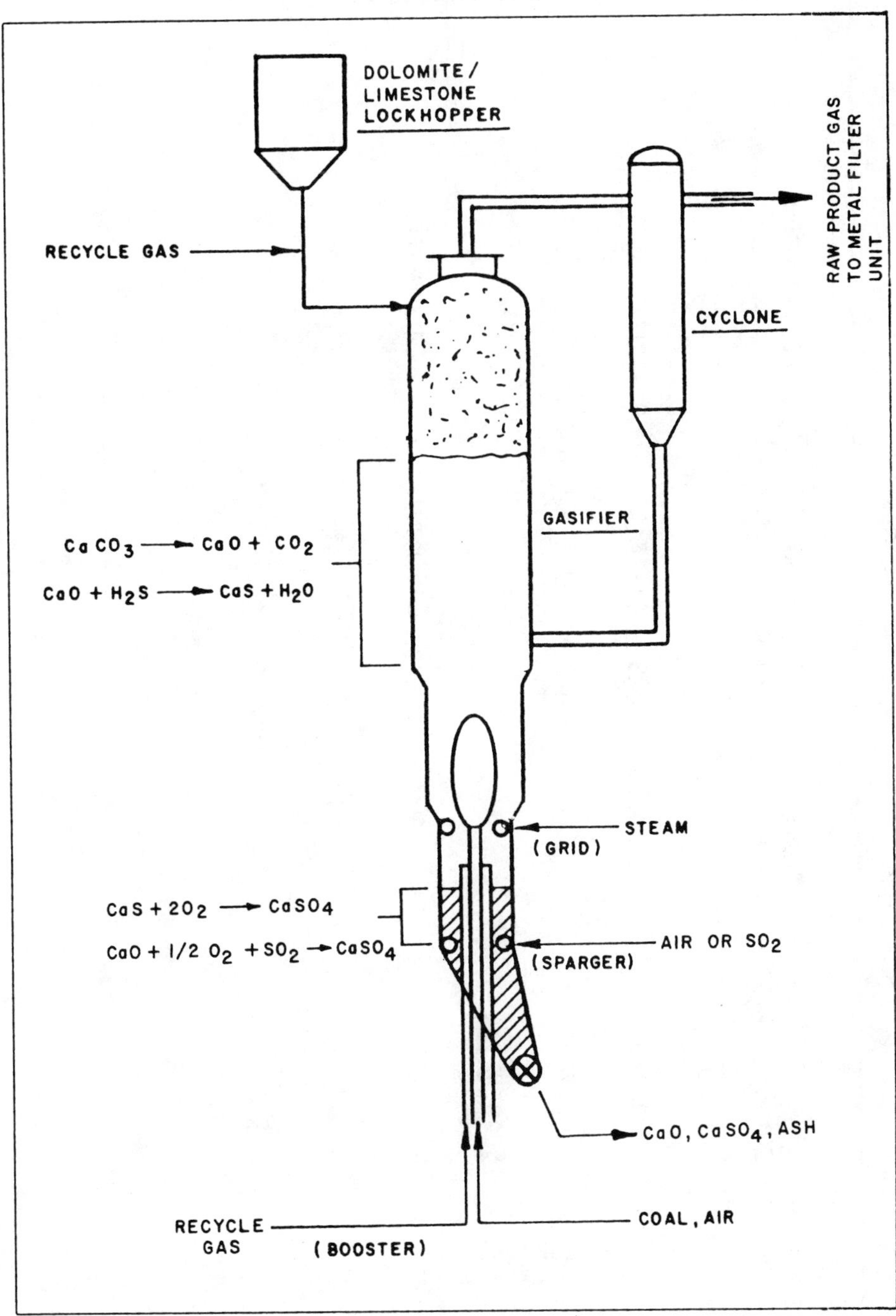

Figure 2. In-Bed Desulfurization Scheme

Dolomite ($CaCO_3 \cdot MgCO_3$) or Limestone ($CaCO_3$) with the Hydrogen Sulfide (H_2S) in the Gasifier Raw Gas

DIRECT

Dolomite $CaCO_3(MgCO_3) + H_2S \rightarrow CaS(MgO) + H_2O + 2CO_2$

Limestone $CaCO_3 + H_2S \rightarrow CaS + H_2O + CO_2$

SEQUENTIAL

Dolomite $CaCO_3(MgCO_3) \rightarrow CaO(MgO) + 2CO_2$ (Calcination)

$CaO(MgO) + H_2S \rightarrow CaS(MgO) + H_2O$

Limestone $CaCO_3 \rightarrow CaO + CO_2$

$CaO + H_2S \rightarrow CaS + H_2O$

Figure 3. Desulfurization Chemistry

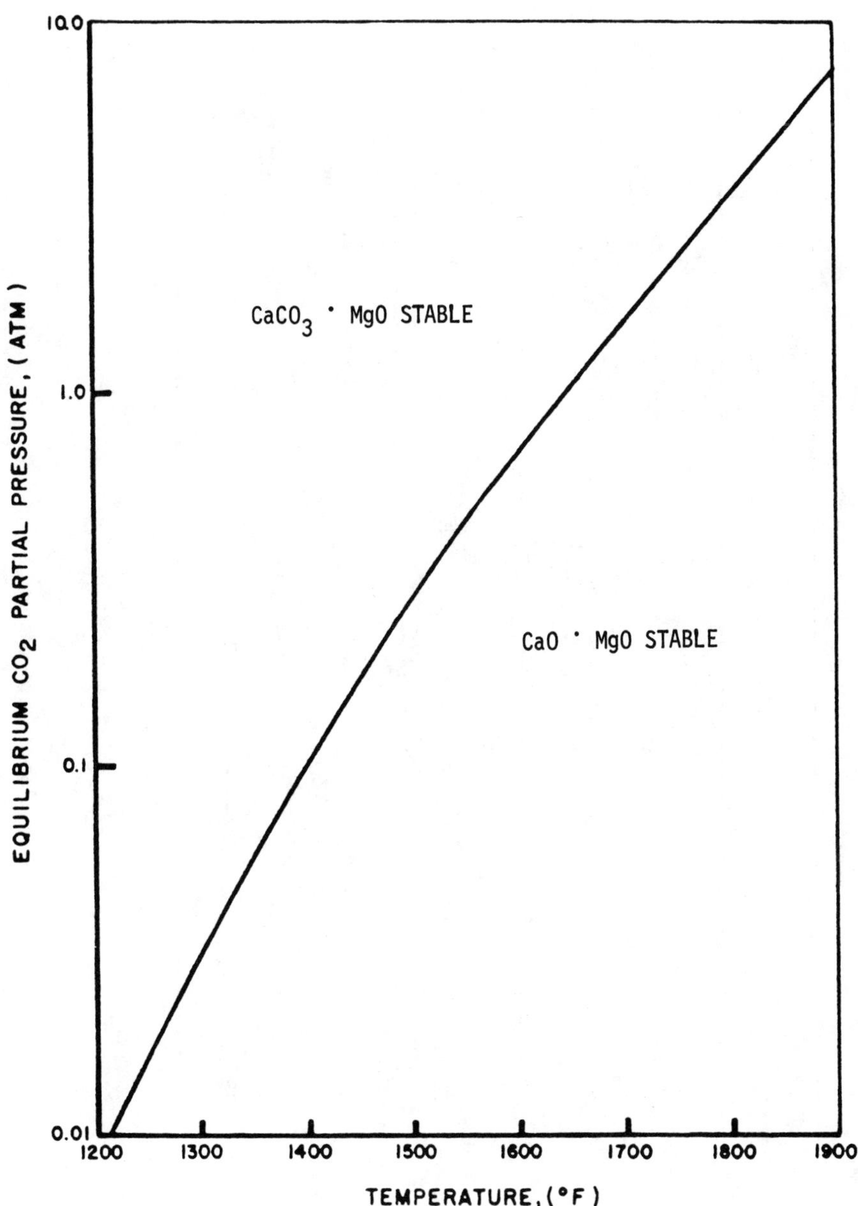

Figure 4. Dolomite Phase Diagram

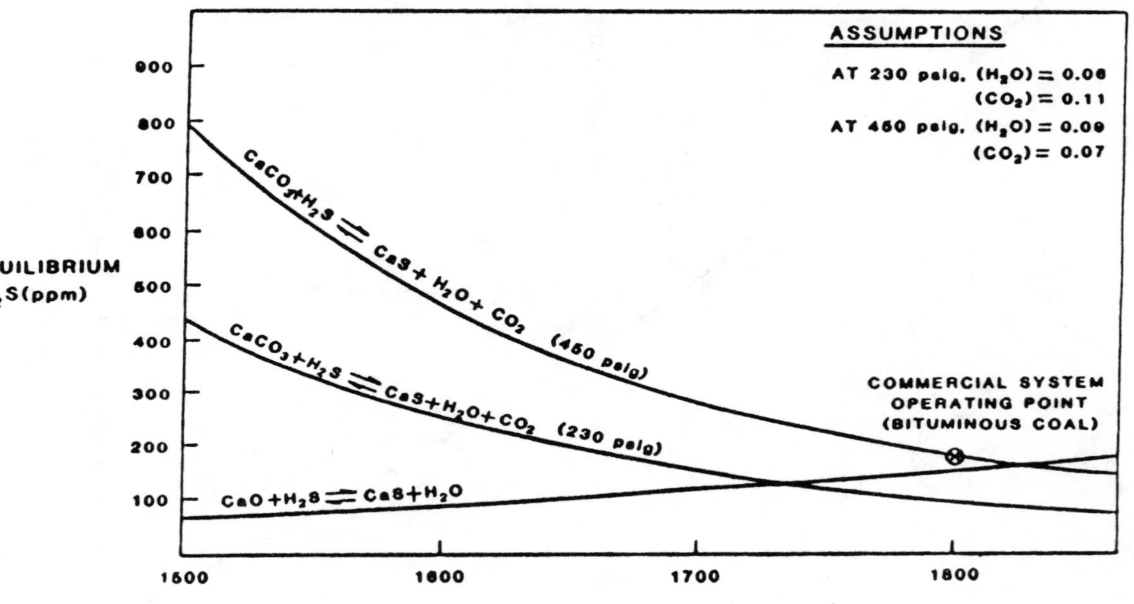

Figure 5. H_2S Equilibrium Concentrations in In-Bed Desulfurization

At the higher temperatures, typical of bituminous coal gasification in the KRW process (∿1800°F), the differences in equilibrium hydrogen sulfide concentrations are small. However, these differences become progressively greater as the temperature is decreased below 1800°F.

In an effort to optimize the gasifier in-bed desulfurization process, KRW has conducted pilot plant testing at the PDU over a wide variety of conditions. The test parameters for this program are shown in Table 1. Both equilibrium and rate mechanism related parameters have been investigated.

Table 1. PDU test parameters for in-bed desulfurization program.

Coals: Pittsburgh #8 (2.4% and 4.5% sulfur), Wyoming (2.0% sulfur)

Dolomite: Plum Run, Glass

Limestone: Greer

Operating Parameters	Ranges
Temperature	1800-1900°F for Pittsburgh Coal 1500-1600°F for Wyoming Coal
Pressure	180, 230 psig
Dolomite Size	8x16, 12x30, 12x0, 10x40 mesh
Limestone Size	12x30 mesh
Sorbent Injection Location	Freeboard, Above Bed, Coaxial Feed with Coal
Ca/S Ratio	1-1.75, 1.76-2.50, 2.51-3.50
Percent Sorbent in Bed	<40, 40-50, >50
Bed Height	10.6-18.6 ft.
Percent H_2O in Product Gas	<6, 6-8, 8-10

Desulfurization results from the KRW pilot plant tests are summarized in Table 2, where it can be noted that the highest degree of desulfurization, 92 percent, was achieved with a Pittsburgh Seam coal containing 4.5 percent sulfur. Test results indicated that the degree of desulfurization or percent sulfur removal increased as the coal sulfur content increased. It was also noted that the calcium to sulfur molar ratios required to achieve high degrees of sulfur removal in the KRW process were typically less than two. These calcium to sulfur ratios can be compared with those required in atmospheric fluidized-bed coal combustion which are typically in the range from three to six in order to achieve over 90 percent sulfur removal.(2,3)

All Pittsburgh coal in-bed tests were conducted in the air-blown gasifier operation mode at temperatures ranging from 1800°F to 1950°F. At these conditions, it was assumed that the calcium carbonate first calcined, forming calcium oxide, before reacting with the sulfur components of the coal gasification product gas. The data in Table 2 indicated that limestone and dolomite were equally effective sulfur removal sorbents.

EXTERNAL BED DESULFURIZATION

The external bed desulfurization system was designed based on the zinc ferrite sorbent developed by the Morgantown Energy Technology Center(4,5,6) in an effort to find a regenerative solid sorbent to remove sulfur components from hot coal gas. The PDU zinc ferrite desulfurizer is a fixed-bed reactor, 6 feet in diameter and 26 feet high, containing a bed of zinc ferrite pellets. It has a sulfur removal capacity of 200 hours of processing 3100 lb/hr of medium-Btu gas or 4800 lb/hr of low-Btu gas produced in the PDU from 1200 lb/hr coal feed with three percent of sulfur content. The desulfurizer vessel is made of refractory-lined carbon steel that is designed for operation at a temperature of 1200°F and a pressure of 300 psig.

The zinc ferrite pellets are made from an equimolar mixture of zinc oxide and iron oxide with some binding material.(7) The physical and chemical properties of zinc ferrite pellets are summarized in Table 3.

Table 2. Summary of PDU in-bed desulfurization results.

Coal Type	Coal Sulfur Content (%)	Sorbent Type	Reaction Path	Ca/S Feed Ratio	Steady State Desulfurization Achieved (%)
Pittsburgh #8	2.3	Dolomite	Sequential	1.67	86
Pittsburgh #8	4.5	Dolomite	Sequential	1.55	92
Pittsburgh #8	4.5	Limestone	Sequential	1.84	91[@]
Wyoming	2.0	Dolomite	Sequential	2.00	85

@ 94% Peak desulfurization achieved

Table 3. Physical and chemical properties of zinc ferrite pellets.

1.	Composition:	ZnO	33 ± 2%
		Fe^2O_3	65 ± 3%
		Binder	2 ± 0.5%
2.	X-Ray Diffraction:	$ZnFe_2O_4$	100%
3.	Physical Properties:		
	Size		3/16 - 5/16 inch, spherical or cylindrical shape pellet
	Bulk Density		82 - 88 lb/ft^3
	Crush Strength		8 pound/pellet
	Surface Area		1 - 3 m^2/g
	Pore Volume (Mercury Porosity)		0.23 - 0.40 cc/g
4.	Theoretical Sorbent Capacity		35.2% Sulfur
	Design Sorbent Capacity		25% Sulfur

This sorbent is regenerative, therefore, multi-cycle operations of sulfidation and regeneration are possible as illustrated by the reaction chemistry shown in Figure 6. During the sulfidation mode, the gasifier product gas is directed downward in the desulfurizer, while the regeneration gas is directed upward during regeneration. The alternative flow path for sulfidation and regeneration will be controlled by four high temperature valves.

Performance of the desulfurizer is predicted by the thermodynamic equilibrium of the gas mixture with the zinc ferrite system,(8) where iron oxide removes the bulk of the sulfur in the gas stream and zinc oxide removes the balance of the sulfur. The level of the H_2S in the desulfurizer exit gas is determined by the reaction equilibrium of hydrogen sulfide with zinc oxide.

$$ZnO + H_2S \rightarrow ZnS + H_2O$$

The reaction equilibrium sulfur content in the exit gas stream is expressed as a function of bed temperature and steam concentration as shown in Figure 7, the shaded area indicates the PDU operating range.

The external bed desulfurization program objectives are as follows, and specific test objectives with their expected results are summarized in Table 4.

- Characterize the zinc ferrite system in the PDU scale both for sulfidation and regeneration operations,
- Demonstrate reliable performance of the zinc ferrite bed through multiple cycle operation,
- Obtain process and operational data for commercial plant design, and
- Demonstrate integrated operation with in-bed desulfurizer/gasifier.

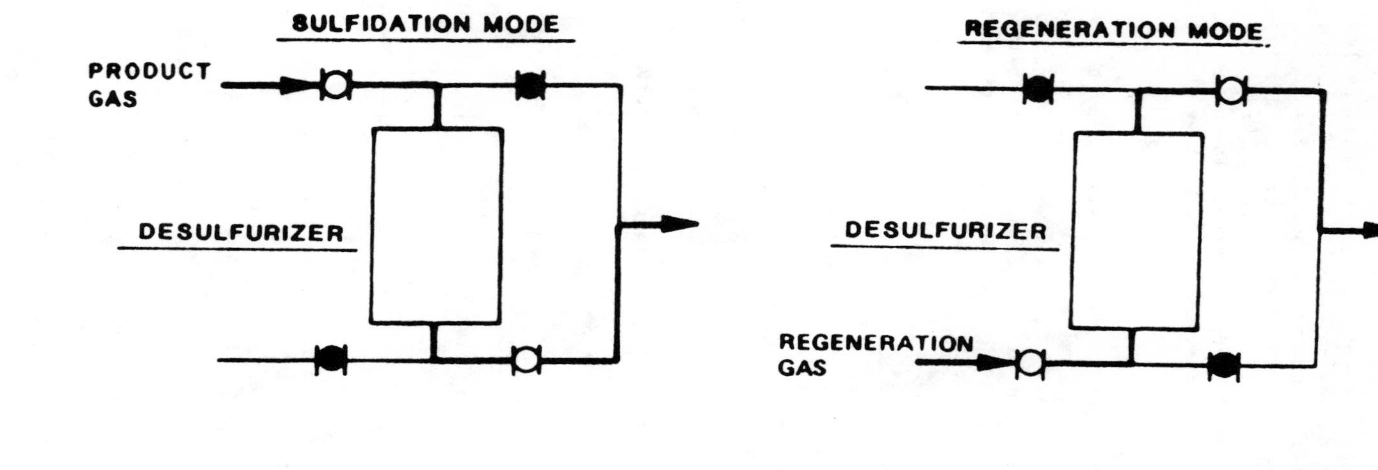

$ZnFe_2O_4 + 3\ H_2S + H_2 \rightarrow ZnS + 2\ FeS + 4\ H_2O$ $ZnS + 2\ FeS + 5\ O_2 \rightarrow ZnFe_2O_4 + 3\ SO_2$

Figure 6. Desulfurization Chemistry of Zinc Ferrite

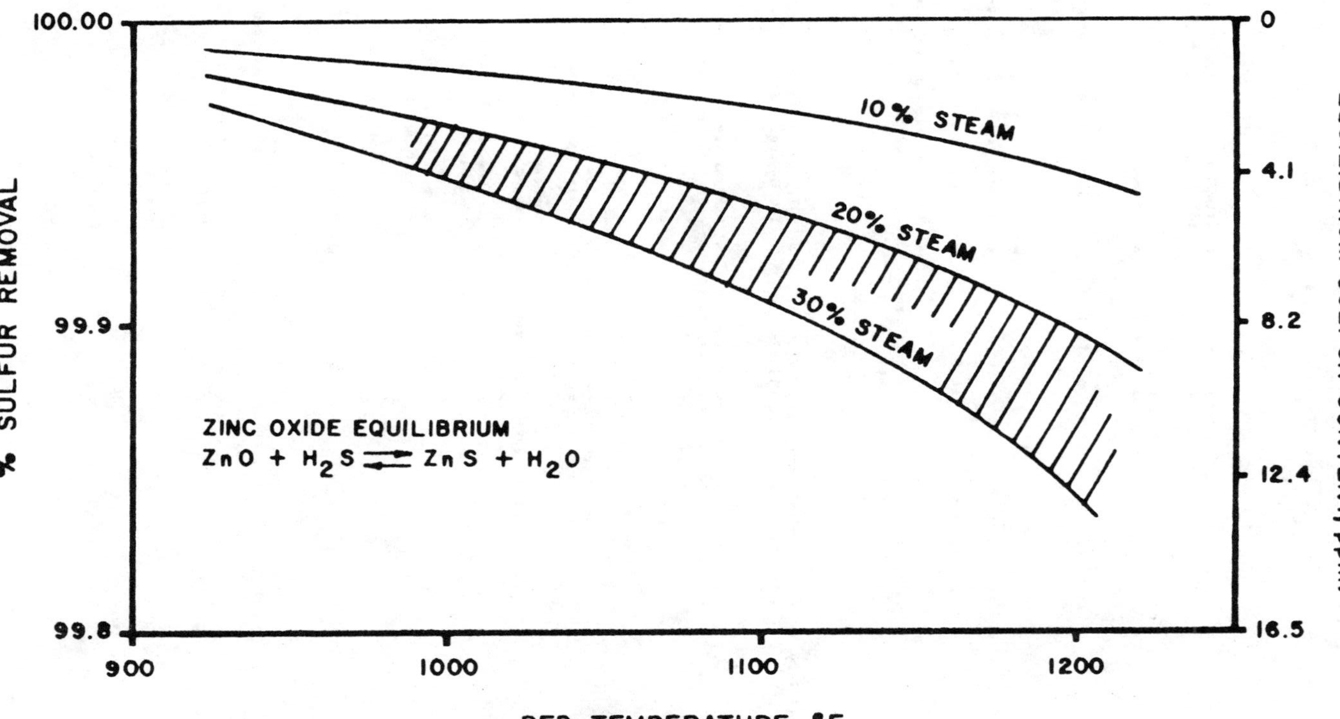

Figure 7. Equilibrium Sulfur Content in Desulfurizer Gas

Table 4. External-bed test program.

PDU Tests	Test Objectives	Expected Results
Shakedown Test	• Commission new PDU subsytems and and external desulfurization equipment. • Obtain sulfur breakthrough curves over 2-3 cycles. • Develop techniques for controlled regeneration of sorbent. • Observe sorbent regenerability.	• Establish operating procedures, diagnostics and system response characterization, • Sorbent characterization, and • Desulfurizer bed performance.
Sorbent Durability Test	• Optimize the operating conditions for sorbent regeneration. (Pre-heat temperature, gas flow rate, O_2 dilution). • Obtain sulfur sorption performance curves over multiple cycles. • Observe sorbent degradation from cycle to cycle.	• Optimum regeneration procedure, • Sorbent durability, and • Degree of sorbent deactivation at each cycle.
Sorbent Durability and Integration Shakedown Test	• Shakedown the integrated operation with gasifier/in-bed desulfurizer. • Observe gasifier performance with excess steam in the annulus simulating regenerated off-gas injection into the gasifier. • Test the integrated operation.	• Sulfur sorption performance for low H_2S concentration, • Operability of integrated operation, and • Recycle of SO_2 to the gasifier.
Desulfurizer Design Test	• Demonstrate the design condition in a long-duration run with zinc ferrite only. • Observe sulfur breakthrough curve with a full bed and low space velocity effect on bed performance.	• Sulfur breakthrough curve at lowest space velocity, • Sorbent bed performance at varied space velocity, • Physical durability of sorbent at increased bed height, • Sorbent sulfur sorption capacity at various bed levels, • Bed pressure drop data, and • Operational data of long duration.
Integrated Regeneration Test	• Demonstrate integrated operation with in-bed desulfurization.	• Evaluation of two sorbent systems, • Sorbent performance at varied sulfur concentration, • Physical durability of sorbent, and • Gasifier performance with SO_2 recycle.

KRW TWO SORBENT PROCESS CONCEPT

In-bed desulfurization with dolomite or limestone in the KRW fluidized-bed gasifier and external bed desulfurization using zinc ferrite sorbent have both been described as a means of sulfur removal from a sulfur-laden gas. Each concept can be used individually for sulfur removal; however, used together, there is a complementary benefit. A conceptual flow diagram showing an integrated in-bed/external-bed desulfurization system applicable to a coal gasification combined cycle plant is shown in Figure 8. With this two sorbent concept, the bulk of the sulfur removal (80 to 90 percent) can be obtained through limestone or dolomite injection into the gasifier fluidized bed, and the remaining sulfur can be removed in the zinc ferrite external bed desulfurizer. The zinc ferrite system in this design will contain two parallel desulfurization units, so that one unit is desulfurizing while the other is being regenerated. In the regeneration of the zinc ferrite bed, the regeneration off-gas which contains an oxidizing gas, including sulfur dioxide, will be injected into the lower section of the gasifier. The following reactions can occur in the gasifier in addition to those in-bed sorbent reactions described in Figure 3. First, the calcium sulfide formed in the gasifier bed will react with oxygen present in the regeneration off-gas as

$$CaS + 2\ O_2 \rightarrow CaSO_4$$

and, the unreacted fraction of calcium oxide will react with SO_2 as

$$CaO + 1/2\ O_2 + SO_2 \rightarrow CaSO_4$$

Not all of the sulfur dioxide will be removed by the above reaction. The unreacted fraction will be converted to hydrogen sulfide in the reducing atmosphere of the upper section of the gasifier and, subsequently, will be removed from the gas by dolomite or limestone in the fluidized bed. In this respect, the two sorbent system is complementary; the in-bed sorbent provides a means of sulfur disposal for the external bed zinc ferrite system, while the external system provides an oxidizing gas for conversion of the calcium sulfide produced from the in-bed process to calcium sulfate.

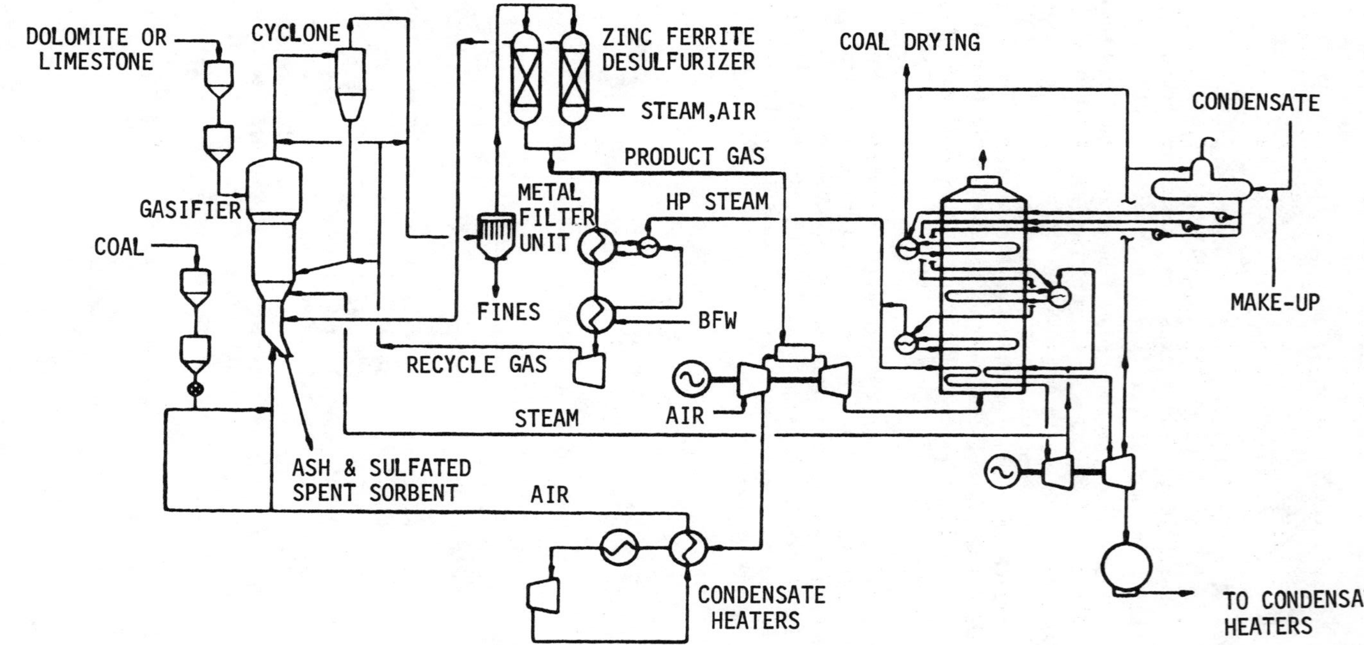

Figure 8. Integrated In-Bed/External-Bed Desulfurization Coal Gasification Combined Cycle Electric Power Generation

CONCLUSIONS

Two processes have been described for removal of sulfur from coal gasification product gas at high-temperature conditions: in-bed desulfurization using dolomite and external desulfurization using zinc ferrite. KRW PDU testing of the in-bed process has been conducted and the design for the external bed process has been completed. Important conclusions drawn from this work are as follows:

- In-bed desulfurization using a calcium carbonate based sorbent is feasible in the KRW gasifier providing as much as 92 percent sulfur removal.
- External bed desulfurization has the potential to provide from 99.85 to 99.95 percent sulfur removal from the process; this will be confirmed in the 1986 PDU testing.
- The in-bed and external bed processes can be integrated into an economical system which will provide high-temperature sulfur removal of virtually all of the sulfur produced from a fluidized-bed coal gasifier and eliminate the need for a sulfur recovery process.

ACKNOWLEDGMENTS

The authors appreciate the support given by the United States Department of Energy and KRW Energy Systems Inc.

REFERENCES

1. Schmidt, D. K., "KRW Process Development Unit," Fifth Annual DOE Gasification Contractors Meeting, DOE/METC-85/6024, pp. 296-304 (1985).

2. Green, O. P., "The Utility Perspective on Dry SO_2 Control Technologies," First Joint Symposium on Dry SO_2 and Simultaneous SO_2/NO_x Control Technologies (1984).

3. Keairns, D. L., et al, "Chemically Active Fluid Bed for SO_x Control: Volume I Process Evaluation Studies," Contract No. EPA-600/7-79-158a (1979).

4. Grindley, T. and G. Steinfeld, "Zinc Ferrite Hydrogen Sulfide Absorbent," in Third Annual Contractors' Meeting on Contaminant Control in Hot Coal-Derived Gas Streams, Report No. DOE/METC/84-6, pp. 145-171 (1983).

5. Grindley, T. and G. Steinfeld, "Testing of Zinc Ferrite Hydrogen Sulfide Absorbent in a Coal Gasifier Side Stream," in Fourth Annual Contractors' Meeting on Contaminant Control in Hot Coal-Derived Gas Streams, Report No. DOE/METC/85-3 (1984).

6. Grindley, T., S. S. Kim, E. E. Gorski, and G. Steinfeld, "Aspects of the Design of a Process Development Unit for High-Temperature Coal Gas Desulfurization," in Fifth Annual Contractors' Meeting on Contaminant Control in Hot Coal-Derived Gas Streams, Report No. DOE/METC-85-6026, pp. 33-62 (1985).

7. Jha, M., "Enhanced Sorbent Durability for Hot Coal Gas Desulfurization," U. S. Department of Energy, Contract No. DE-AC21-84MC21168 (1985).

8. Schrodt, J. T., "Dynamic Simulation of a Fixed-Bed Zinc/Iron Oxide High Temperature Desulfurization Process," Report No. DOE/MC/21582-1500 (1983).

HIGH-TEMPERATURE CONTROL OF CONTAMINANTS IN COAL-DERIVED GASES

D. C. Cicero, S. C. Jain, and J. S. Halow
U.S. Department of Energy
Morgantown Energy Technology Center
P.O. Box 880
Morgantown, West Virginia 26505

The U.S. Department of Energy is currently sponsoring various projects aimed at identification and control of contaminants in hot gas streams produced by both coal gasification or combustion. Contaminants in these streams and the costs associated in controlling them by conventional methods are the main limitations on the commercialization of many advanced coal technologies for power generation. The advanced power generation systems of interest to the Department include integrated gasification combined cycles using gas turbines, direct coal-fired turbines, coal gasification systems using fuel cells, and pressurized fluidized-bed combustion systems. Contaminants which effect these systems or which may result in adverse environmental emissions include sulfur, alkali metals, particulates, chlorides, nitrogen, and various trace metals. This presentation will describe the effects of coal contaminants on these power generating systems and recent advances made in methods of controlling them.

INTRODUCTION

The objective of the U.S. Department of Energy (DOE) hot gas cleanup program is to develop the technology to economically remove hot gas stream contaminants. The projects that comprise the program are aimed at gas streams generated at pressures greater than 6 atmospheres and temperatures greater than 1,000°F. These elevated pressures and temperatures are associated with emerging technologies such as integrated gasification combined-cycle (IGCC) and gasifier/molten carbonate fuel cell power plants, direct coal-fueled turbines, and pressurized fluidized-bed combustors. Control and removal of the contaminants is either a technical limitation or a major cost associated with these advanced systems. Selection and continuation of each cleanup project is based on its ability to meet performance and environmental requirements and to show promise of significant cost reduction over available methods.

DOE carries out development programs besides cleanup projects, for many of the other fuel conversion and power generating components for advanced energy systems based on coal. As an implementing arm of DOE, the Morgantown Energy Technology Center (METC) manages programs in gasification, heat engines, pressurized fluid-bed combustors, and fuel cells. A high degree of coordination among these programs is maintained to insure system compatibility and proper integration of development efforts. This degree of integration has allowed system tradeoffs to be made during the development stage of the individual technologies. Systems such as fixed-bed and fluid-bed gasifiers coupled to hot gas cleanup devices have become centerpieces of the DOE gasification and cleanup programs. In addition to these activities, the DOE programs emphasize laboratory research into the fundamentals and basic mechanisms of gasification, cleanup, and related processes. The intent is to develop a detailed and comprehensive understanding of these processes leading to major improvements in their technological performance, especially with respect to the fate of contaminants.

DISCUSSION

The products of coal gasification or combustion contain contaminants which were part of the coal feedstock. These contaminants include sulfur compounds, chlorides, nitrogen compounds (HCN, NH_3, and NO_x), alkalis, particulates, and trace elements. Removal of the sulfur and nitrogen compounds and the particulates is necessary to meet national air quality standards for SO_x, NO_x, and suspended solids. Removal of the other contaminants is required for some applications to avoid equipment damage or serious degradation in performance (1). In particular, gas turbines have a very low tolerance for alkali compounds and molten carbonate fuel cells (MCFC) for sulfur compounds. The fuel cell sulfur limit is well below the environmental requirements for sulfur control. The effect of sulfur on cell operation has been studied both theoretically and experimentally. The results of system studies conducted by the Institute of Gas Technology (2), United Technologies Corporation (3), and General Electric (4) indicate that sulfur present in either the anode or cathode feed affects the performance of the nickel anode.

Matching Cleanup Subsystems

The DOE Hot Gas Stream Cleanup program has centered on matching of the cleanup subsystem temperature and pressure to that required by the end use process element and generated by the upstream gasifier or combustor. This matching is expected to provide attractive systems by eliminating heat recovery equipment and compressors or expanders. Figure 1 shows a temperature-pressure plot with various subsystems in their operating envelopes. With gasifier-to-turbine power systems, an ideal match is achieved with

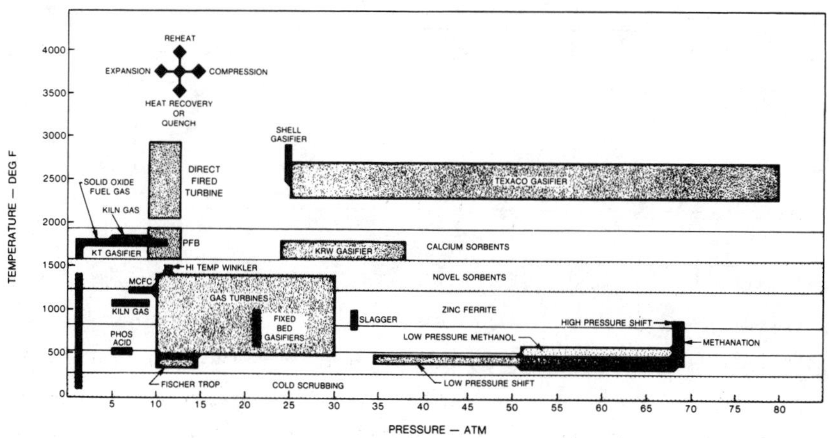

Figure 1 Temperature-Pressure Envelopes

fixed-bed gasifiers and a zinc ferrite desulfurization subsystem. This combination of gasifier and hot cleanup has been entitled the "gasification island" concept by METC and has been receiving high priority in development. Molten carbonate fuel cells operate at about the same temperatures but at lower pressures than gas turbines; both the zinc ferrite subsystem and some novel sorbent subsystems would be an ideal temperature match.

Integration of Hot Gas Cleanup for Electric Power-Generating Systems

METC has proposed the "gasification island" concept for IGCC systems which combines temperature and pressure matching, modular shop fabricated construction, and steam and air integration of the gasifier and turbine. The systems use an air-blown gasifier along with hot gas cleanup. Figure 2 illustrates the simplification of an IGCC system which the gasification island offers. This approach

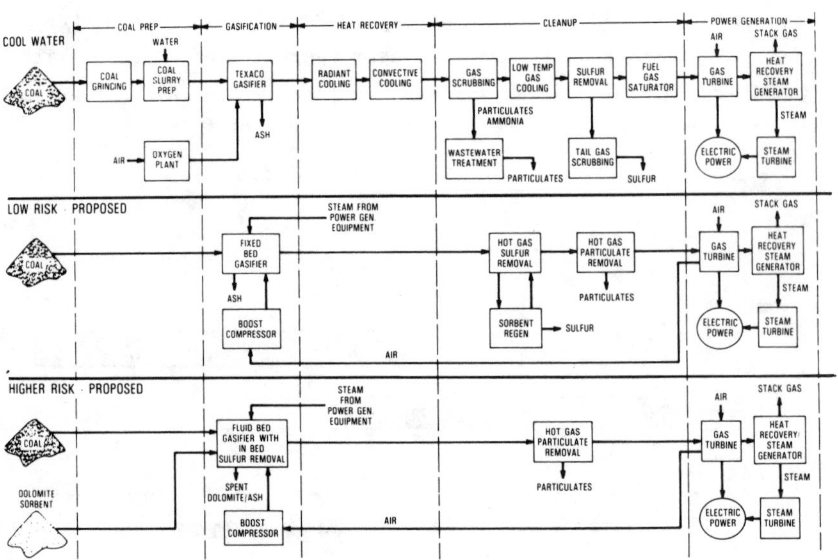

Figure 2
Simplification of IGCC Systems

optimizes the total enthalpy and chemical energy delivered to the turbine combustor while meeting fuel specification requirements with low capital cost. It is important to consider the total system and not just one type of equipment versus another in order that costs are minimized. Using hot gas cleanup, for example, is not a tradeoff of the cost for the hot cleanup equipment and operation versus that of the cold cleanup equipment, because using a cold cleanup system also requires heat exchangers to cool down the gas, reheat equipment, and a more elaborate waste water cleanup system. This type of matched integrated system promises to offer attractive modular power increments and, with the prospects of high-efficiency turbines in the near future, very attractive heat rates and low capital costs.

Table 1 depicts the relationship of the advanced IGCC system with the contaminants to be controlled and the removal devices which may be employed. The first system uses a fixed-bed gasifier which can deliver a dirty fuel gas at temperatures up to 1,000°F. At this temperature, desulfurization down to less than 10 ppm can be achieved by a zinc ferrite hot gas desulfurization process.

TABLE 1

Advanced IGCC Systems and Contaminant Control Processes

Gasifier Type	Contaminant Control Method: Sulfur	Alkali	Particulate	Ammonia
Fixed Bed	Zinc Ferrite	Not Needed	Cyclones	Steam Injection; Staged Combustion
Fluid Bed, Product Gas at 1,200°F	In-Bed Desulfurization and Zinc Ferrite	Not Needed	Cyclones and Advanced Devices	Not Needed if In-Bed Desulfurization is Used
Fluid Bed, Product Gas at 1,400°F	In-Bed Desulfurization and Novel Sorbents	Control Method Not Known	Cyclones and Advanced Devices	Not Needed

Additionally, turbine systems require alkali levels be less than 20 ppb after combustion to avoid turbine corrosion. Alkali appears to be condensed on particulates at gas temperatures less than 1,200°F. This permits alkali to be controlled in lower temperature fuel gas by removing particulates to sufficiently low levels. The quantity of particulates elutriated from fixed-bed gasifiers is relatively minor and can be easily controlled by state-of-the-art cyclone separators. An additional contaminant in this system is fuel bound nitrogen, primarily formed as ammonia. Steam injection, staged combustion, or catalytic decomposition are potential methods for control.

The second advanced IGCC system would employ a fluid-bed gasifier. This type of gasifier can produce a fuel gas in the 1,200°F and higher range, which can also be desulfurized by the zinc ferrite process. A calcium-based sorbent can be added directly to the fluid-bed gasifier to capture sulfur as it is released from the coal. Capture efficiencies have been shown to be as high as 90 percent in a fluid-bed gasifier using a high-sulfur coal. This system variation would be useful to meet new source performance standards (NSPS) for coal-fired utilities. A zinc ferrite unit could be added as a polishing step in order to achieve very high sulfur removal efficiency. In this second system, sufficient particulate removal does not appear to be possible with state-of-the-art cyclone separators due to the high level of elutriation of fine ash and char particles. Advanced, high efficiency, particulate removal devices are being developed for these high-temperature and high-pressure operating conditions. Fluidized-bed gasifiers produce little ammonia or other nitrogen compound contaminants formed from the fuel-bound nitrogen. Therefore, nitrogen control techniques should not be necessary for this system.

The third advanced power generating system represents a possible improvement in efficiency and economy to the system by using higher temperature fuel gas from a fluidized-bed gasifier without significant quenching of the gas to a lower temperature. The combination of high fuel feed temperatures and control of contaminants in order to simultaneously protect equipment and meet environmental emissions requirements is the primary problem in developing this system. Novel sorbents which can operate effectively at 1,400°F and higher are necessary to provide desulfurized gas. These advanced sorbents are being developed at the laboratory scale.

Advanced designs of particulate collection devices will also be required because of the higher operating temperatures.

Advances Made in Hot Gas Cleanup Technologies

Sulfur Control

Sulfur removal is important for all coal gas applications and mandatory in order to comply with environmental standards. Certain applications, such as molten carbonate fuel cell (MCFC), require far more stringent sulfur removal capabilities than the environmental standards dictate. To prevent poisoning of the electrodes, an MCFC requires a fuel gas with no greater than 1 part per million (ppm) of sulfur. Sulfur removal to below 10 ppm could be additionally beneficial to turbine applications, since sulfur and alkali compounds interact to cause corrosion and/or deposition on turbine blades. Thus, reducing sulfur may allow less stringent alkali removal requirements.

Figure 3 shows sulfur removal efficiencies for various processes that operate within the fuel gas temperature ranges of interest to DOE. For fuel gas temperatures ranging from about 850° to 1,200°F, two systems have shown promise: an iron oxide sorbent and a zinc oxide-based sorbent. Iron oxide was tested at METC in the 1970's on fuel gas from a fixed-bed gasifier (5). These tests showed feasibility of the sorbent for achieving up to 90 percent sulfur removal. A Japanese company, IHI, has been testing this sorbent in a fluid-bed process on a pilot scale (i.e., 40 tons/day of coal feed). The sulfur removal processes being evaluated by DOE are zinc ferrite and advanced novel sorbents. The following is a discussion of the development status of these processes.

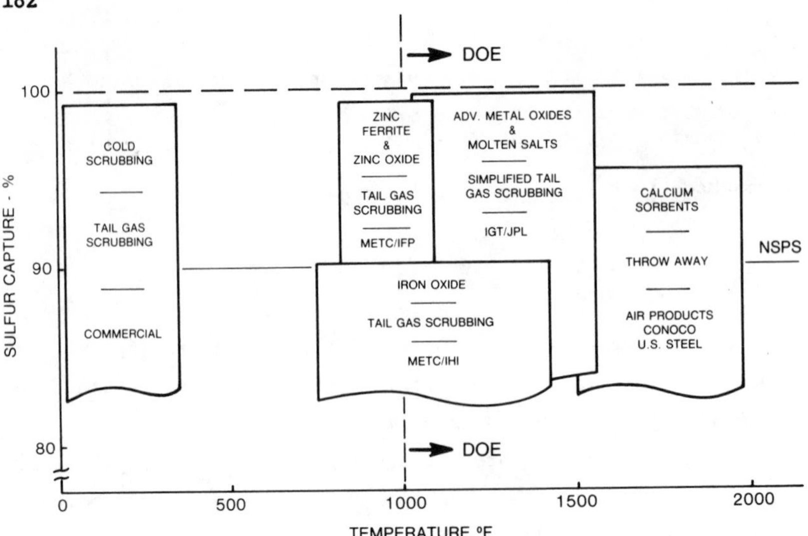

Figure 3
Fuel Gas Desulfurization

Zinc Ferrite

Zinc oxide-based sorbents were under development and testing at METC in the late 1970's and early 1980's. The results indicated that zinc ferrite ($ZnFe_2O_4$) sorbent proved to have the greatest sulfur absorption capability in the temperature range of 1,000° to 1,200°F. Above this range, the absorption capacity of all the sorbents tested declined (see Figure 4). Those containing only zinc oxide tended to be incompletely sulfidized because of the formation of a shell of zinc sulfide on the particle surface. The drop in sulfur absorption capacity of zinc ferrite above 1,200°F was attributed to the formation of wustite (FeO) in the sorbent. Figure 5 is a plot of the exit hydrogen sulfide (H_2S) level attained over three cycles of absorption and regeneration. The H_2S level was about 1 to 5 ppm before the sorbent's sulfur absorption capacity was reached, allowing breakthrough of the hydrogen sulfide. There was no significant drop in the sorbent sulfur loading following regeneration. Upon regeneration,

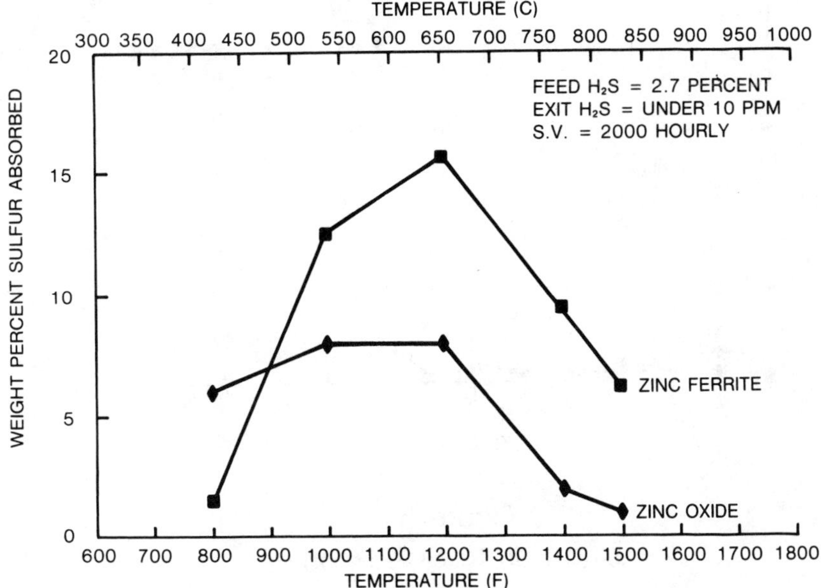

Figure 4. Sulfur Loading
Zinc Ferrite and Zinc Oxide

most of this sulfur was burned out leaving a residue of not more than about 1 weight percent in the sorbent as sulfate. A number of uncertainties were uncovered by the testing and several projects were initiated to address each of these.

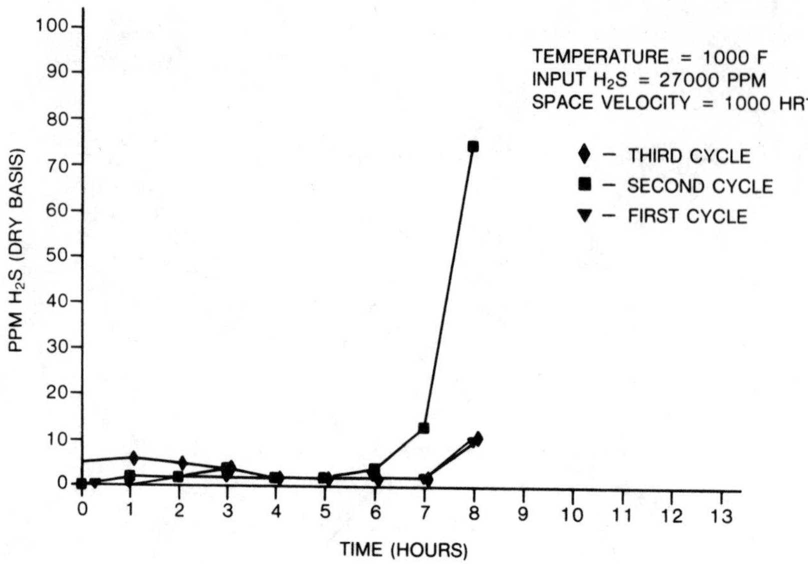

Figure 5
Zinc Ferrite Sulfidation

- Solid Phase Boundaries -- Because the gasifier was operated at a low-steam/air ratio, the high reducing power of the product fuel gas caused the formation of iron carbides in the sorbent, with a consequent weakening and deteriorating effect. To delineate the range of gas compositions which would avoid this reduction, an investigation was contracted to SRI International. SRI International has completed the thermodynamic computations and experimental studies for the phase boundaries of Fe/wustite, wustite/Fe_3O_4, Fe_3O_4/Fe_2O_3, and carbide formation for total pressures of 1 to 20 atmospheres at temperatures of 930° to 1,300°F. These studies will be used to establish the carbon-hydrogen-oxygen (C-H-O) gas phase compositions that delineate the limits of stability of solid phases in equilibrium with the gas.

- Sulfate Formation During Regeneration -- A small amount of residual sulfur in the form of sulfate remains in the sorbent after regeneration. It decomposes to sulfur dioxide and escape from the sorbent in the initial stages of subsequent sulfur absorptions. A study was contracted to SRI International to identify the conditions required to minimize sulfate formation during regeneration of sulfided zinc ferrite sorbent and to determine the chemical behavior of sulfates present in the regenerated sorbents when exposed to hot fuel gas.

  SRI conducted fixed-bed desulfurization reactor experiments to determine the amount of sulfate formed as a function of several process variables. The results indicated that sulfate formation is a function of the duration of the regeneration cycle, the temperature of the sorbent, and the amount of oxygen in the regeneration feed gas. Longer regeneration cycle duration decreases the amount of residual sulfate, as does an increase in space velocity. In the range of particle size from 0.5 to 5 mm, the sulfate formation was not affected by the particle size. The presence of SO_2 in the feed gas, however, promotes significant sulfate formation.

  The sulfates present in the regenerated sorbent decompose when reexposed to the reducing conditions of hot coal gas, releasing sulfur oxides. The sulfur oxides can be reduced and removed by a second bed of fresh zinc ferrite by providing sufficient hydrogen in the gas stream. If these considerations are followed, this process will allow the removal of sulfur oxides to a level below 1 ppm if required. A variety of other methods are possible if the acceptable sulfur level is less stringent.

- Sorbent Structural Changes -- Although the zinc ferrite sorbent was characterized after METC slipstream tests (6), a more detailed knowledge of the structural changes, both chemical and physical, and their relationship to reaction rates was needed. To accomplish this, a contract has been awarded to Louisiana State University (LSU). LSU studied structural changes in hot metal oxide sorbents in addition to zinc ferrite through several adsorption and regeneration cycles.

Experimental effort to date has concentrated upon the sulfidation of $ZnFe_2O_4$. Complete reaction to sulfides is possible over the entire temperature range. The global kinetics exhibit a relatively weak temperature dependence. Early experiments on the regeneration of $ZnFe_2O_4$ have defined the upper temperature limit for regeneration. As indicated in Figure 6, regeneration temperatures in excess of 1,400°F are to be avoided to prevent sintering of the sorbent and loss of reactivity.

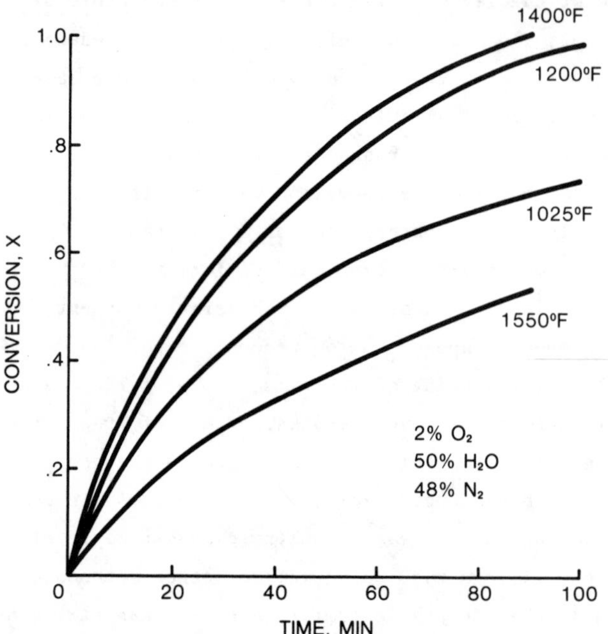

Figure 6. Regeneration
The Effect of Temperature

- Sorbent Durability -- Probably the greatest uncertainty in the development of the high-temperature desulfurization concept concerns the durability of the sorbent over many absorption/regeneration cycles. Not only must the desulfurization sorbent retain its absorption capacity, but it must also retain its strength and not disintegrate. To be economic, durability on the order of over 100 cycles is thought to be necessary. To establish this durability, a contract was awarded to AMAX Extractive Research and Development, Inc., to test several potential methods such as modified chemical composition, agglomerating and hardening procedures, and agglomerate size and shape. The durability is being tested in a 2-inch, fixed-bed reactor using synthetic coal gas doped with H_2S. The test conditions such as gas flow rates, temperature, etc., are essentially the same as used in past METC work. The AMAX sorbent has shown double the physical crush strength and twice the sulfur-bearing capacity of the initial METC sorbent. During the desulfurization testing, the AMAX sorbent has shown a better capacity for retaining its initial surface area. The AMAX sorbent also showed improved resistance to fines attrition during desulfurization.

- Regeneration Offgas Processing -- The sulfided zinc ferrite sorbent must be regenerated, and the sulfur must be recovered in an environmentally acceptable, resalable, or readily disposable form. Zinc ferrite, when regenerated with steam and air, produces an offgas consisting primarily of steam, nitrogen, and sulfur dioxide. Several methods exist to recover sulfur from the offgas in the form of elemental sulfur via state-of-the-art technology (7). An innovative concept being investigated is to recycle this offgas to the gasifier. The recycled stream would be useful as a gasification medium and the sulfur removal would be affected by the addition of a calcium-based sorbent to the gasifier system. Beyond the removal of sulfur dioxide, the concept has shown added potential benefits such as reductions in coal agglomeration tendencies, in tar yields, and increases in coal/char gasification kinetic rates. Laboratory-scale investigations are underway at this time. A similar concept is planned to be explored by KRW Energy Systems, Inc., at the PDU-scale, fluidized-bed gasifier in their Waltz Mill research site during 1986.

- Zinc Vaporization -- The zinc oxide-containing sorbent is limited to a maximum operating temperature of about 1,300°F. Some investigators speculate that zinc vaporization may be unacceptably high above this temperature. Studies by AMAX have shown that at operating conditions of less than 1,200°F, 1,000 reciprocal hours space velocity, and greater than 300 psi, the zinc vapor migration in the bed and vaporization from the bed will be negligible compared to sorbent replacement cycles. The phenomenon, ways of avoiding it, and alternative sorbents are the subject of an investigation at Massachusetts Institute of Technology (MIT).

Novel Sorbents

Novel desulfurization sorbents have the potential for higher temperature applications and simpler offgas treatment schemes than the zinc ferrite desulfurization process. The Battelle solid-supported molten salt (SSMS) system (8), under development since 1974, is at the bench-scale development stage and reduces the concentrations of sulfur compounds and chloride from 6,000 and 200 ppm respectively, to less than 1 ppm each. The SSMS sorbent consists of porous lithium aluminate ceramic pellets loaded with low-calcium salt. An advantage of the SSMS system is that the sorbent can be regenerated with a carbon dioxide-steam mixture to produce a hydrogen sulfide-rich gas from which elemental sulfur could be efficiently recovered.

The sulfur absorption capacity for the SSMS system has been shown to increase with more reducing gases, higher temperatures, and lower pressures. Removal to 1 ppm or less was demonstrated for H_2S, COS, and HCl through multiple cycles, and conditions for continued removal of H_2S to less than 1 ppm have been established. The chloride absorption capacity is high, but the reaction may be too irreversible for regeneration by simple gas purging.

A mixed-metal oxide system under development by the Institute of Gas Technology (IGT) since the late 1970's employs two sorbents (9). One sorbent, cobalt titanate, removes approximately 70 percent of the sulfur and can be regenerated with nitrogen/oxygen/sulfur dioxide mixtures to produce elemental sulfur. A second metal oxide such as zinc oxide, iron oxide, or copper oxide can remove sulfur compounds to low levels and can be regenerated with air and steam or nitrogen to produce sulfur dioxide. Production of elemental sulfur during regeneration would greatly simplify and reduce energy requirements for off-gas treatment. If elemental sulfur represented 20 percent of the total sulfur in the regenerated gas, energy requirements would be reduced from around 10,000 Btu/lb sulfur for conventional processes to 3,000 Btu/lb. In tests on cobalt titanate in a laboratory-scale reactor, elemental sulfur represented up to 35 percent of the total sulfur in regeneration gases.

A research program completed at the Jet Propulsion Laboratory (10) was aimed at synthesizing high-temperature sulfur removal sorbents that combine several desirable physicochemical properties. Mixed-oxide sorbents have been developed that show rapid absorption kinetics and good sorbent regenerability at 1,000° to 1,500°F. The rapid absorption rates are achieved by eliminating or minimizing the resistance associated with solid state diffusion. The mixed-metal oxide sorbents include $ZnFe_2O_4$, $CuFe_2O_4$, $CuFeAlO_x$, and $CuMo_xAl_yO_z$. These sorbents were tested in a laboratory-scale apparatus at 1,000° to 1,500°F for sulfidation and 1,100° to 1,500°F for regeneration. These sorbents are characterized by high sulfidation and regeneration rates. Figure 7 shows the equilibrium H_2S levels prior to breakthrough for the novel desulfurization sorbents. All four sorbents

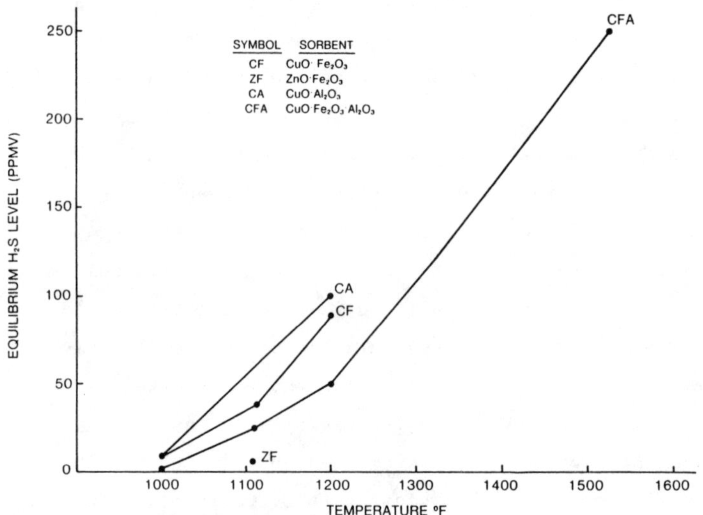

Figure 7
Equilibrium Hydrogen Sulfide Levels for Novel Desulfurization Sorbents

depicted on this figure would be suitable for gas turbine applications in the temperature range of 1,000° to 1,200°F. The combination of $CuO-Fe_2O_3-Al_2O_3$ appears to be superior to either $CuO-Fe_2O_3$ or $CuO-Al_2O_3$ in terms of attaining low equilibrium H_2S levels.

Alkali Control

Turbine systems require that alkali levels be of the order of 20 ppb to prevent corrosion of turbine components. A significant portion of the alkali species will be in the vapor phase at temperatures above approximately 1,400°F. At lower temperatures, the major portion of the alkali present will be condensed on particulates and can, therefore, be removed using conventional particulate removal devices. In order to determine how to approach the problem of alkali removal, more information is needed with regard to the amount of alkali actually present in a coal conversion stream (both as total alkali and as vapor phase alkali alone), and also mechanisms by which alkali can be captured and removed to the extremely low levels required.

Ongoing research is designed to tie together a lot of currently disparate factors -- the total amount of alkali present in coal conversion process streams, the effects of various coal additives, the amount of alkali released (and perhaps subsequently captured by indigenous aluminosilicates) in various coals, the time/temperature dependence of alkali species condensation and the correlation with particle size, and the effectiveness of experimental alkali removal technologies.

METC is conducting in-house studies and is sponsoring fundamental studies of release mechanisms, capture mechanisms, and fate of alkali metals at Midwest Research Institute and the University of Arizona to better understand alkali behavior. Aerodyne Corporation is updating and modifying a computer model for interphase condensation of alkali metals.

Three methods of alkali metal control are being explored by Argonne National Laboratory, Westinghouse R&D Center, and the University of Pittsburgh. The Argonne study has successfully used activated bauxite for alkali gettering in a pressurized fluidized-bed combustor (PFBC) flue gas. The Westinghouse study is directed at alkali gettering in pressurized coal gasification systems using a naturally occurring mineral (emathlite). Investigations have been carried out at a bench scale and have so far shown emathlite's ability to reduce 10 ppm of NaCl in a feed stream to 2 ppm. Figure 8 illustrates the effect of bed depth on using ¼-inch diameter

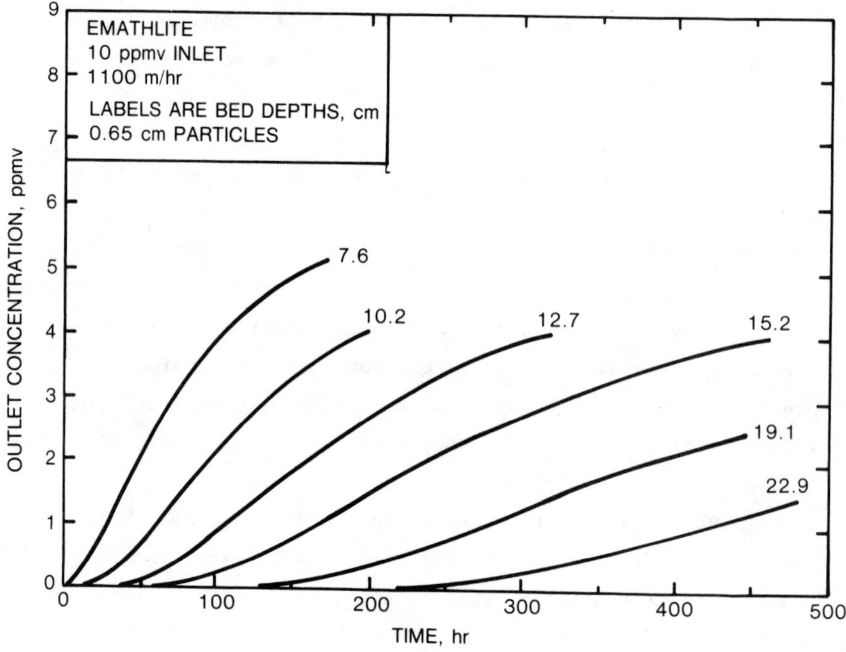

ALKALI BREAKTHROUGH WITH 0.65 cm PELLETS

Figure 8

emathlite pellets for alkali removal. Work in currently being carried out to evaluate the performance of the emathlite packed bed in gasifier conditions. Figure 9 illustrates the demonstrated effectiveness of both activated bauxite and emathlite pellets for removal of vapor phase alkali. The University of Pittsburgh study is concerned with products of entrained gasification using entrained sorbents to trap contaminant species.

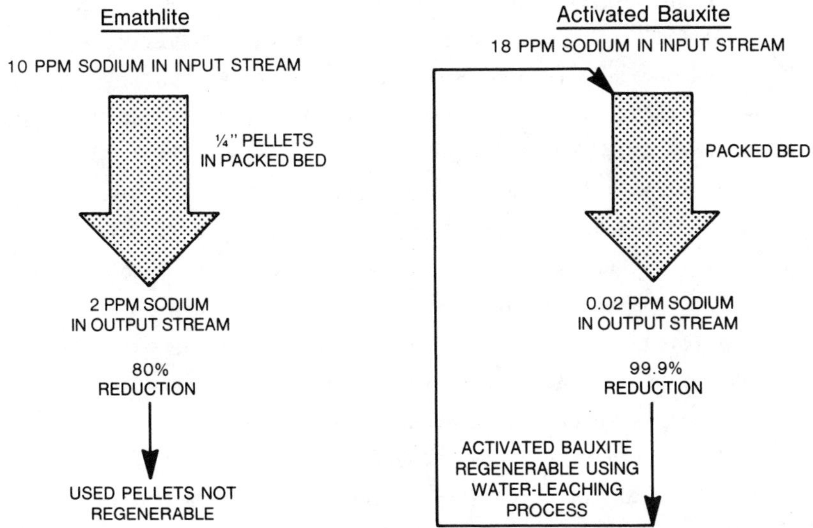

Figure 9
Effectiveness of Experimental Materials for Removal of Vapor Phase Alkali

Particulate Removal

Advanced high-temperature, high-pressure particle control is required to meet environmental and equipment requirements in gasification power generating systems using fluid-bed gasifiers (FBG). Based on data obtained at the KRW Waltz Mill FBG test facility, the elutriated dust cannot be adequately removed by cyclones alone. The large amount of dust, its small average particle size, and the fluctuating gas flow all dictate the need for tertiary particle control.

Tertiary particle control concepts have been investigated by DOE since 1980. Various configurations and combinations of barrier filter, inertial separators, granular beds, and electrostatic precipitators have been evaluated. Testing was carried out over a wide range of scale, from laboratory/bench through subpilot, in the most promising concepts. Through 1984, the work was directed toward meeting the requirements of a PFBC power system. Then an assessment for technical and economic merit narrowed the candidates for further development (11). This was done for PFBC-based and gasification-based power systems. The results showed that several concepts warranted further development. The concepts include crossflow filter (12), ceramic bag filler (13), electrostatic precipitators (14), and granular bag filters (15). A program was devised which included testing under both gasification (bench scale) and PFBC (subpilot scale) conditions to 1,600°F and 10 atmosphere pressure. Which devices were tested under which conditions was determined after considering the work which had been completed up to that point, the significance of a reducing versus an oxidizing atmosphere to the collecting mechanism, and the impact of scale effects.

In addition to the above concepts, two second-generation concepts are being pursued: acoustic agglomeration (AA) and turbulent suppression cyclones (TSC). AA is not a collection but a pretreatment to increase the average size of the dust making it collectable using standard cyclones. It has been shown that 160 dB sound at frequencies between 900 to 3,000 hertz will sufficiently agglomerate the dust. The current work centers on developing a small, economic, reliable sound source. The TSC work is directed toward characterizing and minimizing the boundary layer turbulence which develops at the cyclone wall and limits collection of very small particles. Turbulence has been shown to inject particles back into the main flow stream limiting collection efficiency. Several approaches have been identified to minimize this effect. Both of these concepts offer the potential to increase cyclone collection efficiency to the point where further filtration would not be required.

In summary, a number of tertiary particle collection concepts are being developed and evaluated. Each has the potential for use in gasification power generating systems.

Ammonia and Tar Control

Most of the ammonia, as a contaminant in the hot gas stream, is fuel-bound nitrogen derived from complex nitrogenous compounds in the feed coal released in the gasifier. The major fraction of this ammonia is typically converted to nitrogen oxides in the combustion processes, ultimately passing into the environment with the flue gas. Thus, elimination of the ammonia or prevention of nitrogen oxides formation from the ammonia that is produced is necessary to comply with environmental regulations. Previous experimental studies have explored means of ammonia removal via absorption, catalytic decomposition, or staged combustion of the hot gas. Significant test work at General Electric and Westinghouse was conducted for the formation of NO_x under turbine conditions using stage combustors. The results indicated that nitrogen oxides can be controlled to about 40 ppm.

The tars in the hot gas from a fixed-bed gasifier also contain a significant fraction of sulfur, bound in the complex tar molecules. These tar molecules mostly pass through the zinc ferrite sorbent without being cracked or desulfurized during sulfidation process. The sulfur contained in the tar corresponds to about 100 ppm of the gas. Battelle-Pacific Northwest Laboratories (BPNL) has developed the zeolite catalysts to convert the sulfur in the tar to a form (such as H_2S) which can be removed by a zinc ferrite process being developed by DOE/METC and convert the tars to compounds which will not crack and deposit coke (16). The simplest process is to add a catalyst to convert tars just upstream of the solid zinc ferrite sorbent for sulfur removal in the same reactor. Conversion increased significantly at the higher pressure and improved only slightly with increased temperature.

CONCLUSION

The results obtained from ongoing DOE projects for hot gas stream cleanup are encouraging. Significant advances have been made in developing hot control techniques for sulfur, alkali, particulates, tars, and some trace species to very low levels. These advances open up a spectrum of possible efficient, low-cost systems not only for power generation, but for other applications of coal gasification and combustion. Advanced research efforts will continue to determine the overall technical and economic feasibility of these concepts for possible future commercialization by the private sector. Advances in contaminant control technology will lead to very cost-competitive modular systems to meet future electric power needs in the United States.

REFERENCES

1. Cicero, D. C. and S. C. Jain, "Status and Overview of the DOE-Funded projects Related to Chemical Gas Stream Cleanup Program," presented at the American Chemical Society Symposium on High-Temperature Fuel Gas Cleanup in Chicago, Illinois, September 8 through 13, 1985.

2. Benjamin, T. G., E. H. Camara, and L. G. Marianowski, "Handbook of Fuel Cell Performance," prepared by Institute of Gas Technology for the U.S. Department of Energy, Contract No. EC-77-C-03-1545, May 1980.

3. Healy, H. C., et al., "Development of Molten Carbonate Fuel Cell Power Plant Technology," Quarterly Technical Progress Report No. 2. Prepared by United Technologies Corporation for the Department of Energy, Contract No. DE-AC01-79ET15440, August 1980.

4. Reinstrom, R. M., et al., "Development of Molten Carbonate Fuel Cell Power Plant," Quarterly Technical Progress Report for May through July 1980. Prepared by General Electric Company for the Department of Energy, Contract No. DE-AC02-80ET17019, September 1980.

5. Grindley, T. and G. Steinfeld, "Desulfurization of Hot Coal Gas by Zinc Ferrite," Pg 419. Acid and Sour Gas Processing, Gulf Publishing Company, 1985.

6. Grindley T. and G. Steinfeld, "Testing of Zinc Ferrite Hydrogen Sulfide Absorbent in a Coal Gasifier Sidestream." Published in the Proceedings of the Fourth Annual Contractors' Meeting on Contaminant Control in Hot Coal-Derived Gas Streams," sponsored by U.S. Department of Energy, December 1984. DOE/METC-85/3. NTIS/DE85001954.

7. Technology Status Report -- Gas Stream Cleanup, 1985, by METC (In Press).

8. Lyke, S. E., L. J. Sealock, Jr., and G. L. Roberts, "Development of a Hot Gas Cleanup System for Integrated Coal Gasification/Molten Carbonate Fuel Cell Power Plants," Final Report, January 1985, DOE/MC/19077-1830. NTIS/DE85009691.

9. Anderson, G. L. and Others, "Development of a Hot Gas Cleanup System for Integrated Coal Gasification/Molten Carbonate Fuel Cell Power Plants." Final Report, October 1985. DOE/MC/19403-1816.

10. Flytzani-Stephanopoulos, M. and Others, "Novel Sorbents for High-Temperature Regenerative H_2S Removal." Final Report, October 1985. DOE/MC/20417-1898, JPL Publication 85-80.

11. Rubow, L. N. and R. Zaharchuk, "Technical and Economic Evaluation of Ten High-Temperature, High-Pressure Particulate Cleanup Systems for PFBC." Published in the "Proceedings of the Fourth Annual Contractors' Meeting on Contaminant Control in Hot Coal-Derived Gas Streams," sponsored by U.S. Department of Energy, December 1984. DOE/METC-85/3. NTIS/DE85001954.

12. Ciliberti, D. F. and Others. Hot Gas Cleanup Using Ceramic Cross Flow Membrane Filters. December 1983. DOE/ET/15491-1565. NTIS/DE84003097.

13. Shackleton, M. and Others. High-Temperature, High-Pressure Gas Cleanup with Ceramic Bag Filters. January 1984. DOE/ET/17092-1504. NTIS/DE84003081.

14. Kumar, K. S., P. L. Feldman, and F. S. Jeselsohn. Results of HTHP ESP Testing at Curtiss-Wright. Interim Report. March 1984. DOE/ET/17094-1503. NTIS/DE84009274.

15. Guillory, J. and Others. Granular Bed Filter Development Program, Phase II. May 1983. Final Report under DOE Contract No. DE-AC21-77ET10373.

16. Baker, E. G. and L. K. Midge, "Tar Removal in a Hot Gas Desulfurization Process." April 1986. Final Report under the DOE Contract No. DE-AC21-84MC21290.

SIZE, MINERAL, AND ORGANICS DISTRIBUTION IN CRUSHED OIL SHALES

Ulrich Grimm and Lawrence Shadle
U.S. Department of Energy
Morgantown Energy Technology Center

The Morgantown Energy Technology Center (METC) is investigating the high heating rate pyrolysis of oil shale in an entrained reactor system. As part of this study, an investigation was carried out to find the minimum size range of crushed oil shale particles which still can be considered to be representative of the basic kerogen-containing material. The size should be small and the size range narrow to more easily allow modeling of the entrainment and the pyrolysis effect for this reactor system. A western and an eastern oil shale were processed in a hammer mill in a once-through mode, and the resulting powder was screen analyzed. The different size ranges were rigorously analyzed for their integrity, and it was found that the grain sizes of the mineral matter inter and intra the bedding layers are alike and around 5 μm. The fact that the intra bedding layer particles are bound to a lesser degree by the kerogen causes the mineral material to be increased in the smaller size fractions. It has been determined that a size range of 74 to 88 μm is favorable for comparative studies of both oil shales. Recommendations are made for the classification of shales from other sources.

INTRODUCTION

Research in oil shale retorting has shown that high heating rates can increase oil yields by virtue of reducing secondary cracking and coking reactions (1). The advantages are significantly greater for eastern shales, which are Type II and Type III kerogens, as compared to the Type I kerogen in the western region of the U.S., since the more mature sediments in the eastern basins are more susceptible to coking and cracking. Rapid heat-up of shale is generally achieved using relatively small particle sizes which facilitate the rapid release of volatiles and minimizes intraparticle secondary reactions.

METC is currently constructing an entrained-flow reactor for the study of oil shale pyrolysis mechanisms under rapid heat-up conditions. The facility will be a laboratory-size unit and is intended to operate under atmospheric pressure. In order to achieve rapid entrainment of the injected shale particles and to realize short reaction times and high heating rates, it will be necessary to utilize very small particles in this investigation. Another stipulation is that a size distribution as close to monomodal as possible be used to simplify the mathematical treatment of the test condition. The crushing and grinding of shale to small particle sizes has the additional potential advantage of facilitating beneficiation of shale in which the mineral-rich particles are separated from the fuel-rich particles. Shales are sedimentary rocks which display a laminated structure having separate mineral-rich and kerogen-rich layers. The layers vary in thickness depending upon geological considerations such as age, overburden, depositional environment, as well as the concentration of organic material. Within the kerogen-rich layers, the organic material is intimately associated with the inorganic rock. Thus, there is little advantage for beneficiation to reduce particle sizes much smaller than the lamination thickness unless the size distribution of specific minerals such as pyrite is favorable so that finer grinding would facilitate their removal. Furthermore, in order to describe shale retorting mechanisms, it is advantageous that shale particles be studied which represent the typical kerogen-containing minerals. It is the purpose of this study to determine the criteria by which shale particles representative of kerogen-containing minerals can be rapidly pyrolyzed with minimal resistance to mass and heat transfer processes.

The samples were crushed/ground and subjected to a screen analysis, and size distributions were determined as needed by Coulter counter. The different size fractions were then analyzed by scanning electron microscopy, thermogravimetry, energy dispersive X-ray spectroscopy, and X-ray diffraction. Procedures and results of this investigation are presented in this report. The rationale for selecting a size range for both shale materials from 74 to 88 µm is elucidated. The need to further clean the particles in this size range was established, and several methods were investigated. Optimized cleaning procedures were worked out, and recommendations were made concerning the two size ranges where both shales could be comparably tested. The application of the procedures to other oil shales is discussed.

BACKGROUND

Bradley (2) described the thickness of the varves in shale as a function of shale grade. Each varve represents seasonal deposition having two laminae, one rich in organic matter and one poor. The average varve thickness was 1.8 mm, with the thickest being 9.5 mm in the richest oil shale and the range extending to 0.014 mm.

The size distribution of shales and inorganics has been characterized for western shales (3). Low-temperature oxidation (400°C) was used to remove the organic matter without substantially altering the inorganic minerals. The feed and the residue were characterized by measuring the surface area and size distribution. In rich shale, the particles were all less than 44 µm, while conglomeration in leaner shales resulted in a particle distribution in which only 62 percent of the minerals were smaller than 44 µm. The shale and minerals are generally nonporous, and based on pore volume considerations, the distribution of organic matter in the inorganic matrix is thought to be essentially interparticle, not intraparticle. Estimates suggest that only a small amount of organic matter is bonded either chemically or physically to the mineral constituents. As might be expected from these results, several researchers have found that the small particle size fractions that have been separated by sieve techniques from crushed shales contain less organic content than the larger size fraction based on oil yield (4;5). Although similar trends have been observed for eastern shales, the results for these shales were less consistent and depended on the source of shale and the extent of grinding (6;7).

Several investigators evaluated the segregation of organics from minerals by fine grinding. Even for a relatively rich western shale (49.4 GPT), Larson, et al., (8) found that by crushing to less than 45 μm, more than 85 percent by weight has a specific gravity less than 2.0, so that relatively little separation between mineral matter and kerogen was evident by fine grinding. Lamont and Hanna (7) similarly found that only small quantities of kerogen can be separated by float sink. Robinson (9) was able to reduce the mineral content in shale by 24 to 99 percent from shales of various origins based on wettability differences between kerogen and the mineral matrix. The procedure is to grind the kerogen rock with water and oil phases. The kerogen tends to separate progressively with the oil. Misra, et al. (5), characterized the particles present in small size fractions from eastern oil shale by image analysis to determine the extent of segregation of minerals and organics. In the smaller size fractions (11 x 16 μm), about 32 percent of the particles are either pure mineral matter (13 percent) or pure organic matter (19 percent). In larger size fractions (35 x 45 μm), much less separation is observed, only 5 percent pure mineral matter and less than 1 percent pure organic matter. Jeong and Kobylinski (10) provided the additional information that particle size distribution of specific minerals were different in a Colorado Green River shale by removing minerals successively with acid treatments. Carbonate minerals in 88 to 149 μm shale particles are bimodal, having particles in the size ranges of 3 to 4 μm and 32 to 80 μm. The silicate minerals have a particle size of 4 to 10 μm, and the organic plus pyrite had a bimodal distribution of 3 to 4 μm and 10 to 40 μm particles, with the kerogen being most probably bimodal with 3 to 4 μm and 10 to 25 μm size particles.

Thus, it is apparent that fine grinding of oil shale can facilitate a separation of inorganic and organic fraction due to the interparticle distribution of organic matter in the mineral matrix. However, the amount of separation achieved by grinding to very fine particle sizes is limited by conglomeration of minerals and by the mineral grain size. The richness of oil shales and the nature of the mineral matter affects the distribution of organics and inorganics in various oil shales. Therefore, to assure that a pyrolysis study of finely ground shale is representative of the whole kerogen-containing shale, it is necessary to devise a method for evaluating the size distribution of organics and inorganics.

EXPERIMENTAL

Samples

Two oil shales, a western oil shale from Colorado and an eastern shale from Kentucky, were selected for study to investigate the effect on the oil shale during a grinding operation. The western oil shale is from the Mahogany zone of the Green River formation, 24 km north of Parachute, Colorado. The location is further described as the top 9.14 m mining horizon of the upper portion of the Mahogany zone and 1.52 m below the Mahogany marker. The eastern oil shale sample was obtained from the lower 1.52 m of the Clegg Creek Member of the New Albany shale (upper Devonian) in Bullitt County, Kentucky, which is located south of the city of Louisville. The sample site is located approximately 2.4 km southwest of KCERL core site A-12. A stratigraphic column (1) of A-12 with carbon distribution and the sample interval is presented in Figure 1. The samples were taken as cantaloupe size chunks.

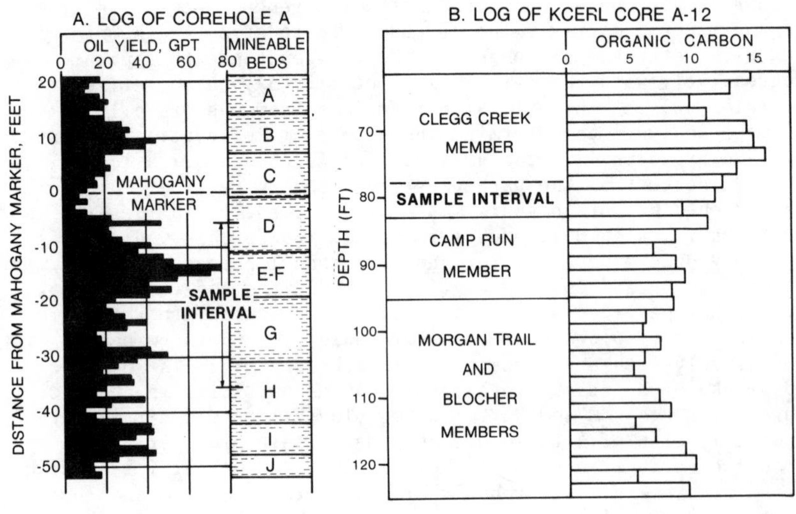

Figure 1. Stratigraphic column and relative organic contents of a) core intervals and mineable beds at Anvil Points Mine, Rifle, Colorado (Larson, 1981) similar to those for the Colorado shale, and b) the KCERL core A-12 with carbon distribution similar to those for Kentucky shale (11).

The chemical analysis and some selected physical properties of the two shales used in this investigation are listed in Table 1, and mineralogical data are shown in Table 2. The Colorado shale is higher in organic carbon than is the Kentucky shale, suggesting that the organic content may be greater in the Colorado shale. This observation is supported by the higher Fischer Assay oil yield for the Colorado shale (Table 3); however, the Fischer Assay is generally considered a low measure of the organic content or oil generating potential from eastern shales. The atomic hydrogen to carbon ratio in both samples is 1.55 ± 0.04 which is not typical for a comparison between eastern and western shales. Eastern shales are usually lower in hydrogen content (12). Although the similarity may exist, one cannot rule out the possibility that a portion of the hydrogen is in a mineral form such as in the water of hydration of clays. Clays are the predominant minerals in the Kentucky shale. The Colorado shale consists principally of carbonates, but it also has significant quantities of quartz and clays. Of course, the water of hydration in clays is highly variable.

The Kentucky shale also contains a large amount of pyrite indicating that the environment of deposition for each shale is substantially different. The high pyrite content in the eastern shale indicates regular influxes of marine water while the low pyrite content of the Colorado shale suggests a fresh water depositional environment. Thus, these two shales represent the diverse character existing in shales.

Table 1. Composition of Colorado and Kentucky oil shale.

	Colorado Oil Shale Mahagony Zone Green River	Kentucky Oil Shale New Albany Clegg Creek Member
Chemical Analysis (Dry) (Percent)		
C (Organic)	17.0	11.7
C (Mineral)	5.33	0.30
H	2.2	1.5
N	0.6	0.4
S (Total)	0.8	4.4
S (Organic)	0.23	0.32
S (SO_4)	0.04	0.06
S (Pyr.)	0.54	4.02
Ash	60.3	82.5
Moisture, As Received (Percent)	0.15	1.46

Table 2. Mineralogical analysis: X-ray diffraction.

	Colorado Oil Shale (%)	New Albany Shale (%)
Quartz (SiO_2)	11.1	5.6
Illite	22.9	79.5
Kaolinite	--	8.9
Calcite ($CaCO_3$)	14.9	--
Bassanite	--	1.1
Pyrite (FeS_2)	5.3	3.4
Plagioclase	9.7	0.7
Siderite ($FeCO_3$)	2.4	0.3
Dolomite/Ankerite	33.7	0.3

Table 3. Fischer assay values.

	Colorado Oil Shale Mahogany Zone Green River	Kentucky Oil Shale New Albany Clegg Creek Member
Oil (Gal/Ton)	32.8	12.4
Water (Gal/Ton)	2.7	5.4
Gas + Loss (Wt %)	2.5	2.5
Spent Shale (Wt %)	83.8	90.4
Oil Density (gm/cc)	0.92	0.945

The photomicrographs presented in Figures 2 and 3 are taken from actual cuts across the bedding layers of larger pieces of shale. They show the laminated structures of both shales very well. The distance between these layers is clearly narrower for New Albany oil shale.

Sample Preparation

The shale samples were taken from sealed drums and crushed to -8 mesh in a nitrogen atmosphere, thoroughly riffled, and resealed in the drum under nitrogen. From these storage drums, 0.45 kg samples were obtained and ground in a nitrogen atmosphere in a Wise, Company, Crusher, which is a small hammer mill, 15.2 cm wide, and has a capacity of approximately 75 L/hr. The resulting powder was subjected to screen analysis. One (1) pound of the crushed shale sample was placed on a sieve having 149 μm size opening (100 mesh), and the sieve screen was agitated mechanically until no further sample passed through. The amount of sample retained on the screen was determined. The material which passed through the sieve was placed on a screen having a smaller size opening, and the process was repeated. The sieve screens mesh

Figure 2. Kentucky oil shale.

Figure 3. Colorado oil shale.

sizes used were 100, 120, 140, 170, 200, 230, 270, 325, and 400 corresponding to sieve openings of 149, 125, 105, 88, 74, 63, 53, 44, and 37 μm, respectively.

Analysis

The different size fractions were tested in a thermogravimetric analyzer, Model TGS-2, manufactured by Perkin Elmer. For the sake of comparison, Fischer Assay heat-up rates (12°C/min) and holding times (20 min at 520°C) were used. Nitrogen was used to provide a measure of the volatile matter of each fraction. In separate tests, an atmosphere of air was used to provide a relative measure of the organic content.

For selected fractions the size distribution was obtained by a Coulter counter, Model TAII, made by Coulter Electronics, Inc. It was generally necessary to treat the oil shale/Isoton suspension in a sonic bath for several hours to separate the fines from the larger particles. The samples were treated for 24 hours or until no further breakup of the agglomerates could be detected. However, reliable data on the oil shale could not be readily obtained until the analytical methods were modified. For example, it was not possible to put the material, especially the Colorado shale, into suspension because the particles were not wetted by the standard Isoton solution. Adding small amounts of isopropanol solved this problem. The wide size distribution of particles produced a second difficulty during Coulter counter analysis; in order to properly sample particles at both ends of this distribution, apertures of 70 and 560 µm were used in the Coulter analysis. Tests are conducted with each aperture individually and particle sizes outside the limits of the respective aperture are ignored. During an analysis, the larger particles blocked the smaller aperture immediately, and it was necessary to wait for several minutes until the larger particles settled out before useful data could be obtained with the smaller aperture. The results presented in Figure 4 demonstrate the sharp break in size. In order to correlate the quantities of particles gained by different apertures, the sample concentration and volume of the dispersion analyzed must be known, and a statistical treatment applied to combine the data from two tests into one single overall particle size distribution.

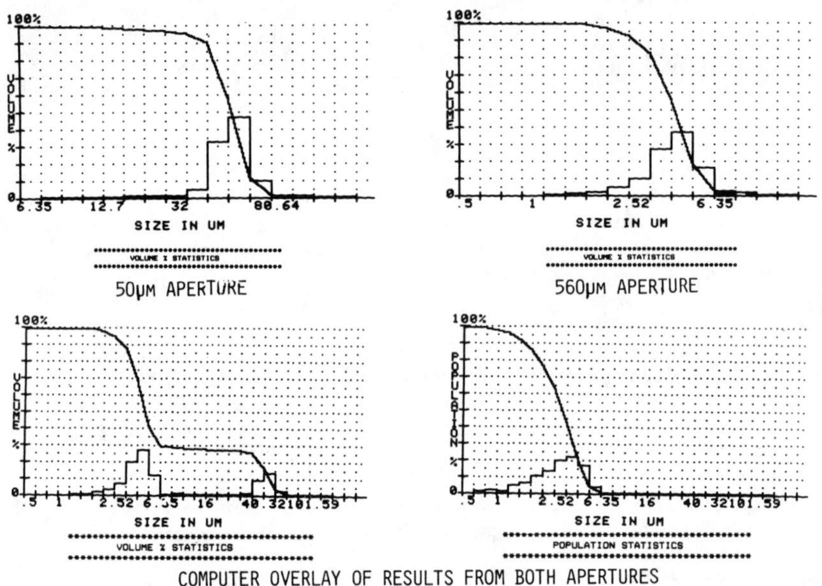

Figure 4. Size distribution of Colorado oil shale 88 x 105 µm size fraction using (a) a 560 µm aperture and (b) a 50 µm aperture.

X-ray diffraction was used to identify and quantify the mixture of crystalline phases in each size fraction. The instrument used in this analysis was a Philips X-ray diffractrometer with a Philips APD 3500 data acquisition system. The procedures and methods in this analysis are described in detail by Renton, et al. (13).

Photomicrographs were obtained to inspect the distribution of particles in each size fraction using a ETEC autoscan Scanning Electron Microscope (SEM). The SEM system was connected with a Tracor Northern TN-5500 Energy Dispersive X-ray Spectrometer (EDS). This allowed the mapping of elemental specimen within the size fractions. The output from the EDS analysis includes a photomicrograph of the respective particle on the right side of the figures, and the smaller pictures are the elemental mapping of the different elements on the surface of this particle. The brightness of the spots on the maps corresponds to the amount of the specific element present.

Separation of Fine Particles

Two methods were developed to separate fine particles (<15 μm) from the bulk of the selected size fractions including (1) a wet-sieving procedure which relies on washing the fines from the selected size fraction or (2) an aerodynamic classifier procedure in which a fluidized-bed is operated in the elutriation mode. The liquids used in the wet-sieving method were water, benzene, acetone, and isopropanol. The tests were carried out on the previously crushed Colorado oil shale in the 75 to 88 μm size range. About 15 grams of sample were taken from this material and suspended in 100 ml of the respective solvents. After these samples were sonified for 100 minutes, they were poured through a 200-mesh sieve and rinsed with another 100 ml of the original solvent. The material retained on the sieve was dried in an oven for 30 minutes at 100°C. The liquids were separated from the fines by evaporation and the solid likewise dried in an oven for 30 minutes at 100°C. The separated fines were typically 20 percent of the original sample (2.9 g from 14 g for isopropanol). TGA data were taken before and after each of the washing processes. The progress in this procedure was monitored by SEM micrographs. This method turned out to be very elaborate and led to some loss of the organic materials in the shale.

The second approach was to separate the fines in an aerodynamic classifier. The apparatus is illustrated in Figure 5. The basic unit consists of a 2-inch diameter fluidized bed, the flow controls, and a vacuum cleaner bag to collect the fines. Parametric studies involved changes in the superficial velocity and test duration. The fluidizing gas was nitrogen. The distributor plate consists of two layers of 200-mesh steel thread net. As the bulk oil shale particles are fluidized by nitrogen gas introduced into the bed, the gas passes through the fluidized particle layer to entrain and carry the fines out of the bed to be collected in the cleaner bag. The bigger particles are not entrained in the fluidizing gas but remain in the bed.

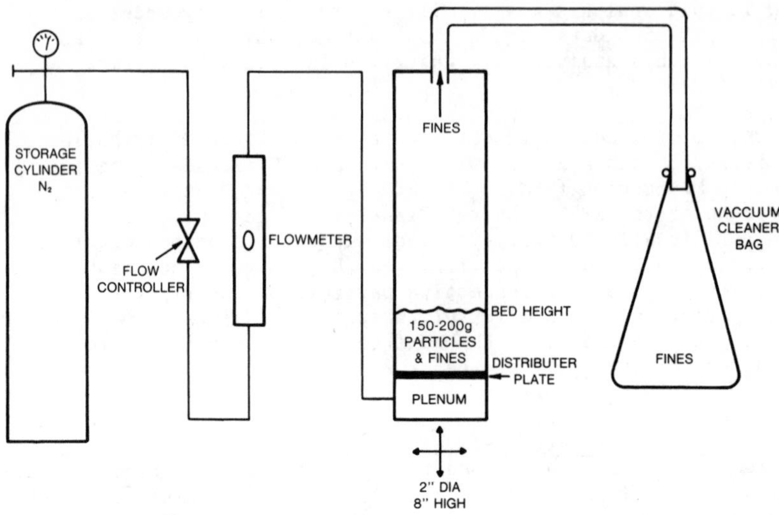

Figure 5. Schematic of the aerodynamic classifier.

Five sets of tests have been performed using 220 grams of Colorado oil shale, 74 to 88 μm size fraction. The parametric studies involved changes in the superficial velocity (or flow rate) and the test duration. Fifteen (15) separated particle samples were obtained under three different flow rates and four different test time intervals. An additional test was conducted on the 88 to 105 μm size fraction at a single set of operating conditions for comparison.

RESULTS AND DISCUSSION

Characterization of Particle Size Fractions

The results of the TGA data and the screen analyses for the eastern and western shales are combined in the graphic presentation of Figures 6 and 7. In these graphs, the mesh sizes and particle diameters of each size fraction are plotted versus the total amount of material (left scale) retained on each screen and versus the TGA weight loss (right scale) of the material retained on each screen. A deviation from the normal distribution is evident in the sieve analysis of both shales. The amount of material retained on the sieves in the 40 to 85 μm particle range was unusually high. Contrary to the general trend of decreasing amount of shale in the smaller size fractions, the Kentucky shale shows a marked increase in the amount of material in size fractions between 50 and 65 μm while the Colorado shale displays a sharp increase in the amount of shale in the 70 to 85 μm size range. In addition, the plot of the TGA-organics content showed an inflection point associated with the same size range. It is apparent that the fines have considerably lower volatile matter content and organic material than the larger size fractions. If intraparticle mass transport limitations were important, then the larger particles would yield less volatile matter due to secondary reactions; however, this was not observed. Since it was the intent of this investigation to find the smallest particle size in the crushed shale material which was still characteristic for the original shale with respect to the kerogen pyrolysis, size fractions in the 52 to 88 μm size range (shaded regions of Figures 6 and 7) were analyzed in further detail.

Differences in particle sizes can easily be distinguished by comparing photomicrographs of narrow size fractions (74 x 88 μm and 63 x 74 μm) from the Colorado shale (Figures 8 and 9). While distinguishable particles in the size range of the sieve opening are observed in the 74 x 88 μm size fraction (Figure 8), masses of agglomerated smaller particles are observed in the 63 x 74 μm size fraction (Figure 9). These fine particles can also be seen as they adhere to the large particles and in the background. All photomicrographs of particles retained on sieve sizes with openings larger than 88 μm appeared to be similar to Figure 8 while all particles on the smaller sieve sizes looked similar to Figure 9. Coulter counter analysis of the same two Colorado shale sieve fractions confirms the appearance of a distribution of particles smaller than the sieve openings; however, a third peak corresponding to particles larger than the sieve opening is also observed. Based on the SEM evidence, this third peak is probably anomaly due to aggregates of fine particles which are not dispersed under the test conditions. Similar behavior was observed when analyzing the Kentucky shale size fractions in the range of 88 to 37 μm. Sieve fractions obtained with openings less than 53 μm consisted primarily of aggregates of small particles while particles in fractions obtained on the larger sieve sizes were typical of the sieve openings. In contrast to the Colorado shale, aggregates of small particles were not observed during the Coulter counter analysis of Kentucky.

A further question to be addressed is the possible segregation in the mineral content of the different size fractions. The results of a semi-quantitative analysis by X-ray diffraction are presented in Table 4. Mineral analysis indicates that there are no major differences in the mineral composition for the Colorado oil shale fractions. For the New Albany shale, a trend can be seen where the illite decreases and the pyrite increases for the smaller size fractions. However, the differences are not significantly larger than the error associated with the analytical technique. Thus, no pronounced mineralogical differences were detected in the different size fractions.

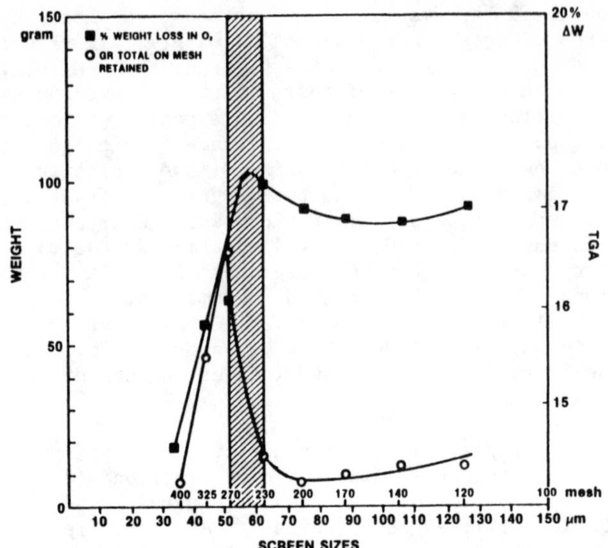

Figure 6. Sieve analysis of the Kentucky oil shale, New Albany, Clegg Creek, and weight loss in a TGA.

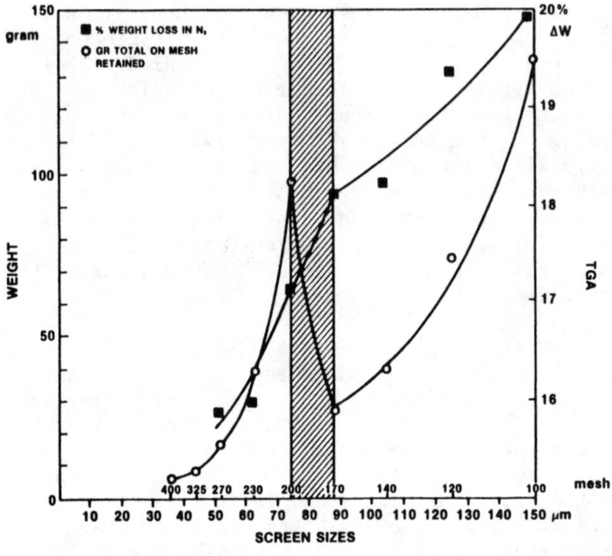

Figure 7. Sieve analysis of the Colorado oil shale, Mahogany zone, Green River, and weight loss in a TGA.

Table 4. Mineralogical analysis of different size fractions of the Kentucky and Colorado oil shales (as weight percent).

Mineral	Kentucky New Albany Shale			Colorado Green River Shale	
	88 x 105 μm	53 x 63 μm	37 x 44 μm	74 x 88 μm	37 x 44 μm
Quartz	7	8	9	11	9
Illite	80	73	67	23	22
Kaolinite	9	9	10	--	--
Calcite	--	<1	1	15	15
Bassanite	1	1	2	--	--
Pyrite	3	5	8	5	6
Plagioclase	<1	1	<1	10	13
Siderite	<1	<1	1	--	--
Dolomite/ Ankerite	<1	<1	1	34	34

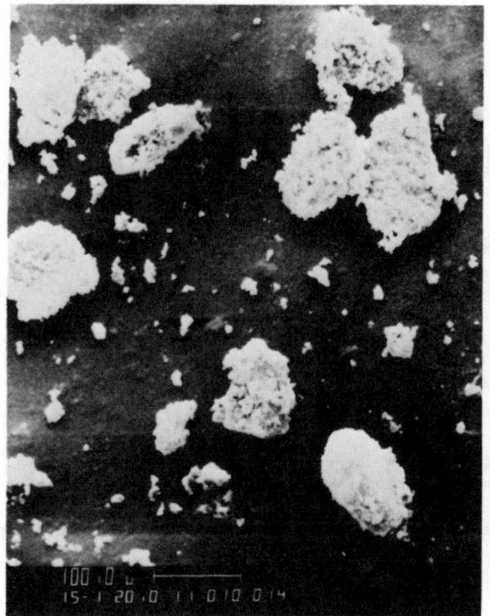

Figure 8. Colorado shale sieve size fraction of 74 x 88 μm particles.

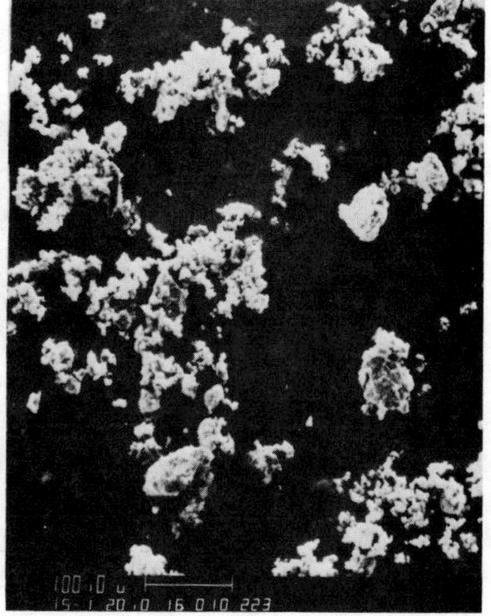

Figure 9. Colorado shale sieve size fraction of 63 x 74 μm particles.

Separation of Fine Particles

It became obvious during this investigation that simply shaking of the sieves was not enough to separate fines from the larger particles because most fractions appeared to be bimodal when suspended in a liquid for Coulter counter analysis. The two methods used to separate fine particles from larger particles in the selected sieve size range were wet-sieving and aerodynamic classification. Wet sieving resulted in a much cleaner particle than the original as displayed in Figure 10 for the 74 x 88 μm particle size Colorado fraction. However, a fraction of the organic material is extracted by the organic solvents (Table 5) and so the measurement of the amount of fines separated can be inaccurate and the nature of the sample can be altered. To determine the extent of these effects, TGA data for weight loss in a nitrogen atmosphere (Table 5) were obtained from the original sample, the washed solvent material, and the fines. The higher weight loss of the fines is attributed to organic material washed out of the shale and dried with the fines as evidenced by a waxy deposit observed on the evaporation vessel. The mineral residue of the materials used in the TGA tests was again subjected to a SEM investigation. These particles do not appear to be decrepitating into smaller material (compare Figures 8 and 10). Analysis by energy dispersive X-ray spectroscopy could not detect difference in the elemental composition of the surface of treated and untreated material.

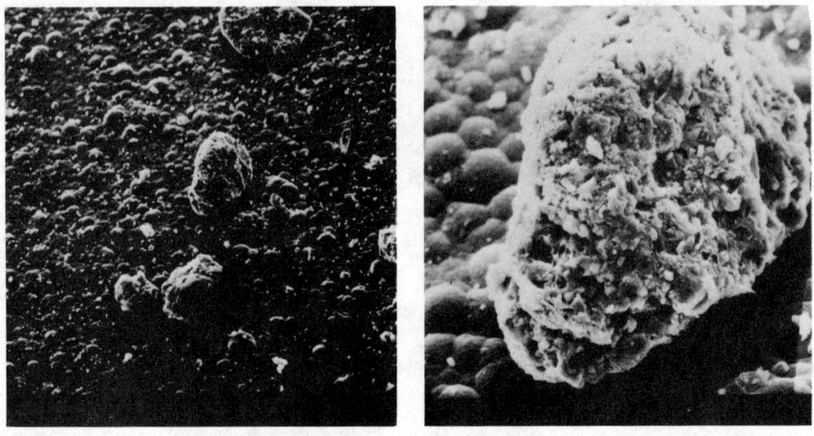

Figure 10. Colorado oil shale on +200 mesh screen after benzene rinse.

Table 5. Weight loss of treated Colorado oil shale in TGA as weight percent.

	Original Shale	Benzene	Acetone	Isopropanol
Sample	16.5	15.6	16.6	17.6
Fines	--	20.14	17.8	16.0

The results from aerodynamic particle size classification for the Colorado shale 74 to 88 μm size fraction are shown graphically in Figure 11. It is very clear that the amount of dust trapped out increases as flow rate of N_2 increases. The greatest portion of dust removed is separated in the initial first hour. Also, in an additional test on larger size fractions (88 x 105 μm) under the same conditions, the amount of dust removed was significantly smaller than the smaller size fraction (e.g., at N = 2.55 40 scfh and 2 hours, fines removed were 4.5 percent compared to 9.9 percent, respectively). Separation of fines by the aerodynamic classifier was also found to be successful in removing a large number of fines from selected size fractions.

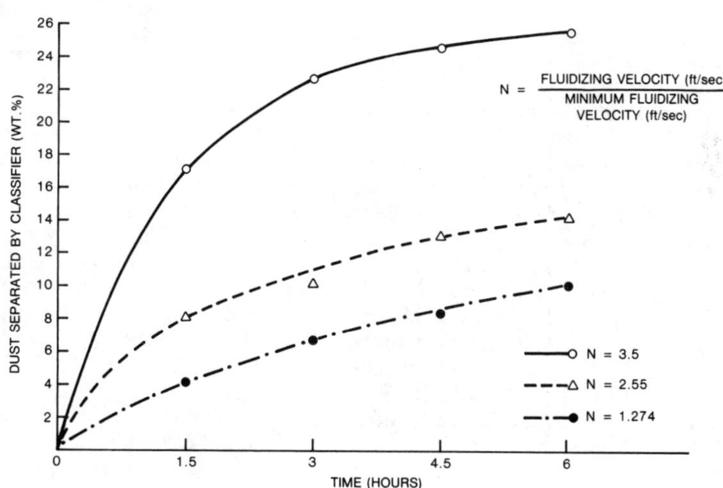

Figure 11. Dust separated from Colorado shale (74 x 88 μm) by classifier as a function of time and N_2 flow rate.

The particles retained in the fluidized bed are of the proper dimensions for particles in this size fraction, and there is very little or no interference from fines and agglomerates of fines having dimensions smaller than the selected sieve sizes (see photomicrograph, Figure 12). The fines (Figure 13) are obviously much smaller in size. The photomicrographs clearly show the degree of separation which was achieved. The enlarged photomicrograph of the aerodynamically cleaned size fraction (Figure 12) shows that the highly clean surface of the solvent-treated material (Figure 10) is not achieved. However, the surface of this material has a freshly broken appearance and the particles show more of the laminated features characteristic of shale. Particle size analysis of the aerodynamically cleaned size fraction (Figure 14) indicates that the fines have been essentially removed. The bimodal distribution of the original shale has been converted to a single normal distribution in the particle size range of the sieve opening. Also, the fines removed consist only of small particles less than 25 μm (Figure 15). Qualitative analysis of SEM indicates, as might be expected, that the number of fines was apparently reduced as the separation time increased. The conditions for separating the fines from a load of 220 grams crushed and sieved Colorado oil shale, 74 x 88 μm, are using 40 scfh nitrogen as fluidization gas, and a run duration of 2 hours.

CONCLUSIONS

Oil shales contain a substantial amount of inorganic material consisting largely of quartz, silt, and clay minerals. Some oil shales are really organic-rich siltstones and mud rocks, while others are organic-rich limestones. The common sedimentary feature of many oil shales is a distinct lamination, on a millimeter scale, of alternating clastic and organic laminae. As a consequence forces acting on the shale material will not lead to random cracking pattern, but will follow structural formations of the rock. The internal weaknesses are of three orders of magnitude -- macrocracks, microcracks, and discontinuities between adjacent structural units, each of which has an effect on the apparent strength of any piece of shale. The contribution of each type of weakness varies with the size of the piece. During grinding, fractures tend to propagate along the bedding layers and through cracks and interfaces between the laminae.

An effort is currently underway to allow a more accurate imaging of the kerogen-rich laminae to determine whether the smallest particles representative of the whole kerogen-containing shale can be correlated with the size of laminations. It is proposed that upon grinding, the shale material crushes initially along the bedding layers with subsequent breakup of the kerogen-rich layers in particles with an approximate thickness of this layer or larger. These particles appear as tightly packed agglomerates. The finer material is assumed to be the less

Figure 12. Colorado oil shale, 74 x 88 µm size fraction retained in fluid bed.

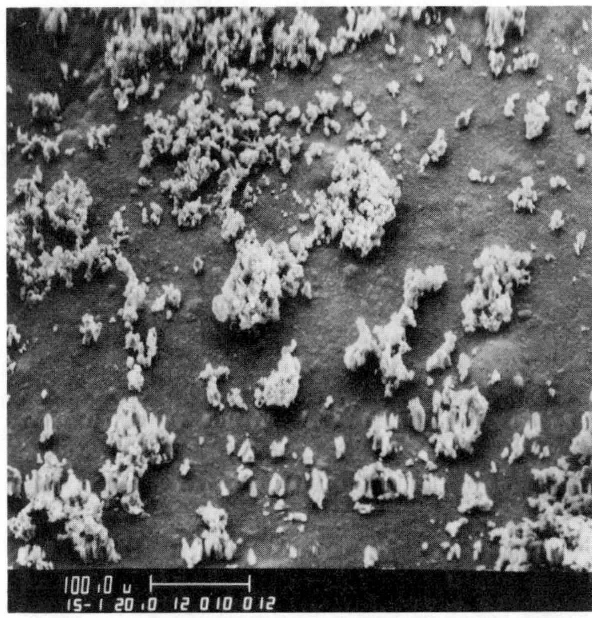

Figure 13. Colorado oil shale, separated fines from 74 x 88 µm sample.

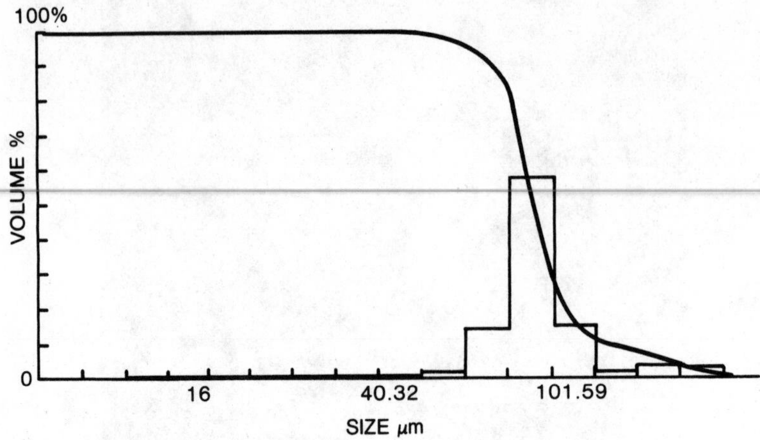

Figure 14. Particle size distribution for aerodynamically cleaned Colorado shale 88 x 105 μm fraction.

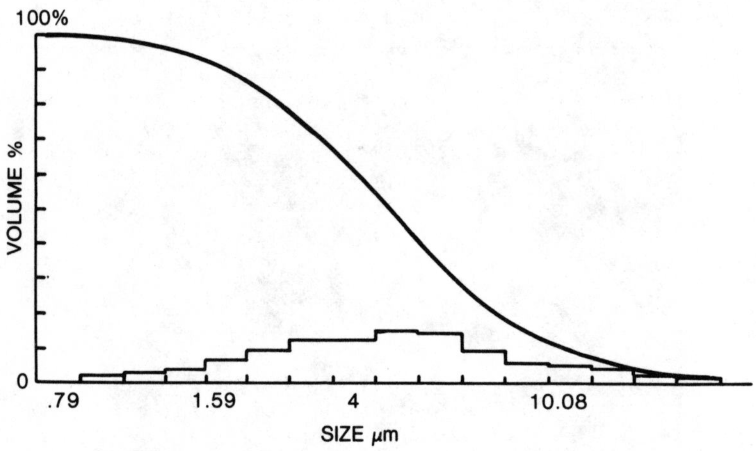

Figure 15. Particle size distribution for fine material separated from Colorado shale 88 x 105 μm size fraction using aerodynamic classifier.

strongly bound minerals from between the kerogen layers and are observed in their natural grain size or as loosely formed aggregates. As a consequence of their origin, these smaller particles carry a lesser amount of kerogen. Grinding over extended periods or repeated grinding procedures result in a product which shows large clusters of fine particles. It appears that forcing the "natural thickness particles" to break up exposes more and more bitumen- and kerogen-rich surfaces which, by virtue of the sticky nature of the organic matter, leads to the formation of bigger agglomerates.

The experimental results lead to the recommendation that the 74 to 88 µm shale particles are most suitable for the proposed studies in the entrained reactor. This size range seems to be acceptable for both the Colorado and Kentucky oil shales investigated using the described crushing method. The separation of the fines from the desired size fraction can be satisfactorily achieved using a fluidized-bed classifier. The wet-sieving technique is sufficient to rinse the fines of ~75 µm particles; however, the effects of the solvents can present a problem and solvent removal is time consuming. The use of isopropanol is apparently the best of those tested since the fines, in this case, have less volatiles carried over.

The simplest way to find the size of the smallest particle still representative of the original rock seems to be a once-through crushing of the material with subsequent screen analysis. Washing the desired size fraction with solvents to remove fines worked in both shales, but the better method is by size classification in a fluidized bed. It was determined that the particles undergoing aerodynamic separation in a fluidized bed must be treated for 2 to 3 hours to remove the last remnants of fines retained on their surface. Microscopic inspection of the aerodynamically cleaned particles shows the surface appearing as freshly opened breaks and not the "weathered" surfaces of the solvent-treated material. The entrained-flow reactor tests shall consequently be initiated with this material.

ACKNOWLEDGMENTS

The authors acknowledge Alan Weakly, of the Exxon Company, Grand Junction, Colorado, and Lance Barron of the Kentucky Center for Energy Research Laboratory, Lexington, Kentucky, for providing the samples of the Colorado and Kentucky shales, respectively.

REFERENCES

1. T. Coburn, et al., Retorting Kentucky Oil Shale Yield Optimization at Moderate and Rapid Heating Rates, Sixteenth Oil Shale Symposium Proceedings. 1983, Golden, Colorado.

2. H. W. Bradley, U.S. Geological Survey Progress Paper 158 (1938).

3. P. R. Tisot and W. R. Murphy, Physical Structure of Green River Oil Shale, Hydrocarbons from Oil Shale, Oil Sands, and Coal, AIChE Symposium Series 54, 61, p. 25 (1965).

4. R. S. Datta and C. A. Salotti, Coarse Beneficiation of Green River Shale, 16th Oil Shale Symposium Proceedings, Colorado School of Mines, p. 413 (1983).

5. C. L. Lin Misra and J. D. Miller, Concentration of Eastern Oil Shales by Froth Floatation, Eastern Oil Shale Symposium Proceedings, Lexington, Kentucky, p. 377 (1983).

6. R. S. Datta and C. A. Salotti, Beneficiation of Eastern Oil Shale, Eastern Oil Shale Symposium, Lexington, Kentucky, p. 227 (1982).

7. W. E. Lamont and H. S. Hanna, Upgrading of Eastern Oil Shale by Beneficiation, Eastern Oil Shale Symposium, Lexington, Kentucky, p. 175 (1982).

8. O. A. Larson, C. W. Schultz, and E. L. Michaels, Beneficiation of Green River Oil Shale by Density Methods, ACS Symposium Series 163, ed. H. C. Stauffer, Oil Shale, Tar Sands, and Related Materials, p. 39 (1981).

9. W. E. Robinson, Isolation Procedures for Kerogen and Associated Soluble Organic Materials, Organic Chemistry, p. 183, Springer Verlag, New York (1969).

10. K. M. Jeong and T. P. Kobylinski, Organic Mineral Matter Interaction in Green River Oil Shale, Geochemistry and Chemistry of Oil Shale, ed. M. Kuis and McKay, ACS Symposium Series 230, p. 493 (1983).

11. L. S. Barron, Institute for Mining and Minerals Research, Lexington, Kentucky, private communication.

12. R. M. Cluff, et al., Geologic and Geochemical Studies of the New Albany Shale Group (Devonian-Mississippian) in Illinois.

13. J. J. Renton, Use of Weighted X-Ray Diffraction Data for Semi-Quantitative Estimation of Minerals in Low-Temperature Ashes of Bituminous Coals and in Shale, METC/CR-79/5 (A02) (1979).

SEPARATION CHALLENGES AND OPPORTUNITIES IN THE OIL AND PETROCHEMICAL INDUSTRY

G. C. Blytas
Shell Development Company
P. O. Box 1380
Houston, Texas 77001

In the oil and petrochemical industry, novel separations will be implemented only if they can satisfy the needs dictated by the changing scenarios: offshore oil production, enhanced oil recovery (including thermal), and refining of heavier crudes. New separation problems are related to the nature of the product itself, and/or to the methods of its production and refining. Special separation problems are also involved, in a critical way, in the production and purification of natural gas, and of the CO_2 used in enhanced oil recovery.

In the petrochemical complex, we focus more on value-added, end-use of refinery products and hence purity requirements become important. Finally the disposal of waste streams introduces environmental problems of increasing importance. In this paper, examples are cited in which new or improved capabilities are needed in phase separations, distillation, solvent extraction (including supercritical), adsorption, and various types of membrane processes.

INTRODUCTION

In this paper, the challenges which face the separations scientist and technologist in the oil industry, and some of the opportunities which exist for technological improvement and innovation in the eighties will be outlined.

The oil industry in the USA is a mature industry. Thus, it is unlikely that new refineries will be constructed in the near future in the USA. Still, over 40% of the capital in petrochemical complexes is dedicated to separation units. Furthermore, during the last decade, the energy industry has witnessed dramatic changes, both economic and technical.

In addition to facing rapid changes in economic scenarios, in seeking opportunities for innovation in the oil industry, the technologist is also faced with the conservative attitude of an industry which has been in existence for over 80 years. This cautious attitude is understandable when one considers the staggering costs of unplanned downtime in a typical petrochemical complex.

Nevertheless, technological strength, and the ability to adapt and to innovate as appropriate, are important for the oil industry of the eighties. This was probably not the case before the energy crisis of 1973. Thus, a 1970 study of the petroleum industry (1) compared several oil companies to conclude that "in the mature and homogeneous oil industry, technologically innovative oil companies are not more profitable than non-innovative companies". However, a comparison of these same companies in 1986 shows that the technical leaders are currently the leaders in profitability and are better equipped to handle the current decline in oil prices than the less innovative companies. So, we can state that technologically sound innovation correlates positively with the long-term profitability and longevity of an oil company.

Technological innovation in separations is prompted by the new needs and scenarios of the mature oil industry. These include, for example:

- Changes in crude oil prices
- New types and sources of oil; offshore, thermal, and enhanced oil recovery
- Production and treatment of sour gases
- New conversion processes in the petrochemical complex
- Higher purity requirements in process and product streams
- Environmental constraints; water cleanup, limitations in deep welling, unleaded gasoline.

An important fact to keep in mind is that separation processes are dependent processes. They are generally preceded by a conversion process. The conversion could take place in a reactor in a few minutes or seconds, or in a geologic formation over millions of years.

In this paper, trends in separations technology will be delineated as they impact on the various operating divisions of an integrated oil company. Specifically, separations will be discussed as they impact on oil and gas production, on refining and petrochemicals, and on environmental activities.

I. OIL AND GAS PRODUCTION

The last decade was characterized by increased offshore oil production, by increased use of thermal recovery methods and by the initiation of carbon dioxide flooding. Chemical flooding has been studied and piloted, but will probably not play a significant role in the near future.

(a) In carbon dioxide flooding for Enhanced Oil Recovery (EOR), opportunities for new separations exist both upstream and downstream of the flooding step. The separating medium, supercritical fluid carbon dioxide, may require various purification steps, because it usually has to be pipelined from one location to the other, preferably in the supercritical state. For pipeline transportation, carbon dioxide must be relatively free of methane and nitrogen which impair the miscibility behavior of carbon dioxide and complicate metering in custody transfer; it must be essentially free of hydrogen sulfide (specs 4 to 16 PPMV) to avoid corrosion and safety problems, and it should contain less than 60% saturation of water at minimum pressure and temperature. Many designs call for 30% water saturation (2).

At the source of carbon dioxide (CO_2), we are often faced with the problem of hydrate formation. In order to avoid this problem, we use hydrate inhibitors, which are essentially antifreeze solutions such as ethylene glycol (EG) or diethylene glycol (DEG). During service, these solutions take up electrolyte and water from the carbon dioxide and the formation and eventually lose their effectiveness. In principle, the reclamation of hydrate inhibitor (EG or DEG) from spent solution can be accomplished by (a) evaporating the water, (b) distilling the EG or DEG under vacuum and (c) reconstituting the solution by mixing the desired amounts of DEG and water. However, this approach entails vacuum distillation of a heavy organic in the presence of scaling salts. Under these conditions heat transfer can be impaired by scaling. Furthermore, overheading of DEG results in losses by thermal decomposition and as bottoms.

A more effective approach is to desalinate the inhibitor/water solution to an acceptable level, and then to reduce the water to the desired level by evaporation. One approach towards desalination is by electrodialysis. We have found this approach to be satisfactory in our commercial operations.

The glycols can also be used as a drying agent for the CO_2, which simplifies the management of both hydrate inhibitor and dehydration systems.

In connection with drying supercritical CO_2, a potential problem is the solubility losses of typical agents, such as DEG or TEG. In some cases, these losses can be kept low by using glycerine solutions for dehydration (3). The <u>removal of hydrogen sulfide</u> from high pressure or supercritical CO_2 at the source is another technical challenge. Even selective amines, such as the hindered secondary amines used in the Flexsorb (4) process, do not maintain their selectivity in the presence of very high CO_2 pressures.

Shell is currently studying an H_2S removal process which can be used with supercritical fluid CO_2. The process uses a proprietary iron-redox solution. The conceptualized process is shown in Figure 1. The redox reactions are summarized, again in conceptual form, in Figure 2 (5,6). Our studies have shown that under certain proprietary conditions we can operate such a process at excellent volumetric capacities and with very low consumption of chemicals.

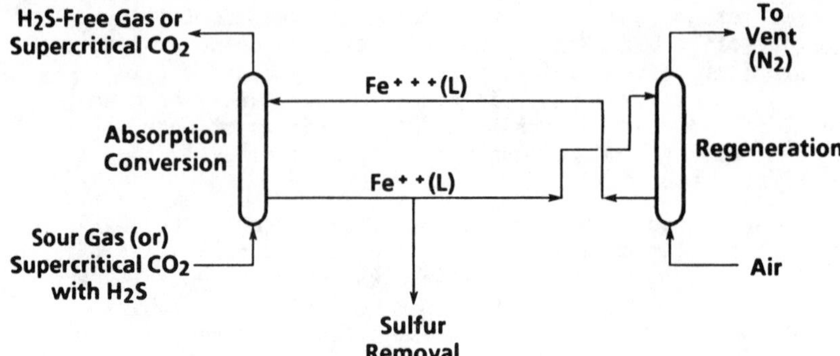

Figure 1. Process for removal of H_2S from sour gases or high pressure and supercritical CO_2.

$$H_2S + 2Fe^{+++}(L) \longrightarrow 2H^+ + S^0 + 2Fe^{++}(L) \qquad \text{Conversion}$$

$$2Fe^{++}(L) + 1/2\,O_2 + 2H^+ \longrightarrow Fe^{+++}(L) + H_2O \qquad \text{Regeneration}$$

$$H_2S + 1/2\,O_2 \longrightarrow S^0 + H_2O \qquad \text{Overall}$$

Figure 2. Oxidation of H_2S to sulfur in an iron redox system.

The removal of H_2S from lower pressures of CO_2 gas is not as difficult as the separation of H_2S from supercritical CO_2. A two-stage Selexol process is planned by Exxon for the removal of 4.5% H_2S from a source which contains 66% CO_2, 22% CH_4, 7% N_2, and 0.5% He. The project, at the La Barge anticline, generates an H_2S-rich Claus feed, high purity CO_2, and a methane/nitrogen/helium stream. This stream is dehydrated and its components are segregated by a cryogenic process (7).

After carbon dioxide flooding, we must provide processes for the recovery of CO_2 from the produced gases for recycle. An interesting engineering problem at this point is the variability in composition of recovered gas. Thus, at the early stages of an EOR project, the CO_2 returning with hydrocarbon gases will be low, under 10% mole. However, 4 to 6 years after CO_2 flood startup the level of CO_2 may increase to 80% or 90%. The optimal process for CO_2 recovery during the early phases of the project could be different from the optimal process at a later time. In the early phase of a project, if separation facilities are not yet available, the gaseous mixture can be reinjected.

Several recent studies have addressed the subject of gas separations in EOR projects. The conclusion from these studies is that the selection of a given process depends on the composition, pressure, temperature and other conditions of gaseous mixtures and of the field (8-12).

Figure 3, from a 1982 study, summarizes the optimum areas of applicability of various types of processes for CO_2 recovery from methane (9). Chemical solvents, physical solvents, and cryogenic and membrane processes were compared. Figure 3 shows that membranes and physical solvents are advantageous at high CO_2 partial pressures, whereas chemical solvents are preferable

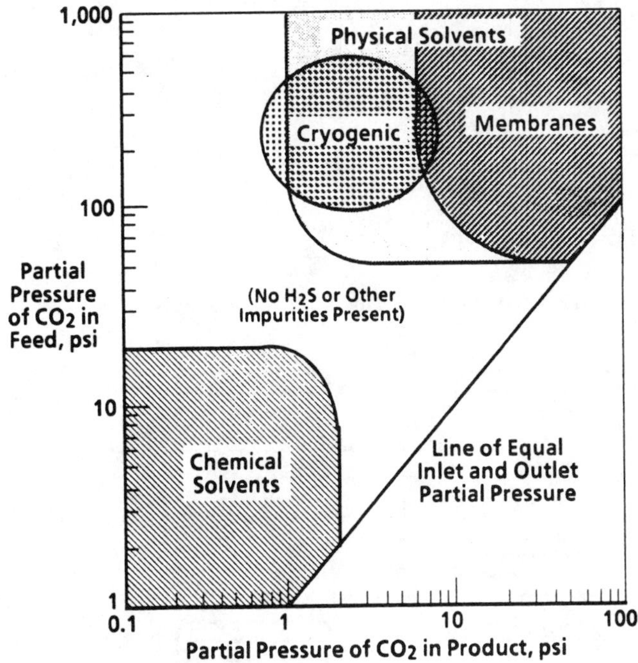

Figure 3. Gas treating for CO_2 removal.

at low CO_2 partial pressures. Therefore, in principle, during the early stages of a project, economics favor an amine process. At the later stages, a membrane process could be retrofited. In membrane-amine processes, the CO_2/CH_4 feed would go to the membrane process first. After recovering the bulk of CO_2, the reject from the membrane process would be treated with the amine process. The modular nature of membrane plants introduces a flexibility which could simplify the design of such plants.

A recent (1986) report suggests that if small amounts of H_2S can be reinjected with the CO_2, then a membrane-only process can be very economical with high CO_2 EOR gases. The report refers to CYNARA membrane systems in their application at SACROC Project (Snyder, Texas) (10).

The separation of CO_2/CH_4 gas for various uses is illustrated by Figures 4, 5 and 6, which show various configurations of one, two and three membrane modules (9). These modules can be hollow fiber, or spiral wound.

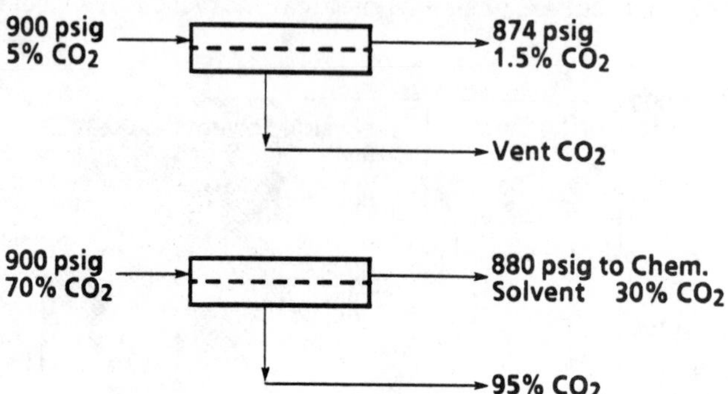

Figure 4. Cleanup and topping units (one stage).

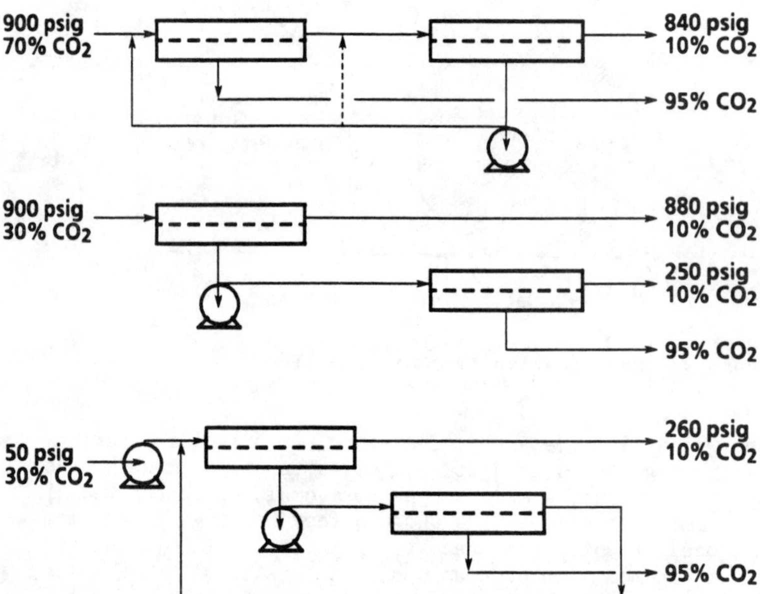

Figure 5. CO_2 and field fuel gas (10% CO_2) separation (two stages).

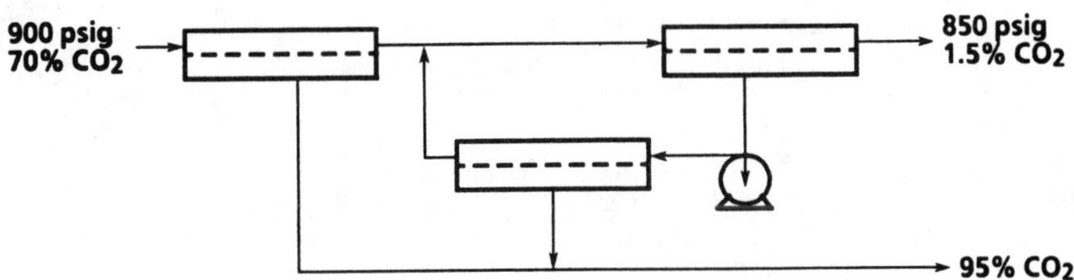

Figure 6. CO_2 and pipeline gas (1.5%) separation (three stages).

The presence of H_2S complicates separations in the EOR produced gas also. This is particularly true when we contemplate cryogenic separations for CO_2 removal from C_2+ hydrocarbons. Shell in its Wasson Denver Unit will employ MDEA (methyldiethanolamine) to remove H_2S prior to dehydration and fractionation. A block diagram of the process is given in Figure 7 (11). Processes which can be considered for the removal of H_2S from CO_2 include the Split-Flow Claus, Selectox,(12,13) Stretford, Lo-Cat,(14) Flexsorb,(4) and Sulfinol-M.

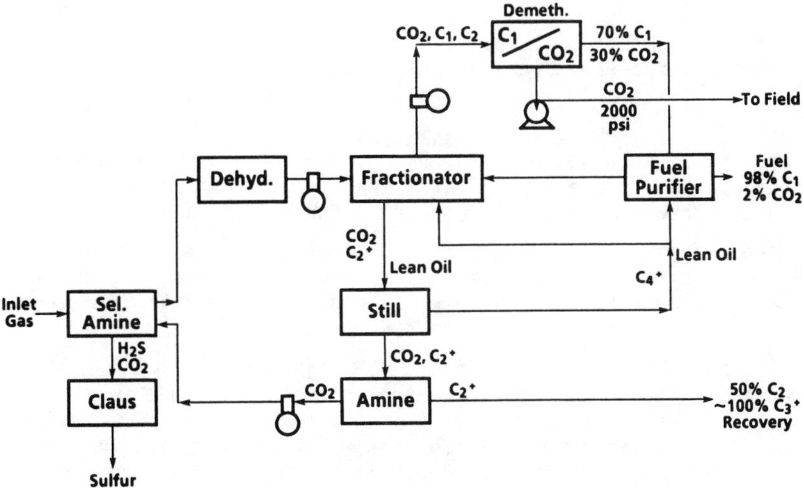

Figure 7. Process scheme for Wasson Denver Unit.

Finally, it should be pointed out that combination "amine-membrane" processes can sometimes be used to advantage to tailor not only the produced gases to purity specifications but also the recycled CO_2 gas for improved miscibility properties. Thus, in order to increase the miscibility region of CO_2 we can allow some C_2+ gases and some H_2S gas to be recycled with the CO_2. This is accomplished by the process shown in Figure 8 (2).

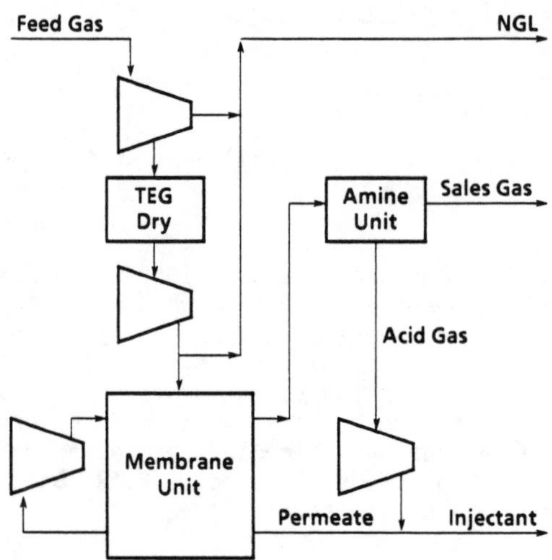

Figure 8. Membrane/amine combination allowing increased miscibility for recycled CO_2 (ref. 2).

(b) Thermal Recovery by steaming is an enhanced recovery process already in wide use. Over 500M barrels per day are produced in this manner. In a typical steaming operation we may need up to nine barrels of steam per barrel of produced crude.

Two separations are important in this operation: Firstly, the recovery of dry crude (less than 3%w water) from very large volumes of crude/water mixtures; secondly, the recovery and recycle of pure water for further steam production.

Since thermally produced crudes are generally heavy, with a density of about 1.0, the oil/water separation is difficult. Thus, dehydration requires very large settling tanks which allow many hours of residence time for phase separations. Furthermore, surfactants are used for the destabilization of emulsions and reverse emulsions. The problem of dehydration is further complicated by the presence of suspended solids, sands and clays, in the production (oil/water) mixture. Solids, and especially clays, stabilize interphases and emulsions. Clays become oleophilic and tend to stay suspended in the oil, thus creating potential problems in downstream processing. A better understanding of the surface phenomena involved in the oil/water/solids separation is needed.

The volumes of water recovered with the thermally produced crude are very large. Recycle of this water to steam generators requires purification and desalting. Optimization of the ion exchange or other desalting or softening processes involved is an important design problem.

(c) Offshore wells account for about 15% of the domestic oil production. Due to the high cost of platform space ($8M to $15M per ft^2) only the most necessary separations are carried out on the platforms. Oil/water separations are carried out in heater-treaters which take less space than the settling tanks used in thermal crude production. Even though the water produced in offshore platforms is generally less than the water produced from steam generated wells, improvements which will reduce the space requirements of offshore oil/water separations would be worthwhile. In fact, water flooding is already used extensively as a means of increasing production. This further underlines the need for compact oil/water separation processes.

A typical oil/water separator system for offshore use is shown in Figure 9. This figure shows that the space taken by oil/water separation equipment is considerable. It also shows that we need chemical additives, antifoams, emulsion breakers, flocculants and biocides, a practice which is under some scrutiny by the EPA.

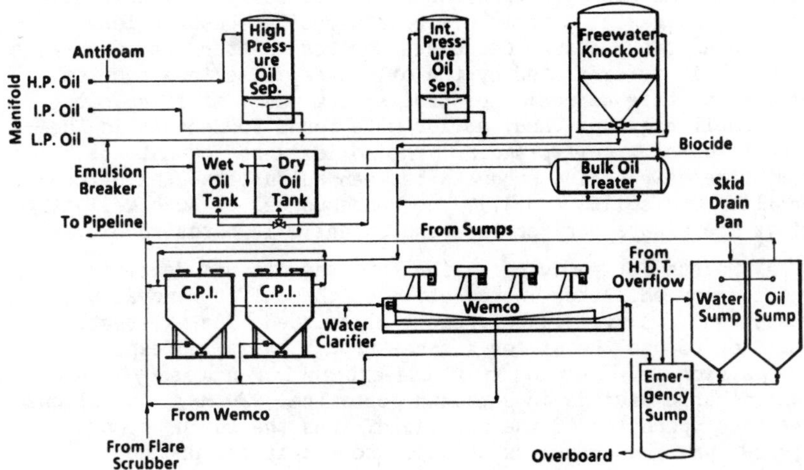

Figure 9. Oil/water separation on an offshore platform.

Other separations of importance to offshore applications are gas separations. Membrane systems, especially if compact and light can become an important tool in the offshore environment.

Thus, Monsanto PRISM systems have found several applications on platforms. They are used in removing oxygen from air to produce dry nitrogen gas, with less than 5% O_2. The gas is used for blanketing equipment and controls, during shutdowns. Units that produce up to 40,000 SCF/hr are available. The systems operate at temperatures from 0°C to 100°C and pressures up to 2000 psi. Conoco (UK) Ltd, and Statoil (Norway) are using such skid-mounted units on their platforms (15).

(d) Gas Production. Gas separations have already been discussed in connection with EOR. In this section an example is discussed in which a solvent extraction process holds the key to the successful operation of several deep sour gas wells. Such wells present many challenges to the gas technologist (16).

Shell operates a profitable gas field in Thomasville, Mississippi, in which sour gas is produced at 21,000 feet and contains over 35% H_2S by volume. The Thomasville gas plant has several gas wells which are subject to very high pressures (20,000 psia initially) and temperatures (400°F).

In order to protect the metal tubing used in these wells against corrosion, a specially formulated oil phase containing corrosion inhibitors is circulated in the well. The flow direction is shown in Figure 10. The oil/inhibitor mixture goes to the bottom of the well through the annular space and returns to the surface through the center tubing, along with the produced sour gas. The sour gas contains dissolved sulfur which is soluble in the gas at bottom hole conditions but becomes less soluble as the pressure decreases towards the surface. Consequently, it is extracted by the oil phase. A serious problem related to this corrosion control system was identified soon after well startup. Thus, solids were found to deposit in the annuli, causing well plugging. The solids were found to be a composite of reaction products between sulfur, the inhibitor and drilling mud solids (16,17). The consequences of such well pluggings are always very costly, and potentially catastrophic.

A process for removing sulfur from the oil was developed (18) and introduced. After introduction of the sulfur removal process, annuli plugging was greatly alleviated. The alternative would be to use the oil on a once-through basis, an expensive proposition. Implementation of the extraction process yielded considerable savings in shutdown costs (ca. \$20 MM), and allowed the safe operation of the gas field. Thus the sulfur removal process was the unobvious solution to a critical production problem.

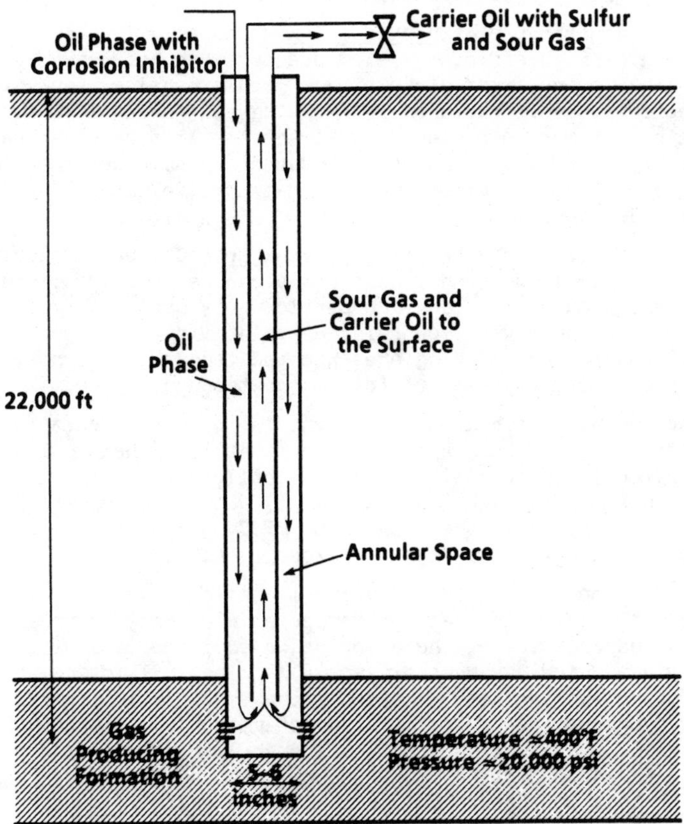

Figure 10. Schematic of deep sour gas well.

II. THE PETROCHEMICAL COMPLEX

(a) Distillation

In the refinery, crude distillation is one of the major users of energy. The distillation process has received renewed attention in the past few years, mostly because of the increased cost of energy.

Even though distillation is 75 years old, there are still research areas in which more effort is justified. In a recent paper (19), Fair and Humphrey emphasize the need for more work on fundamentals of mass transfer as they impact on tray efficiencies, and on thermodynamic analysis and experimental testing as they relate to energy consumption in distillation.

Some important trends observed recently in distillation practice are discussed in the following paragraphs.

(i) The increased use of packed columns, especially the introduction of high efficiency, low pressure drop structured-type packings is one trend. Even though more expensive than dumped packings, the structured packings have high specific surface area and promote efficient mass transfer. This reduces the HETP. The higher efficiency and the reduced pressure drop can justify their cost, especially when heat-sensitive materials are distilled under vacuum.

(ii) The increased use of energy saving methods, particularly heat pumps, is a second trend. Heat pumps can be used in columns with low temperature drops and are particularly useful in distillations of systems with low relative volatility and high reflux requirements. By reducing pressure and temperature drops, structured packings are compatible with heat pumps.

For more details on heat pump designs the reader is referred to the literature (20-23). A notable application of heat pumps is in propylene/propane splitters. This application has been discussed in some detail by Quadri (24,25). Continued growth in the use of heat pumps in the separation of C_3-hydrocarbons, C_4-hydrocarbons and C_5-hydrocarbons is expected (19).

Most heat pump arrangements involve mechanical vapor recompression. With mechanical vapor recompression we have to balance the capital expenditure of the compression equipment against the energy savings which can be realized. A less capital-intensive concept for recovering heat in hydrocarbon distillation is (iii) multistage condensation (26). In a two-stage condensation, the first stage of condensation supplies enough liquid to reflux the column and the second stage provides the top product of the column. For this concept to be applicable the overhead product must be a multi-component mixture with some temperature spread between the dew and the bubble points. Furthermore, the column top temperature should be high enough to be useful as heat for another process.

A typical example is the overheading of gasoline hydrocarbons (400-430°F) from crude oil. Bannon and Marple show that a 75% increase in the heat which is reusable at 300°F can be achieved by going from one stage to two stages of condensation. This is shown in Figures 11 and 12. In a crude column in Shell's Deer Park Manufacturing Complex, two-stage distillation saved 80MM Btu/hr (Figure 13).

Before closing this section it should be mentioned that energy consumption can also be decreased through application of closer process control, for instance on-line analysis of streams. Also improved vapor-liquid equilibrium data, when available, should be incorporated into the design/operation basis as soon as possible. In many cases, these approaches would permit significant reduction of operating reflux.

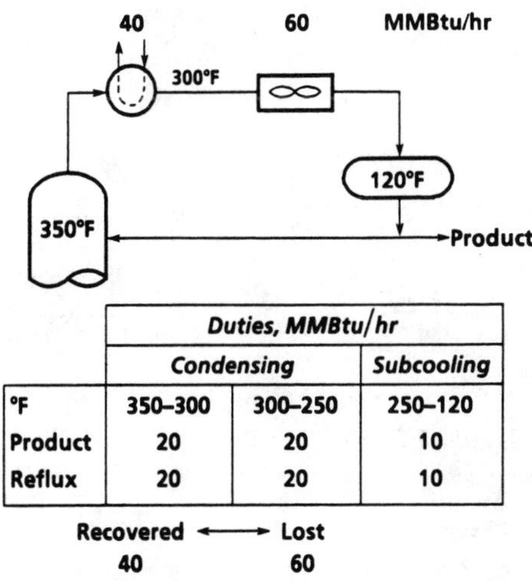

	Duties, MMBtu/hr		
	Condensing		Subcooling
°F	350–300	300–250	250–120
Product	20	20	10
Reflux	20	20	10

Figure 11. Conventional overhead system.

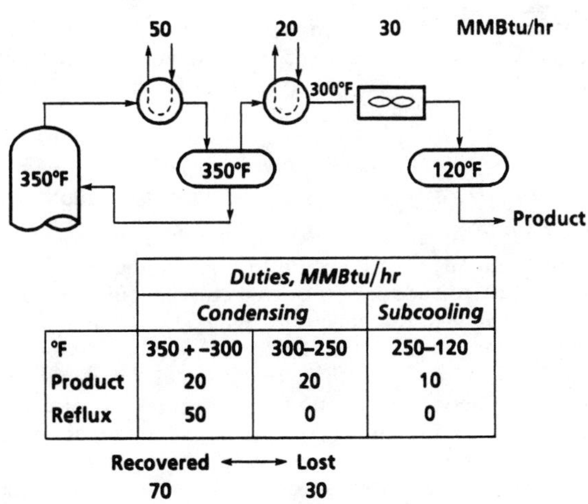

	Duties, MMBtu/hr		
	Condensing		Subcooling
°F	350 + -300	300–250	250–120
Product	20	20	10
Reflux	50	0	0

Figure 12. Two-stage condensation.

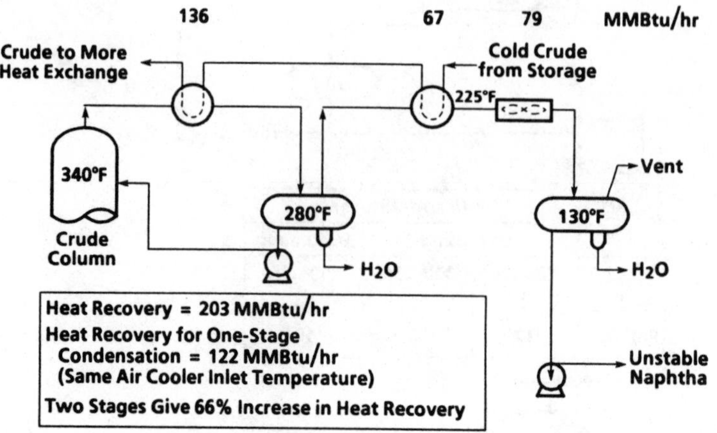

Figure 13. Crude column No. 2 at Deer Park

(b) Novel Concepts in Distillation

(i) Catalytic Distillation. In the last few years methyl tertiary butyl ether (MTBE) has become important as an octane number booster, filling the need left by the removal of tetraethyl lead from gasoline. It is generally made by the liquid-phase reaction of isobutylene (IB) and methanol, using a fixed-bed acidic ion exchange resin catalyst. Steam-cracked mixed butylenes (raffinate from the butadiene extraction plant) and catalytically-cracked C_4 fractions can be used as feedstocks.

Up until 1980 there were four companies offering MTBE processes for license: Huels, Snam Progetti, Arco and Sun. The differences between these processes are in reactor configurations, numbers of reactor stages, isobutylene conversion levels (80 to 99.8%) and purity of produced MTBE (97% to 99.7%). Temperatures from 40°C to 150°C are used (27,28). A generalized flowscheme of the conventional MTBE process is shown in Figure 14. In this process MTBE is produced in the two reactor stages shown in the left side of the flowscheme. Since the reaction between isobutylene and methanol is exothermic, inter- and intrastage cooling is provided to allow completion of the reaction. The mixture of products and unreacted methanol are then separated in the remainder of the process and the recovered methanol is recycled for use.

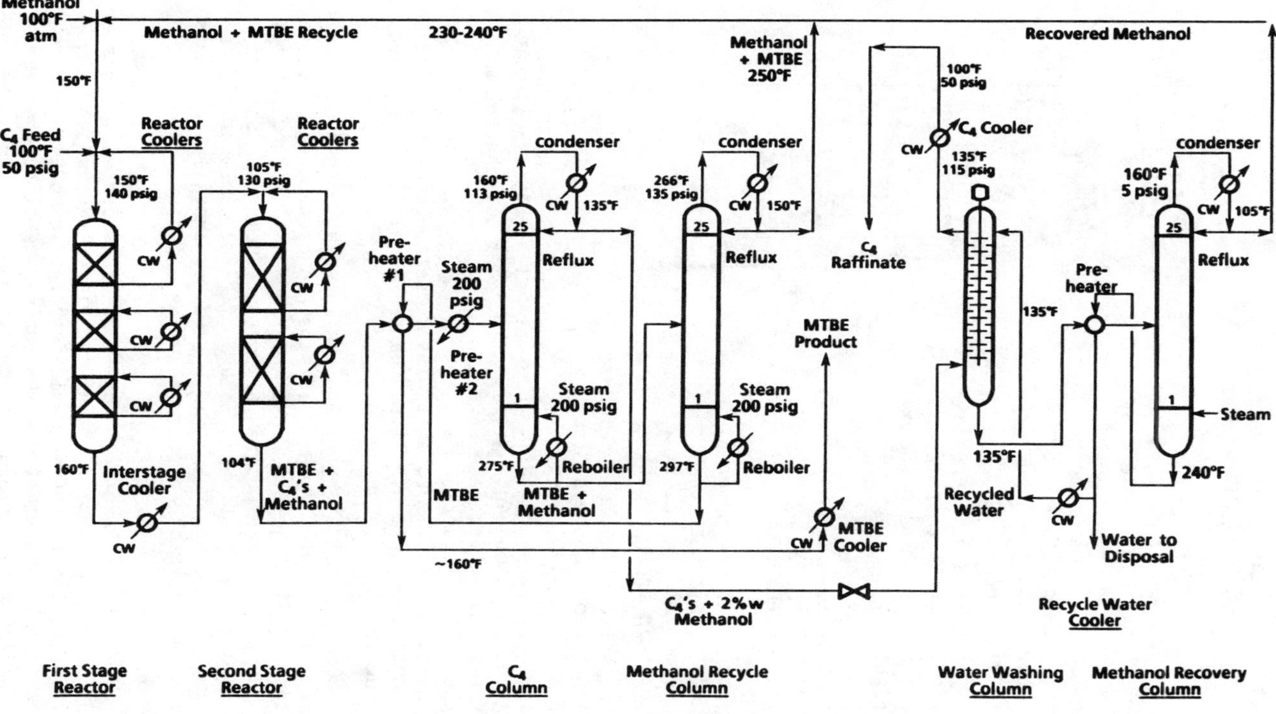

Figure 14. Conventional methyl tertiary-butyl process.

An interesting version of the MTBE process has been recently developed (29). In this version, distillation and catalyzed reaction steps are carried out in the same distillation unit as in Figure 15. The catalyst in this column is packaged in pockets of filter clothing held in place by stainless steel mesh. With this arrangement, there is no contraction and expansion of the resin catalyst bed, thus leading to a longer catalyst life.

This approach takes advantage of the LeChatelier's Principle to push the equilibrium shown below to the right.

$$IB + MeOH \underset{K_2}{\overset{K_1}{\rightleftharpoons}} MTBE$$

The developers of the process, CRL/Neochem, claim that they can approach "multiple-equilibrium stages" by this arrangement.

Also, whereas in the conventional process the heat of reaction has to be removed in interstage coolers to avoid increasing the rate constant K_2 of the back reaction, in the Neochem system the heat of reaction is used to distill the C_4-raffinate. The temperature of the reaction is controlled by controlling the pressure. The integrated MTBE-Neochem process is shown in Figure 16.

Figure 15. Catalytic distillation column.

(ii) Centrifugal Distillation. An interesting new distillation device is the HIGEE, which is a packed contactor rotated at high speed (30,31). The HIGEE was invented and developed at ICI (UK), as part of an R&D program aimed at reducing the size of processing units without sacrificing efficiency (process intensification). Compactness can lead to reduced installation costs and to smaller inventories, thus allowing efficient batch operations.

By using special packing, the HIGEE increases mass transfer rates. By using gravitation, it allows higher throughputs. On the debit side, the use of mechanical energy is costly, and rotating equipment can lead to vibrational instability problems. Also due to the gravitational fields, vacuum operations are not practical.

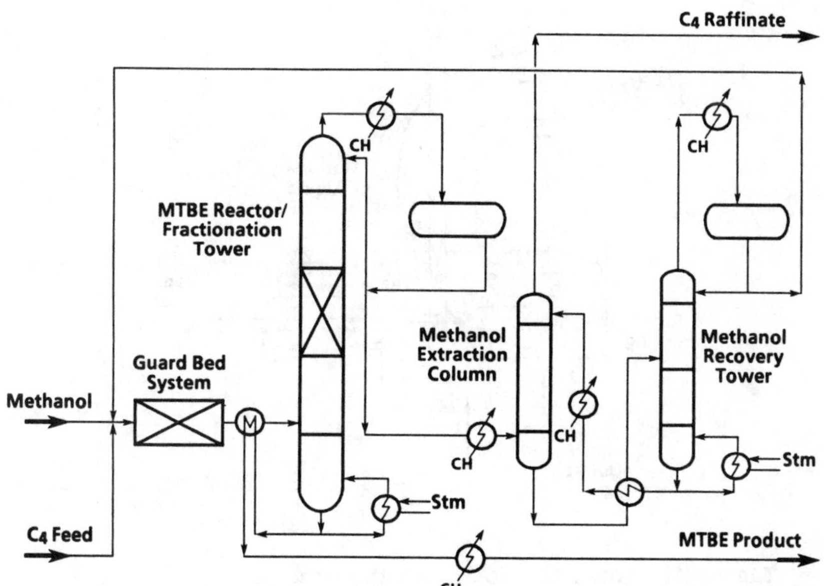

Figure 16. Simplified process flow diagram CR&L/Neochem MTBE process.

The HIGEE unit is illustrated in Figure 17. Figure 18 shows how HIGEE units can be integrated to simulate various sections of a distillation unit.

(c) Solvent Extraction

The most important solvent extraction process in the petrochemical complex is the extraction of aromatics from paraffins. A well known process for this separation is Shell's Sulfolane process, available for licensing through UOP. This process has been optimized further by improved design procedures and by the use of better equilibrium data. However, with the possible exception of the ROSE process (discussed below) no major breakthroughs have taken place in solvent extraction in recent years. The work carried out in the 60's and 70's with cuprous complexes towards separation of olefins (32-34) has not been commercialized, because advances in conventional separation methods and new sources of α-olefins have reduced the incentives for radical innovation. Yet, it is interesting to note that a centrifugal extraction device, which could be the forerunner of the HIGEE distillation, has been in operation for over 25 years. This is the Podbelniak Extractor which is briefly described in the following section.

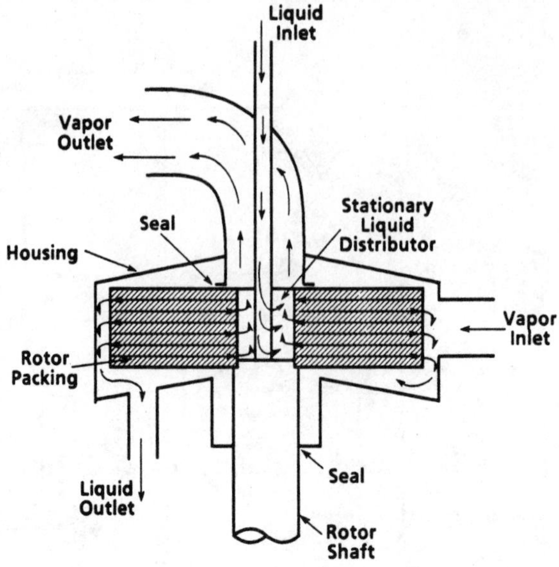

Figure 17. Rotating HIGEE packed contactor.

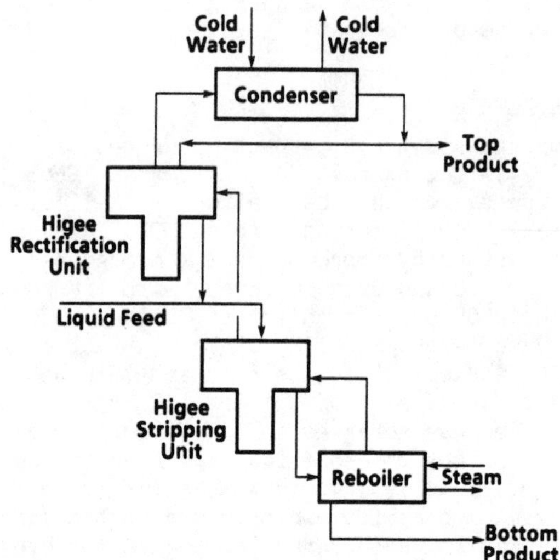

Figure 18. Integrated HIGEE distillation system.

(i) Lube Extraction Using Podbelniak Centrifugal Extractors. Shell has used centrifugal Podbelniak extractors (PODS) for many years to extract lube with phenol or n-methylpyrolidone (NMP). It is estimated that three of these units, 4 ft in diameter and 2 ft in depth (Figure 19), have a throughput equal to that of a 20 ft by 110 ft extraction tower. The vibrational instability problems associated with equipment rotating at over 1000 RPM have been solved successfully, but not without some effort.

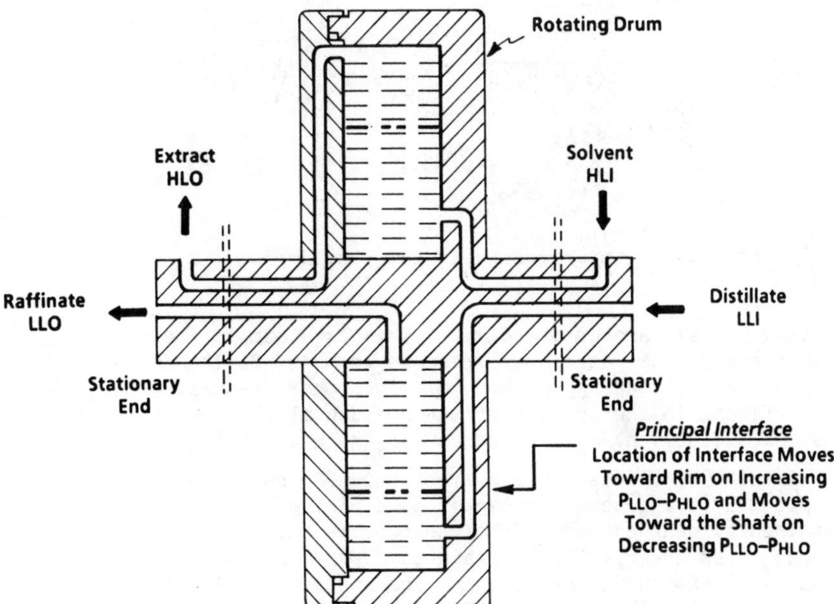

Figure 19. Cross section of Podbelniak extractor.

Each contacting unit provides 1 1/2 to 2 theoretical stages. These units are particularly suitable when the density difference between the two phases is small, as when the solvent is NMP. The PODs facilitate phase coalescence, thus alleviating the need for large diameter columns. They are particularly suitable for use in the lube plant, where batch operation is normal practice. These units, like the HIGEE, have very small inventory.

(ii) Supercritical Extraction in the Refinery. The ROSE Process. The principle utilized in this process is summarized in Figure 20. In conventional deasphalting the solvent plus deasphalted oil (DAO) leaves the extractor as liquid at low temperature and pressure at point A. This mixture is heated and flashed to point C. At C virtually all solvent is vaporized

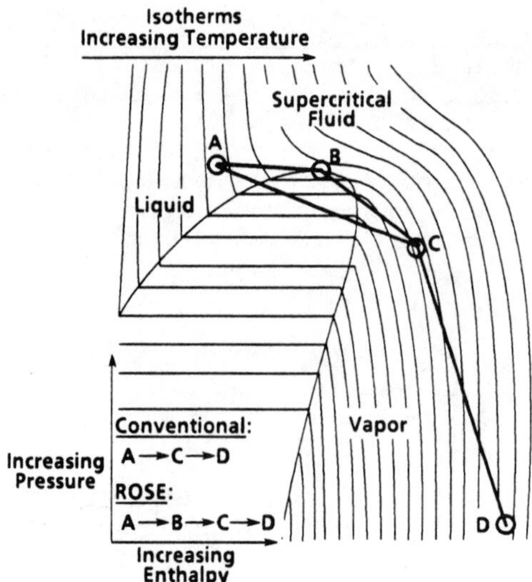

Figure 20. Solvent recovery enthalpy diagram.

leaving a small amount of solvent. This is steam stripped at a higher temperature and lower pressure to point D. The enthalpy change,i.e., the energy needed to effect these separations, is proportional to the product of the horizontal distance times the mass of the solvent and of the DAO at any given step.

In the ROSE Process, the solvent plus DAO solution is heated to point B (using sensible heat). Some of the sensible heat is obtained by heat exchange from the recycle solvent, which of course saves energy. The big gain in energy, however, takes place at B. At the higher temperature at B, the solvent and oil are immiscible. Thus, 85 to 93% of the solvent phases out, and is recovered for recycle to the extraction. From point B, the oil with 7% to 15% of the solvent is taken to point C and eventually to D. In this case, because the solvent was removed by phasing out at B, the total enthalpy needed in the process is at least 40% less than in the conventional deasphalting process.

(d) Adsorptive Separations

In the petrochemical plant adsorptive separations are used both for purification of streams and for bulk separations. For the removal of trace amounts of impurities one needs either molecular sieving action, or a chemical reaction at the surface, chemisorption. Some streams which need high degree of purification are, for example, monomer streams which are feeds to polymerization reactors employing very active catalysts.

In some cases the molecular structures of the impurities to be removed are not known. The first step in solving such a problem is to identify the structure of those impurities and then to find a suitable adsorbent. Fortunately, some adsorbent manufacturers (36) are becoming more involved in studying the adsorptive and physical and chemical properties of their products. This should help the industrial researcher considerably.

Whole new classes of molecular sieves (silicalites, aluminophosphates, etc.) have become available in the last ten years. The adsorption characteristics of these materials and their potential use in separations have to be explored further.

For bulk separations, the UOP SORBEX process, although employing fixed bed adsorbents, simulates solid-liquid countercurrent contacting quite satisfactorily. In fact, the UOP process is presently the most advanced concept for large scale liquid phase separation with sieves. Nevertheless, the engineering of a real solid-liquid countercurrent adsorption process remains a challenge.

Large scale bulk separations can also be carried out by gas phase adsorption. In this case countercurrent contacting is less essential. Commercial scale molecular sieve units recovering detergent range paraffins in a gas phase adsorption process have operated for over 20 years (37). In developing novel adsorption processes for gas phase separations, it is essential to minimize undesirable side reactions which may be catalyzed by the adsorbent surface. Thus, inertness of the sorbate in the presence of the adsorbent, especially at regeneration temperatures, must be ensured.

(e) Membrane Separations

With the exception of water desalination by reverse osmosis and electrodialysis, gas phase separations are being commercialized more rapidly than liquid phase separations. The use of enriched air in petrochemical processing could be facilitated by the availability of oxygen-selective membranes. Hydrogen recovery from refinery gases should also become more attractive as fuel costs decline.

A factor for the ready acceptance of gaseous membrane separations is that gases are more forgiving to membranes than liquids, especially organic liquids. Plasticization of membranes by solutes is being studied extensively. An important component of the Separations Research Program at the University of Texas, Austin, is devoted to the understanding of the solute-membrane-solvent interaction relevant to plasticization (38-40).

Reverse osmosis (RO) membranes would find more uses in industrial separations if they were resistant to attack by organic components. Thus, aqueous solutions of organics, especially when electrolytes are also present, can cause deterioration of RO membranes by differential swelling of the various membrane layers. Work towards avoiding such differential swelling in RO systems is needed. Homogeneous RO systems may be promising in this regard.

Another approach towards making stable membranes is the development of dynamic membranes. These are formed by passing membrane-forming materials over and through a porous metallic support (41). The selectivity of such systems is still inadequate, however. Polyacrylic acid and hydrous zirconia mixtures have been used to make dynamic membranes.

III. ENVIRONMENTAL EXAMPLES

The pressure for a cleaner environment has persisted over the last two decades despite economic cycles. Biotreating has been established as a major process during the 70's. Nevertheless, some water pollutants are not readily biodegradable. The last case study in this paper will be a discussion of two effluents with very different properties, and how they can be handled cost effectively.

The first case is an aqueous effluent containing byproducts of an insecticide. The effluent contains chlorinated and nitrilic compounds of very low solubility in water. The first route considered for the disposal of this toxic stream was the Zimpro wet oxidation process. The high capital and energy requirements of this process, however, prompted a search for a better alternate. The low solubility of the pollutants in water suggested solvent extraction with a light hydrocarbon. This approach, solvent extraction, resulted in a far more economic disposal scheme. In the commercial process, the pollutants are extracted in the light solvent, the solvent is vaporized for recycle, and the pollutants are incinerated (42).

The second case is an aqueous effluent containing a saleable by-product, glycerine, along with NaCl. The selective removal of glycerine from water is not practical. Thus, we opted for vapor compression, which is an energy-efficient method of removing more than 95% of the water, which results in the precipitation of over 95% of the salt. This process is shown in Figure 21. The glycerine concentrate must be further desalted in order to yield salt-free polyglycerol bottoms. This can be accomplished by electrodialysis or by adding a monoalcohol to the concentrate at high temperature, as shown in Figure 22. The NaCl is precipitated from the glycerol/water mixture when pentanol is added, because a pentanol/glycerol/ water mixture has much lower solubility for NaCl than the glycerol/water phase (43). A high temperature is used in the pentanol process because at low temperatures the ternary mixture forms two phases as in Figure 23.

The remaining water is evaporated and the glycerine is overheaded to yield a salt-free polyglycerols bottoms. The salt-free polyglycerols can be incinerated in existing incineration equipment.

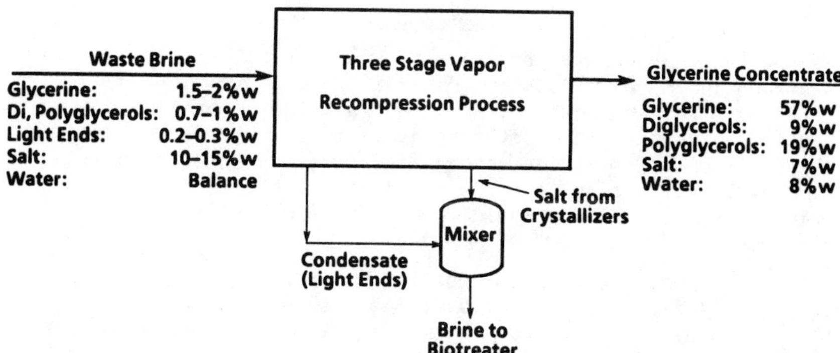

Figure 21. Vapor compression/crystallization process for concentrating glycerine.

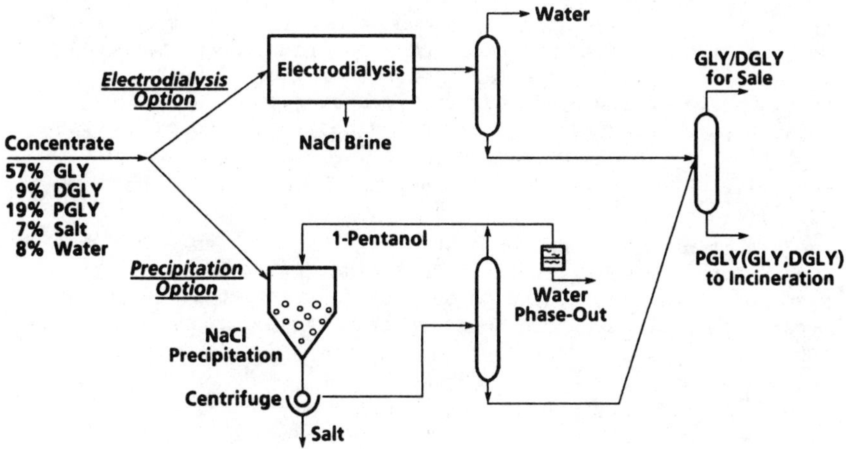

Figure 22. Options for desalting concentrate and glycerine recovery.

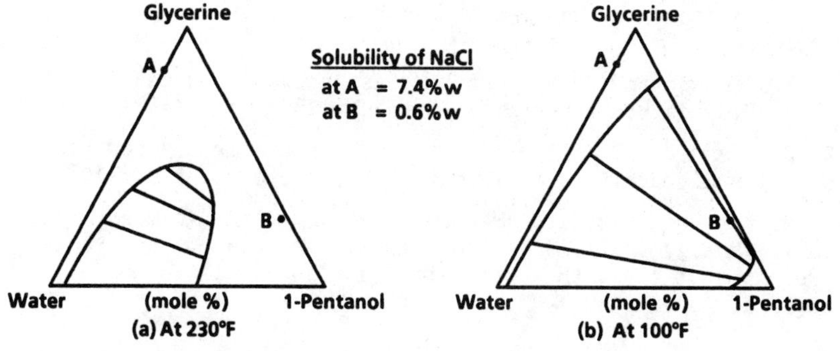

Figure 23. Phase relationships in pentanol precipitation route.

CONCLUSION

The last decade has presented many opportunities for new separations in the oil industry. Many of these innovations were related to the production of oil and gas. The separation advances in the refinery and petrochemical complex were more along the lines of optimization, energy savings, and environmental conservation. This scenario is compatible with the fact that oil prices have been rising rapidly and environmental concerns have persisted.

However, while this paper was being written, the price of crude world-wide declined by over 50%. It is not clear whether this trend will continue or reverse itself. If low crude prices do persist, then the emphasis in the next few years may shift back towards better processes for "value-added" products. That means more profitable operations in the refinery and the petrochemical complex, and hence more opportunities in these areas.

In any case, the technological trends that were started in the last decade have acquired a momentum which will tend to lead to more technological progress. Advances in membrane applications, EOR, and environmental concerns will continue to be important even at low crude prices. According to a University of Texas study, energy savings alone are not enough to push implementation of separation processes. Reliability, productivity, control, safety, environmental, and space factors are important as well (44).

In conclusion, it can be stated that the challenge for the separations technologist in the oil industry is to identify the needs of his industry, and to develop relevant, cost-effective separation processes, and to do so in a rapidly changing economic scenario.

REFERENCES

1. G. Foster, "Process Innovation in Petroleum Refining", Alfred P. Sloan School of Management, Working Paper #490-70, September 1970.

2. J. E. Johnson and F. B. Walter, "Gas Processing Needs for EOR", Hydrocarbon Processing, October 1985, pp. 62-66.

3. Z. Diaz and J. H. Miller, "Drying Substantially Supercritical CO_2 with Glycerol", U.S. Patent 4,478,612, October 27, 1984.

4. H. N. Weinberg, et al., "New Gas Treating Alternatives for Saving Energy in Refining and Natural Gas Production", Paper presented at the 11th World Petroleum Congress, London, England, August 31, 1983.

5. G. C. Blytas and Z. Diaz, Sulfur Process, U.S. Patent 4,356,155, October 26, 1982.

6. G. C. Blytas, Sulfur Separation, U.S. Patent, 4,390,516, June 28, 1983.

7. G. A. Petzet, Editor, "Exxon Outlines Big LaBarge CO_2 Development", Oil and Gas Journal, October 22, 1984, p. 72-74.

8. C. S. Goddin, "Comparison of Processes for Treating Gases with High CO_2 Content", Proceedings of Sixty-First Annual Convention, GPA, p. 60-68, March 15-17, 1982, Dallas, Texas.

9. K. C. Youn, H. Wall, G. C. Blytas, "Role of Membrane Technology in the Recovery of Carbon Dioxide, Economics and Future Prospects", paper presented to GPA Regional Meeting, Houston, November 18, 1982.

10. "Low Cost Membrane Process Handles EOR Associated Gas Streams", Oil and Gas Journal, January 6, 1986, p. 70.

11. A. J. Flynn, "Wasson Denver Unit - CO_2 Treatment", Proceedings of Sixty Second Annual GPA Convention, p. 142-145, San Fransico, California, March 14-16, 1983.

12. G. Goar, "CO_2 Removal from EOR Gases and Attendant Sulfur Handling Problems", Sixty-Fourth Annual convention, GPA, March 18-20, 1985, Houston, Texas.

13. R. V. Bertram, et al., "Selectx Process: Versatile, Economical Technology for Sulfur Recovery", paper presented to the Sufur 85 International Conference, November 13, 1985, London, England.

14. L. C. Hardison, "Catalytic Gas-Sweetening Process Selectively Converts H_2S to Sulfur, Treats Acid Gas", Oil and Gas Journal, June 4, 1984, p. 60-62.

15. "Membrane Based Nitrogen Generator Moves Offshore", Oil and Gas Journal, July 1985, p. 116.

16. T. W. Hamby, Jr., "Development of High Pressure Sour Gas Technology", presented at the 54th Annual Fall Conference of the Society of Petroleum Engineers, September 23-26, 1979.

17. M. C. Place, "Corrosion Control - Deep Sour Gas Production", presented at the 54th Annual Fall Conference of the Society of Petroleum Engineers", September 23-26, 1979.

18. G. C. Blytas, "Sulfur Extraction Method", U.S. Patent, 4,230,184, October 28, 1980.

19. J. R. Fair and J. L. Humphrey, "Distillation Research Needs", Separation Science and Technology, 19 (13-15), pp. 943-961, 1984-85.

20. H. R. Null, "Heat Pumps in Distillation", Chem. Eng. Prog., 72 (7), July 1976, pp. 58-64.

21. F. E. Becker and A. I. Zakak, "Recoverying Energy by Mechanical Vapor Recompression", Chemical Engineering Progress, July 1985, p. 45-49.

22. J. Flores, F. Castells and J. A. Ferre', "Recompression Saves Energy", Hydroc. Processing, July p. 59, 1984.

23. P. S. O'Neil, et al., "Vapor Recompression Systems with High Efficiency Components", Chem. Eng. Progress, July 1985, p. 57-62.

24. G. P. Quadri, "Use of Heat Pump for P-P Splitter. 1-Design", Hydroc. Proc., February 1981, p. 119-126.

25. G. P. Quadri, "Use Heat Pump for P-P Splitter. 2-Optimization", Hydroc. Proc., March 1981, p. 143-151.

26. R. P. Bannon, S. Marple, "Heat Recovery in Hydrocarbon Distillation", Chem. Eng. Prog., July 1978, p. 41-45.

27. Methyl Tertiary-Butyl Ether and Tertiary Butyl Alcohol", SRI Report #131, August 1978.

28. L. S. Bitar, E. A. Hazbun and W. J. Piel, MTBE Production and Economics, Hydrocarbon Proc., October 1984, p. 63-66.

29. E. P. Lander, J. Nathan Hubbard, and L. A. Smith, "Reving-up Refining Profits with Catalytic Distillation", Chem. Eng., April 18, 1983, p. 36-39.

30. C. Ramshaw, HIGEE Distillation", The Chemical Engineer, February 1983, p. 13-14.

31. "New Mass Transfer Find is a Matter of Gravity", Chem. Eng., February 21, 1983, p. 23-29.

32. G. C. Blytas, "Decomposing Cuprous Trifluoroacetate - Olefin Complex with a Paraffin Countersolvent", U.S. Patent 3,546,106, December 8, 1970.

33. G. C. Blytas, "Styrene Separation from Ethylbenzene via Cuprous Complexing", U.S. Patent 3,801,664, April 2, 1974.

34. R. B. Long, et al., "Cuprous Halide Slurry Process for Liquid Purification", U.S. Patent 3,481,971, December 2, 1969.

35. Chem. and Eng. News, November 26, 1984, p. 35-37.

36. K. P. Goodboy and H. L. Fleming, "Trends in Adsorption with Aluminas", Chem. Eng. Prog., November 1984, p. 63-68.

37. E. E. Young and G. C. Blytas, "n-Paraffin Separation Process", U.S. Patent 3,309,415, March 14, 1967.

39. D. R. Lloyd and T. B. Meluch, "Material Science and Evaluation for Liquid Separation Membranes", in Materials Science of Synthetic Membranes, D. R. Lloyd Editor, ACS Symposium Series, No. 269, Washington, D.C. (1985).38.J. M. Dickson, M. Babai-Pirouz, and D. R. Lloyd, "Aromatic-Water Separations by a Pressure Driven Membrane Separation Process", Industrial and Engineering Chemistry, Process Design and Development, 22, p. 625-632 (1983).

40. J. S. Chiou and D. R. Paul, "Sorption and Transport of CO_2 in PVF_2/PMMA Blends", submitted to the Journals of Applied Polymer Science.

41. C. H. Gooding, Chem Tech, June 1985, p. 348-354.

42. G. C. Blytas and R. K. June, "Removal of Organic Contaminants from Waste Water", U.S. patent 4,401,570, August 30, 1983.

43. G. C. Blytas, "Recovery of Glycerine from Saline Waters", U.S. Patent 4,560,819, December 24, 1985.

44. J. L. Bravo, et al., "Assessment of Potential Energy Savings in Fluid Separation Technologies: Technology Review and Recommended Research Areas", The University of Texas at Austin, December 31, 1984.

ETHANOL RECOVERY FROM LOW-GRADE FERMENTATES BY SOLVENT EXTRACTION AND EXTRACTIVE DISTILLATION: THE SEED PROCESS

D.W. Tedder, W.Y. Tawfik, and S.R. Poehlein
School of Chemical Engineering
Chemical Process Design Institute
Georgia Institute of Technology
Atlanta, GA 30332--0100 U.S.A

Solvent extraction may be coupled with extractive distillation to recover ethanol from low-grade fermentates. Water coextracts with ethanol from dilute aqueous mixtures, but the resulting extracts can be dehydrated by extractive distillation. Such dehydrated extracts can then be selectively stripped to produce fuel grade ethanol and regenerated solvent.

INTRODUCTION

Liquid/liquid extraction may be used to separate ethanol from water. Earlier measurements were reported by several investigators (1-8). When the relative effects of various molecular functionalities are considered, water immiscible alcohols exhibit higher distribution coefficients. However, solvents containing phosphate groups, ester, and carbonyl groups are also usable. On the other hand, those solvents with higher distribution coefficients (e.g. 2-ethylhexanol) also exhibit lower selectivities. Moreover, the selectivities of all solvents examined tend to decrease with increasing aqueous ethanol concentrations.

Solvents that are capable of achieving high ethanol recoveries from dilute mixtures, yield selectivities high enough to produce fuel-grade ethanol from dilute aqueous mixtures, and also exhibit adequate solute loadings have not been found. Typically, extracts contain 3 to 5 wt% ethanol and about equal amounts of water. In such cases a single cycle (i.e. solvent extraction followed by regeneration) will not yield a fuel grade product (i.e. 98-99 wt% ethanol).

However, the higher boiling solvents (e.g. tridecyl alcohol) also radically change the ethanol/water relative volatility and can be used in extractive distillation where water is selectively removed from the liquid/liquid extract. In this case, those solvents with higher capacities but lower selectivities can still be used to recover a fuel-grade product.

In the case of the higher boiling solvents, Solvent Extraction and Extractive Distillation (SEED) can be coupled as indicated in Fig. 1 to achieve the desired separation. The resulting process consists of three columns: a liquid/liquid extraction column, an extractive distillation column, and a solvent regeneration column. The last column recovers the desired fuel-grade ethanol product and provides regenerated solvent for the liquid/liquid extraction and the extractive distillation columns as indicated.

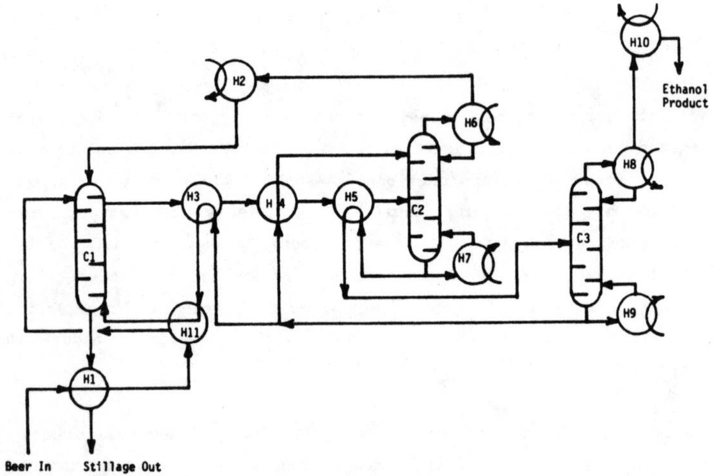

Fig. 1. SEED Process for ethanol recovery from dilute beers. C1 is liquid/liquid extraction; C2 is extractive distillation; C3 is solvent regeneration. Only one solvent is used throughout.

THERMODYNAMIC MEASUREMENTS

The equilibrium compositions of vapor and liquid phases were measured using gas chromatography and internal standards. The modifier/diluent ratio for blended solvents was assumed not to change appreciably due to equilibration. For selected solvent blends, the liquid/liquid mutual solubility curve was also titrated.

Liquid-Liquid Equilibria

Fermentates typically contain small amounts of dissolved salts and species like sugars. Synthetic feeds containing dextrose as a substitute for all nonionic, inextractable dissolved fermentate solids were used to measure liquid/liquid equilibria and to develop the given correlations.

Ethanol and water distribution coefficients in the solvent were measured as a function of ethanol and dextrose weight fractions, the temperature, and the volume fraction of the modifier (either TBP or tridecyl alcohol) initially in the organic diluent (Isopar M). Data analysis then led to the following empirical correlations which are valid over the indicated ranges.

Tridecyl Alcohol/Isopar-M Blends

$$\ln D_e = 7.644 - 3.23\, X_e^2 + 1.76\, X_e + 0.849\, X_D + 2.89\, V_{TDOH} - 3153.4/T$$

$$\ln D_W = 2.664 - 18.941\, X_e^2 + 9.26\, X_e + 1.86\, X_D + 3.73\, V_{TDOH} - 1383.3/T$$

$X_e < 0.33$ $\quad$ $301 < T < 342$ K

$X_D < 0.60$ $\quad$ $V_{TDOH} < 1.0$

These correlations for ethanol and water extraction are based upon 23 equilibrations in which the independent variables were adjusted over the range indicated.

Tri-n-butyl phosphate/Isopar-M Blends

$$\ln D_e = 1.6+58.5\ X_e^2 - 19.3\ X_e + 0.985\ X_D + 3.78\ V_{TBP} - 1007.5/T$$

$$\ln D_W = -5.31-6.96\ X_e^2 + 5.36\ X_e + 1.18\ X_D + 4.95\ V_{TBP} - 465.1/T$$

$X_e < 0.33$ $\qquad$ $301 < T < 342$ K

$X_D < 0.6$ $\qquad$ $V_{TBP} < 0.5$

These correlations are based upon 27 equilibrations.

Vapor Liquid Equilibria

Vapor/liquid equilibria were measured for all binary pairs. Analysis of the data using the Uniquac (9) model led to the parameter estimates that are summarized in Table 1.

$$\text{where } \ln \tau_{ij} = -A_{ij}/RT$$

Table 1. Uniquac binary interaction parameters for the ETOH/H20/TDOH/ISOPAR M system estimated from binary VLE data.

i	H20	ETOH	ISOPAR M	TDOH
H20	0	329.49	348.66	326.1
ETOH	-29.43	0	-117.10	332.80
ISOPAR M	762.37	1839.5	0	887.14
TDOH	326.10	35.2	-348.89	0

LABORATORY APPARATUS

Laboratory extraction tests were carried out using a 2.54 cm diameter glass reciprocating plate column. This unit included 92 stainless steel reciprocating plates and had an active contact height of 244 cm. It is sold by the Chem Pro Corp., New Jersey, U.S.A. as Model KC 1-8.

Two glass vapor/liquid bubble cap columns were built to test the SEED process. Both columns had 7.6 cm diameters. The extractive distillation column consisted of three bubble cap trays above the feed and three below. The solvent regeneration column included four bubble cap trays below the feed and two above.

RECIPROCATING PLATE COLUMN MODEL

A generalized correlation for the height equivalent to a theoretical stage to diameter ratio (HETS/D) was developed using experimental and literature (10,11) data. This ratio is correlated in terms of several dimensionless groups as indicated below. Additional details about the model are given elsewhere (12).

$$\frac{HETS}{D} = 1.03\left(\frac{D\,\Delta\rho g}{\rho_c U_T^2}\right)^{-0.075}\left(\frac{AFt_m}{H\,U_T}\right)^{-1.30}\left(\frac{\sigma}{D\rho_c U_T^2}\right)^{0.625} \quad \pm 16.4\%$$

TOXICITY TESTS

Earlier studies (13) on solvent toxicity suggested that solvent species with at least a ten carbon chain exhibit sufficiently low aqueous solubilities that they no longer interfere with the fermentation process. Studies at Georiga Tech confirmed this observation. For example, 2-ethylhexanol strongly suppresses biological activity while species such as decyl and tridecyl alcohols exhibit very little effect.

FERMENTATE EXTRACTION

Actual fermentates were produced using sucrose, media, and brewers' yeast. These fermentates typically contained about 5 wt% ethanol and 10 wt% inextractable solids. Tri-n-butyl phosphate and diluent (Isopar-M) blends were used as the solvent.

In these tests, whole fermentate was pumped into the top of the reciprocating plate column which was operated organic continuous (i.e. with the liquid/liquid interface below the contact zone). Fermentate droplets containing biomass passed downward through the column and were observed to coalesce at the bottom of the column. Extract was passed through a coalescer, preheater, and into the middle of a bubble cap column containing 6 trays.

Steady state with solvent recycle could be achieved in about 2 hours of continuous operation. Actual ethanol product recovered from the solvent regeneration column averaged to about 45 wt% ethanol and the column performance was comparable to operation with synthetic feed mixtures.

During these tests no significant accumulation of interfacial debris was observed to accumulate. The biomass passed through the column without accumulation.

SOLVENT EXTRACTION/EXTRACTIVE DISTILLATION TESTS

In these cases the reciprocating plate column and two bubble cap VLE columns were used. Several hours of continuous operation with solvent recycle were required to achieve steady state. The aqueous solvent extraction feed contained from 3 to 9 wt% ethanol. Solvents consisted of tridecyl alcohol/Isopar-M mixtures. Product recovered from the top of the extractive distillation column typically contained from 1 to 20 wt% ethanol. Product recovered from the solvent regeneration column ranged from 85 to 98 wt% ethanol.

COMPUTER AND ECONOMIC ASSESSMENT

Computations using both a rating simulator (PROCESS) and a design estimator (RUNOPT) were completed. The rating simulator was developed by Simulation Sciences, Inc., in Fullerton, CA. The design estimator is available at Georgia Tech on the CYBER.

Generally, the computer models and the actual process runs agree. For example, an analysis of the second test described above indicates that the plates below the feed in the EDC exhibited about 60% overall efficiency. The predicted temperatures and pressures also agreed well with the actual experiment. At specified pressures, the temperatures of EDC and SRC bottoms and distillates were within $5_{o}C$ of the predicted values.

Cost analysis was completed in which three beers (5.15, 1.9, and 0.57 wt%) were processed using the SEED process. Comparisions were then made with Berkeley optimized distillation (14) concept using their net energy balances, cooling water, and theoretical tray estimates. Azeotropic distillation costs were modelled using the data provided by Black (15). All cases assumed a 99% ethanol recovery and purity (mole basis).

Computer modelling studies suggest that energy savings are possible compared to conventional distillation. Economic comparisons suggest that the SEED process may save 3 U.S. cents/liter when processing to 10 wt% beer up to 13 U.S. cents/liter for a 3 wt% fermentate feed. Ethanol recovery via this route become more attractive as the mixture quality decreases.

DISCUSSION

The solvent polarity and the ethanol/water loading both affect the water/ethanol relative volatility in the presence of the solvent. In the presence of solvents, the water/ethanol relative volatility ranges from 10 to 30. Pure diluent (Isopar-M) yields the highest water/ethanol relative volatility whereas 2-ethylhexanol yields lower values. Blended mixtures of tridecyl alcohol and Isopar-M exhibit intermediate volatilities, but water is always more volatile than ethanol in the presence of high-boiling solvents.

The cell viability after solvent extraction was not studied in this work. Hence, these tests do not prove that the raffinate could be recycled to the fermenter while still maintaining the culture. However, they do suggest that fermentates can be effectively processed by solvent extraction without adverse effects from biomass accumulation in the column.

Optimization studies in which the solvent blend was treated as a variable led to a solvent consisting of pure tridecyl alcohol. This result suggests that solvent loading is more economically important than selectivity during solvent extraction and relative volatility during extractive distillation. Further solvent improvements may be achieved through the use of high-boiling mixtures that exhibit even higher loadings than pure tridecyl alcohol.

REFERENCES

1. D. F. Othermer, R. F. White and E. Trueger, Ind. Eng. Chem., 33, 1240 (1941).
2. A. V. Brancker and T. G. Hunter, ibid, 32, 38 (1940).
3. I. Bachman, ibid, 32, 35 (1940).
4. J. L. Schweepe and J. R. Lorach, ibid, 46, 2391 (1941).
5. D. F. Othmer and P. E. Tobias, ibid, 34, 696 (1942).
6. C. M. Qualline and Van Winkle, ibid, 44, 1668 (1952).
7. J. W. Roddy, Ind. Eng. Chem. Proc. Des. Dev., 20, 104 (1981).
8. J. W. Roddy and C. F. Coleman, Ind. Eng. Chem. Fund., 20, 250 (1981).
9. J. Prausnitz et al., Computer Calculations for Multicomponent Vapor Liquid and Liquid Liquid Equilibria, Prentice Hall (1980).
10. A. E. Karr and T. C. Lo, Proc. of Inst. Sol. Ext. Conf., London (1971).
11. A. E. Karr, AIChE J., 15, 232 (1959).
12. W. Y. Tawfik, Design of Optimal Fuel Grade Ethanol Recovery Systems Using Solvent Extraction, Ph.D. Thesis, School of Chemical Engineering, Georgia Institute of Technology, Atlanta, GA (1986).
13. J. Ribuad, Design of Liquid-Liquid Extractants for the Recovery of Fuel-Grade Ethanol, M.S. Thesis, Chemical and Biochemical Engineering Dept., University of Pennsylvania (1980).
14. T. K. Murphy, H. W. Blanch, and C. R. Wilke, "Recovery of Fermentation Products from Dilute Aqueous Solution," LBL-17979, Lawrence Berkeley Laboratory (April 1984).
15. C. Black, CEP, 78 (September 1980).

NOMENCLATURE

A	=	Amplitude of reiprocation
A_{ij}	=	Uniquac binary interaction parameter
D	=	column diameter
D_e	=	ethanol distribution coefficient (ratio of weight fractions)
D_w	=	water distribution coefficient (ratio of weight fractions)
F	=	frequency of reciprocation
g	=	gravity constant
H	=	plate spacing
R	=	gas constant
t_m	=	plate thickness
T	=	temperature, K
U_T	=	phase relative velocity
V_{TDOH}	=	volume fraction of tridecyl alcohol in solvent
V_{TBP}	=	volume fraction of TBP in solvent
X_D	=	weight fraction dextrose in equilibrated aqueous phase
X_e	=	weight fraction ethanol in equilibrated aqueous phase
Δp	=	phase density difference
p_c	=	continuous phase density
δ	=	liquid/liquid interfacial tension

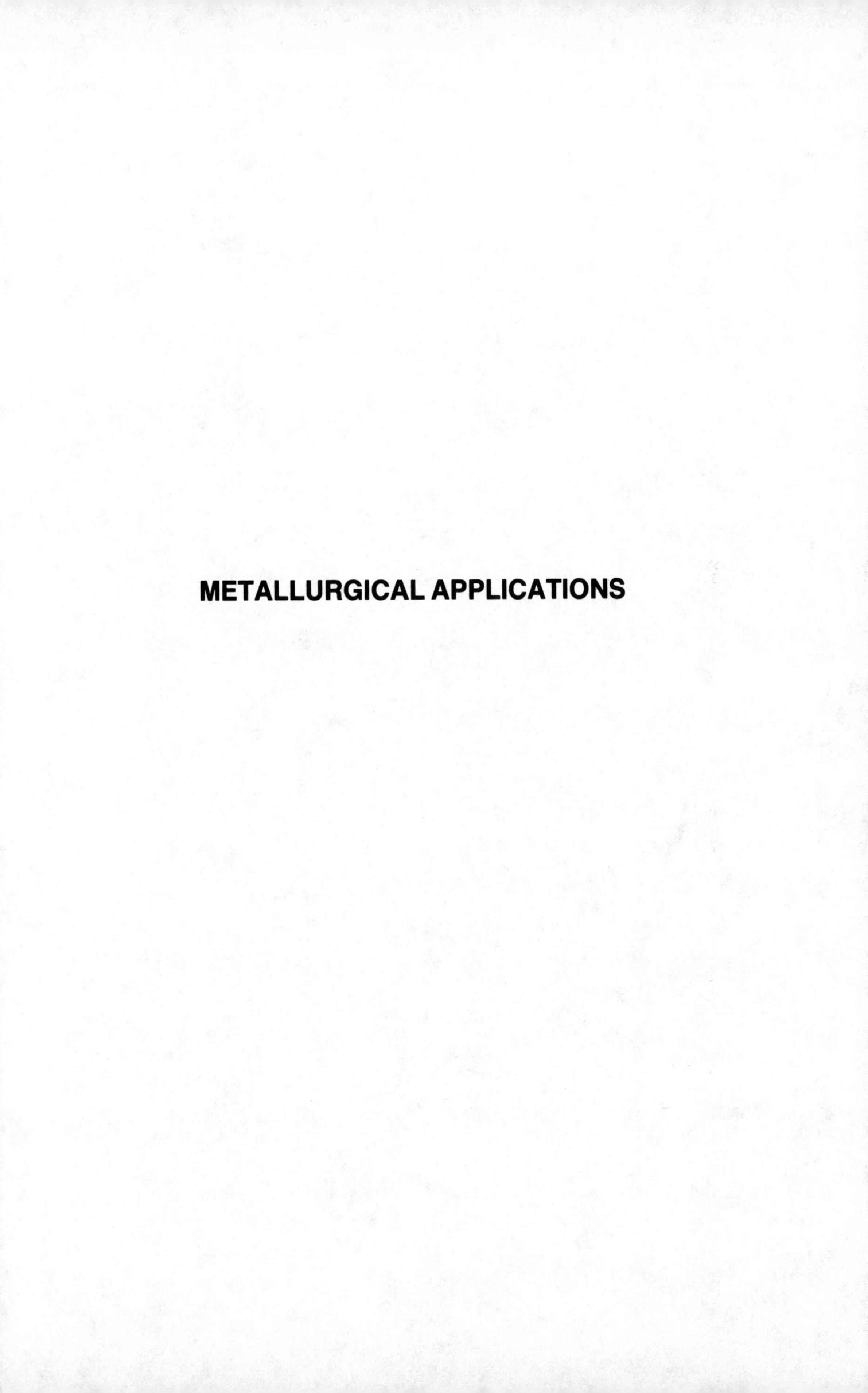

METALLURGICAL APPLICATIONS

HYDROMETALLURGY - ITS DEVELOPMENT AND FUTURE

G.M. Ritcey
CANMET, Energy, Mines & Resources Canada
Ottawa, Canada

Hydrometallurgy, first as an alternative to pyrometallurgy because of noxious emission by the latter and then as a combined process operation of pyro-hydrometallurgy, has seen significant developments in the past 25 years. These developments have resulted because of constraints imposed on the process. These include economics, the requirement to treat low grade or complex fine-grained ores difficult to selectively float; the requirement to recover valuable by-products; increased energy costs and the environmental aspects as reflected by the various governmental agencies and pressure groups.

Because of these various constraints, there have been numerous developments in using chemistry to the best advantage to achieve the objectives. Together with chemistry has been the development of corrosion resistant materials to withstand the developing new technology, and the subsequent interface with chemical engineering to design and produce equipment that can run with minimum maintenance and provide the necessary mass transfer as well as mass transport.

This plenary discussion will briefly cover the developments and examine the present and future application of hydrometallurgy.

INTRODUCTION

The definition for Hydrometallurgy, according to the U.S. Bureau of Mines "Dictionary of Mining, Mineral and Related Terms" (1) is "...the treatment of ores, concentrates, and other metal-bearing materials by wet processes, usually involving the solution of some component, and its understanding and application has evolved considerably, particularly during the past 20 to 30 years. Although it has been used (certainly some of its unit processes) for much longer, it is only in recent times that this processing technique has become generally widely accepted. The processing of gold ores was one of the early **areas** treated by hydrometallurgy, and of course the recovery of copper and other metals sulphides that had been leached in the open by sulphide-oxidation and subsequent acid generation and leaching has been going on for centuries. But this paper is dedicated to the present practice and future technology and applications.

The rise of the popularity of hydromerallurgy and its technology started probably with the nickel (2) and uranium plants (3) in the early 1950's and spread with the development of the copper (4) industry in the 1960's as environmental regulations for SO_2 emissions by smelters were imposed on those pyrometallurgical operations. Subsequently other metal recovery plants have incorporated or changed to hydrometallurgical processing, such as base metals (Cu, Ni, Co, Pb, Zn) (5-8) and light metals (Al, Ti) (9). Also its use in gold recovery (10) and the developments in the overall nuclear industry have resulted in considerable technology for the recovery of thorium (11), zirconium, hafnium (12) and its use in the processing of spent nuclear fuels. And in more recent years, hydrometallurgy processes are being applied to waste effluents (13).

The Flowsheet

A typical integrated hydrometallurgy circuit is shown schematically in Fig. 1 for the conventional acid-leach uranium mill.

An appreciation of the "upstream and downstream" of the process is necessary if any one or more of the unit processes is to be a viable component of the integrated plant. Some of the questions to ask before a choice of the process route is made include the following:

- what is the nature of the ore (mineralogy, values, gangue)?
- will beneficiation be viable?
- based on mineralogy, what are the possible lixiviants?
- what impurities would be expected in the leach solution?
- what metal or metals are to be recovered?
- what type of purification process is possible?
- are by-products possible?
- what are the environmental constraints?
- energy requirements?
- corrosion problems?
- overall costs?

The choice, design and optimization of the process involve the three E's - economics, energy and environment.

In Fig. 2 is shown a block diagram of the various unit processes in the hydrometallurgical flowsheet, together with notations for most of the units regarding present technology. Briefly each of the unit processes will be covered as to the present, and later the future possibilities in the development of technologies will be discussed.

Crushing, Grinding, Beneficiation and Mineralogy

The grind of the ore must be optimized with respect to the leach, roast-leach, or possible beneficiation circuits such as tabling, flotation, high intensity magnetic separation, heavy media, etc. Liberation of the mineral by the process is governed by the grain size and the complexity of the mineralization (e.g., complex fine grained sulphides). Also the gangue material is important as it may provide a clue to possible excessive reagent consumption during the leaching process. It is therefore essential to have a good and thorough mineralogical exami-

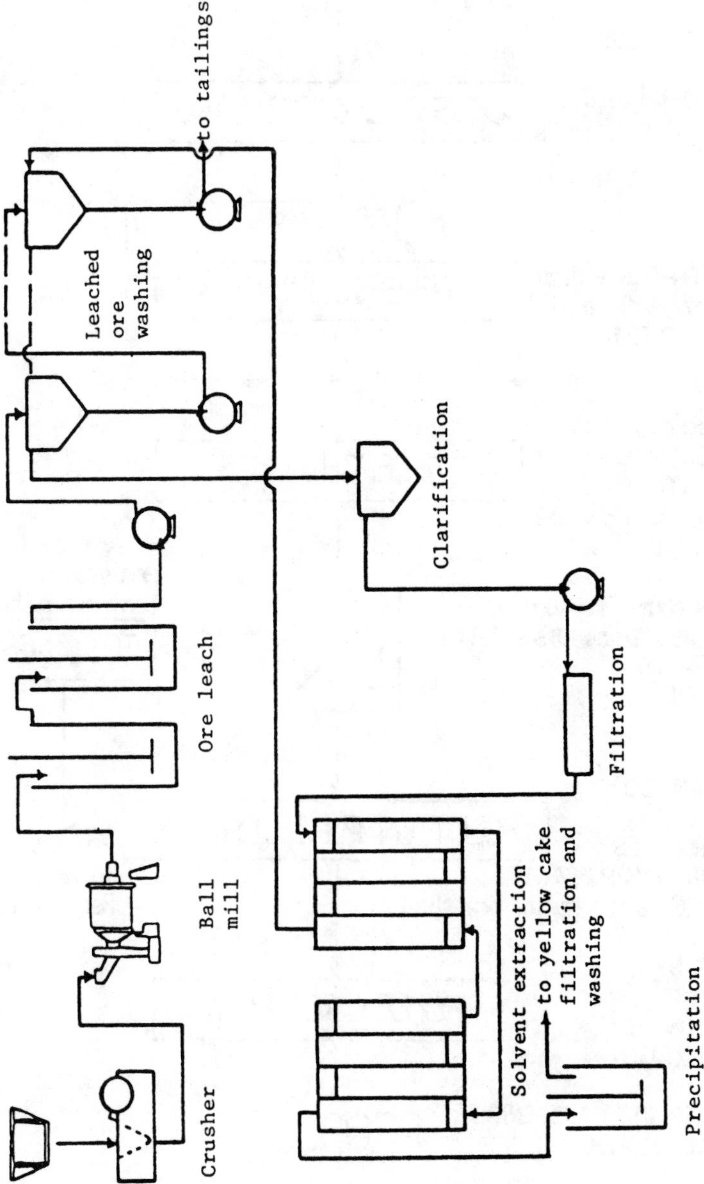

Fig. 1. Conventional acid-leach uranium mill

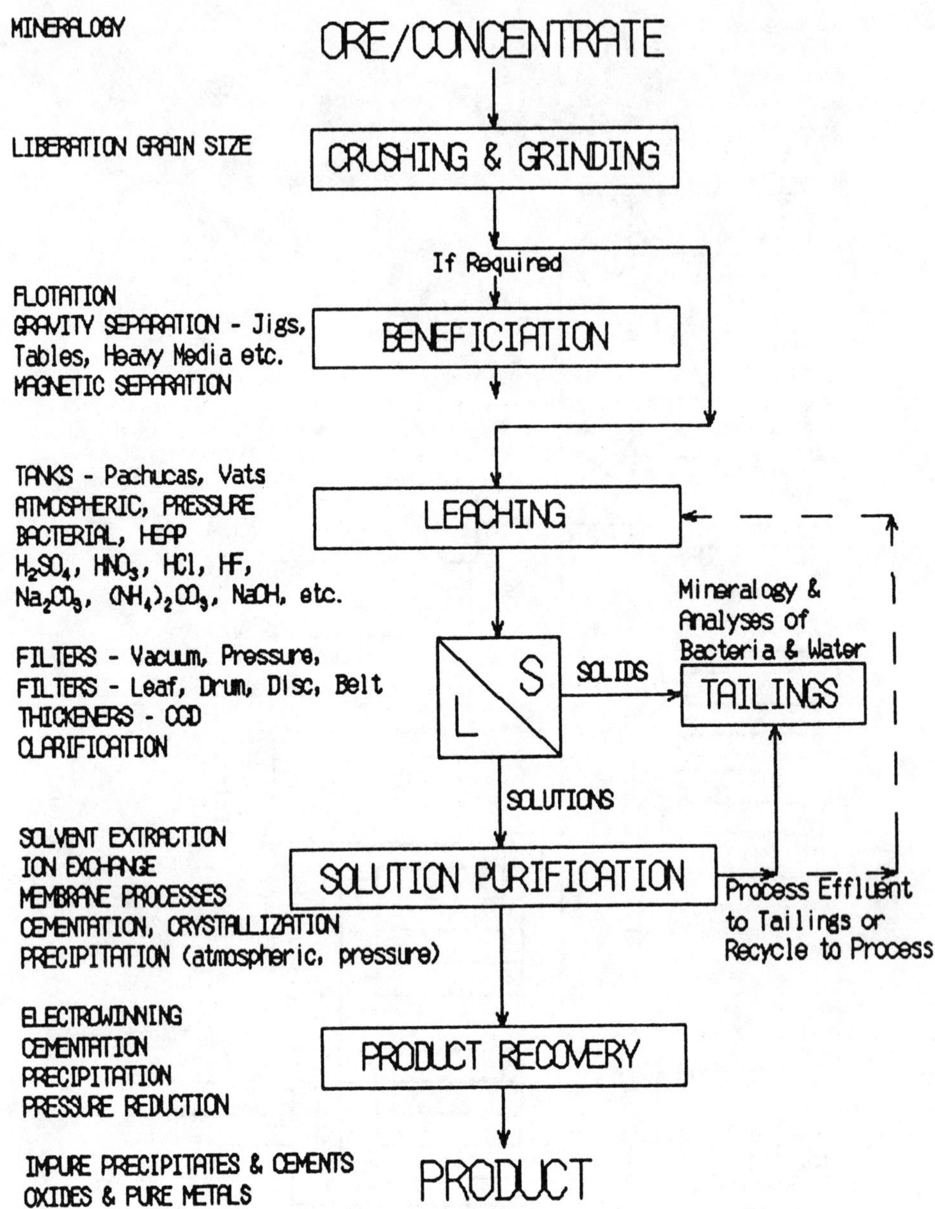
MINERALOGY
ORE/CONCENTRATE
LIBERATION GRAIN SIZE
CRUSHING & GRINDING
If Required
FLOTATION
GRAVITY SEPARATION - Jigs, Tables, Heavy Media etc.
MAGNETIC SEPARATION
BENEFICIATION
TANKS - Pachucas, Vats
ATMOSPHERIC, PRESSURE
BACTERIAL, HEAP
H_2SO_4, HNO_3, HCl, HF,
Na_2CO_3, $(NH_4)_2CO_3$, NaOH, etc.
LEACHING
Mineralogy & Analyses of Bacteria & Water
FILTERS - Vacuum, Pressure,
FILTERS - Leaf, Drum, Disc, Belt
THICKENERS - CCD
CLARIFICATION
L
S
SOLIDS
TAILINGS
SOLUTIONS
SOLVENT EXTRACTION
ION EXCHANGE
MEMBRANE PROCESSES
CEMENTATION, CRYSTALLIZATION
PRECIPITATION (atmospheric, pressure)
SOLUTION PURIFICATION
Process Effluent to Tailings or Recycle to Process
ELECTROWINNING
CEMENTATION
PRECIPITATION
PRESSURE REDUCTION
PRODUCT RECOVERY
IMPURE PRECIPITATES & CEMENTS
OXIDES & PURE METALS
PRODUCT

Fig. 2. Schematic of hydrometallurgical flowsheet

nation performed to determine not only the minerals of economic interest but also the possible reagent consumers. Such an examination may strongly influence the type of process that is used to recover the values. Once a process has been selected, the mineralogy of the ore before and after treatment must be considered, as well as the mineralogy of the solids going to tailings impoundment.

Leaching

The material fed to the leaching stage may be a low grade ore, concentrate or a complex ore. In recent years there has been a renewed interest in the reprocessing of very low grade tailings for recovery of metals such as copper (14), uranium, silver or gold (15). Leaching takes place under many situations. The methods employing vat leaching, stirred tanks or pachucas are still common. With the decreasing amount of higher grade ores and the viability of reprocessing tailings, we now have heap leaching, in-situ leaching and in-place leaching of broken rock underground. While atmospheric leaching was for many years the normal method, pressure leaching at elevated temperatures has proven extremely useful for some ores and concentrates, particularly those that are refractory or those which, because of their mineralogy, are high acid consumers at normal temperatures and pressures. The use of bacterial leaching, to oxidize the iron sulphides, converting to acid and ferric sulphate has been shown to be effective for the dissolution of uranium ores (16) containing pitchblende, uraninite and brannerite, copper ores (17) as well as some refractory gold ores containing arsenopyrite (18). Because of the varied mineralogy, economics, energy requirements and environmental problems that might be associated with any particular lixiviant, we have seen plants using H_2SO_4, HNO_3, HCL, HF, thiourea, (19) or mixtures of some of these such as H_2SO_4 + HF (20), H_2SO_4 + HNO_3 (21), H_2SO_4 + H_2O_2 (Caro's acid H_2SO_5) (22). Alkaline circuits have included Na_2CO_3 or $(NH_4)_2CO_3$ and alkaline cyanide.

As an alternative to conventional processing, chloride metallurgy has made considerable progress in the laboratory and pilot plant, and even a few commercial plants are now using some aspect of chloride technology (23). Although most of the effort has been devoted to treating complex base metal sulphide bulk flotation concentrates (23, 24), other ores have included uranium, gold and precious metals (25, 27). Corrosion is still a problem, but not to the same extent as formerly with the development of synthetic materials to contain the reaction. Low temperature leaching is presently more viable compared to the high temperature non-aqueous chlorination processes mainly because many engineering problems remain unsolved at the higher temperature (these processes are also usually more costly unless concentrates are being treated).

With the developments and improvements in leaching together with the aid of improved mineralogy the leaching recovery efficiency has increased. Also by-product recovery is now easier to achieve.

Liquid-Solids Separation

The liquid-solids separation is the single most expensive unit operation in the mill, accounting for about 50% of the capital and operating costs in most circuits (28). It is also a very important step in order to minimize the loss of soluble metal values to the tailings. Good washing, settling and clarification are necessary before purification. Filters such as vacuum, pressure, leaf, drum and disk were used for many years. More efficient belt filters have been recently introduced in many mills and have resulted in decreased soluble losses because of increased washing efficiency. Liquid-solids separation has been achieved by cyclones and thickeners. Thickener separation using a continuous countercurrent decantation system has been improved by high capacity thickeners resulting in improved economics to the plant.

Prior to solution purification the solution is clarified to reduce the suspended solids, using pre-coat filters.

Solution Purification and Product Recovery

Solution purification for metals recovery can be accomplished by a number of routes, or a combination of several, depending upon the metal being recovered and the impurities, the grade demanded, and the value of the metal. These options include cementation, precipitation, crystallization, ion exchange, solvent extraction, reverse osmosis, electrowinning. A schematic of the process alternatives is shown in Fig. 3 (29).

The recovery of metals from solution and the demand for high product purity has resulted in more effort in both solvent extraction and ion exchange and in the application of carbon-in-pulp for gold recovery. Continuous ion exchange systems (30-33) have been developed and installed, and are being used mainly to produce a more concentrated solution for feed to solvent extraction purification. Ion exchange processes are used in the treatment of effluents for contaminant removal prior to the discharge to the environment. New solvent extraction reagents have added to the scope by which this unit process can be an asset to the operation. Better metal selecivity is being achieved with the reagents which of course results in an overall decrease in the operating costs as well as the capital costs. Continued engineering design and research has resulted in better understanding of the relation of the mixing shear to stable emulsions and crud (34, 35) in the solvent extraction process, so that better mixer-settler designs (36, 37) have resulted to improve on the overall efficiency of the process. Solvent extraction has also been used in the treatment of effluents prior to disposal (38). Interest in solvent-in-pulp recovery of metals to lower the capital and operating costs of liquid-solids separation has resulted in at least one commercial plant. Research on the production of metal powders by pressure or hydrolytic stripping (37) from the loaded solvent has continued with particular success in the recovery of gold, platinum and palladium (40).

Carbon adsorption processes have been developed primarily for the recovery of gold from alkaline cyanide leach solutions as well as from the leach slurry or in the presence of the pulp after leaching (41). The gold is recovered now by pressure

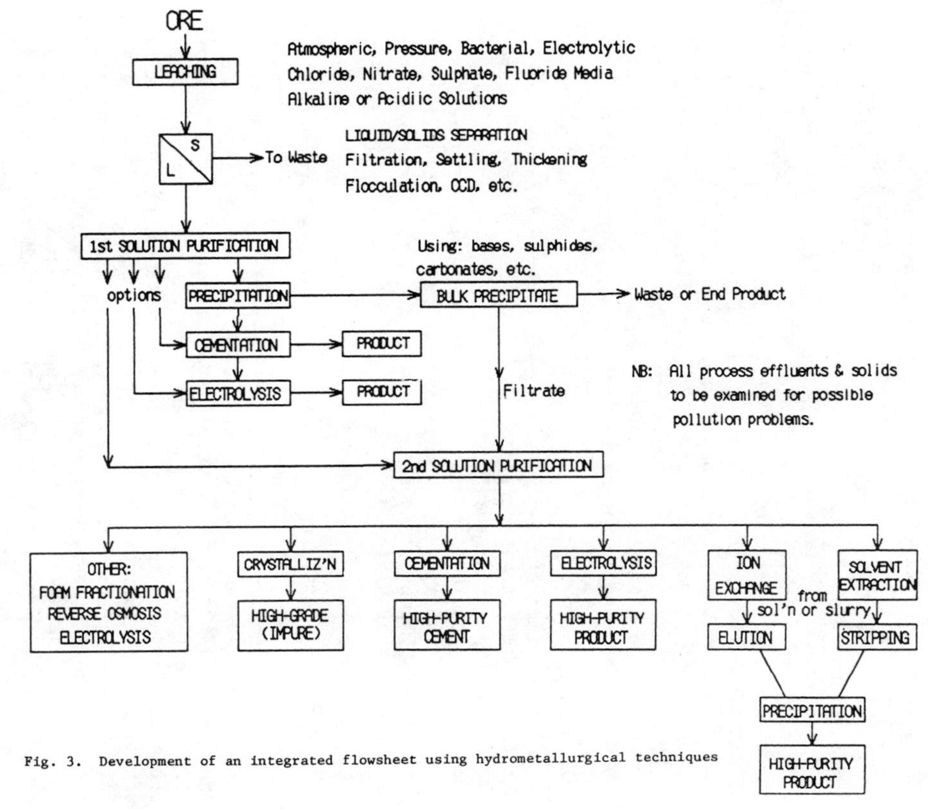

Fig. 3. Development of an integrated flowsheet using hydrometallurgical techniques

elution (42) instead of the previous zinc precipitation from the leach liquor and the many subsequent purification stages required. Carbon adsorption is also presently used for the removal of organics and other contaminants from waste streams.

Electrolytic processing has had a continuing developing period in the leaching and recovery of metals from complex sulphides (43). Direct electrolysis from leach slurries and electrowinning from chloride solvent extraction strip solutions for the recovery of copper, zinc and lead has met with some success (44). This has resulted in a reduction of process operations, and mechanical stripping of cathodes has now become a cost-effective operation. Electroelution of gold from carbon or resins is presently in the early stages of development and optimization (45).

The method of recovery of the final product has traditionally been dependent upon the type of solution and impurities and the grade of product desired. A bulk, impure product can be obtained by cementation, by direct electrolysis or by a precipitation as a sulphide, carbonate, oxalate, hydroxide, etc. and drying. If a prior purification stage has been carried out, such as by solvent extraction, then a considerably higher grade product may be obtained by pressure reduction to produce a metal powder, or by electrowinning to produce high purity cathodes.

Conservation

Conservation of resources has been a major item of concern for the metallurgical operation. Most plants now have incorporated some type of scheme to conserve water, so that now water recycle is up to 100% in many plants (46). Also, many metals which were formerly sent to tailings are now being considered for recovery, both for monetary as well as environmental reasons. In addition there is some reagent recycle practiced.

Environment

Environmentally, the past ten years has seen a very positive approach to improving the quality of the effluents discharged from the process and ultimately to the receiving environment. With the concern and pressure by the public and the regulatory agencies, and the cooperative efforts between industry, regulatory agencies and research laboratories, much has transpired. These developments include a better understanding of the fixing of process precipitates, such as jarosite, goethite and hematite, for disposal (47); improved design of tailings impoundment areas, and of disposal techniques; revegetation; stabilization; covers; dam liners; seepage quality of effluents; contaminant migration patterns through the tailings; and the overall biogeochemical interactions occurring within the tailings as weathering proceeds (48).

FUTURE DEVELOPMENTS

Mineralogy

With improved instrumentation coupled with computer control, significant improvements have already resulted in the last several years in the area of applied mineralogy which is a necessary tool for the successful metallurgical process design. Image analysis is one prominent example and this coupled with other sophisticated analytical and surface chemistry techniques will undoubtedly assist in improving the process and decreasing the overall cost.

Control

The tremendous development of the micro-computer and its impact on the process through modelling and process optimization has already been highly significant. However, only a small portion of the industry is presently taking advantage of currently available technology. Software packages are being improved continually and as the metallurgist becomes acquainted and knowledgeable with some of these powerful tools of the electronic age many significant improvements in the process should result. This will be reflected in a better understanding of the chemistry, metallurgy, biogeochemistry and engineering as applied to the process.

Conservation and By-Product Recovery

The aspect of conservation and utilization of our total resources will become more important through time. Already there is concern in some areas of processing, and in some parts of the world more than others - depending upon the availability or lack of the raw material. Processes will be developed and plants installed for the increased recovery and treatment of secondary sources such as baghouse dusts and metallurgical scrap as well as the treatment of process effluents, tailings and by-products from present process streams.

Processes to recover gold and precious metals and the rarer metals, such as gallium, indium, germanium, etc. which would be used in the electronics industry, will be developed. These will require an increased effort in separations chemistry and in producing high purity chemical precipitates. Rare earth recovery, although still developing relatively slowly, should increase as new industrial uses are found. In addition, continued processing of base metals such as Co, Ni, Cu, Pb, Zn will occur, together with Cr, Mn, V. Other metals will include U, Th, Zr, Hf, Ta, Nb.

Leaching of low grade ores, on surface or in-situ underground, will develop to a greater extent as the general biotechnology efforts to improve the understanding of this discipline develop. Up to the present, the development of cultures of bacteria specific for certain metals has not been successful, and generally the leaching rates are slow. An improvement in the leaching kinetics would be a major accomplishment. However, bacteria leaching can be applied only where the ore is sufficiently porous to permit leaching, so the type of ore and mineralogy, size and the element to be leached are constraints to the development of this technology. The use of bacteria to preoxidize complex gold ores containing arsenopyrite prior to cyanide leaching is presently showing promise, and this technique may well develop. Oxidants use in general, and pressure processes together with EMF control will improve selectivity and recovery efficiency. Heap leaching of ores is well developed in some areas and with additional research and the understanding necessary to carry out the process under arctic conditions will undoubtedly be attractive to many mines. Increased re-use of lixiviants should develop to decrease leaching costs as well as effluent neutralization disposal costs. Finally, the application of chloride metallurgy and its acceptance by the plants will increase the possibility for metals recovery, lixiviants re-use, by-products recovery and pollution-free tailings.

Liquid-Solids Separation

This expensive operation requires the most attention if we are to reduce the total capital and operating costs of the circuit and to reduce the present metal soluble losses going to the tailings. Incorporation of solvent-in-pulp or resin-in-pulp would be major factors in achieving a significant cost reduction.

Solution Purification

In addition to improvements in the chemistry of the applications of ion exchange and solvent extraction together with better engineering design, solution purification will include carbon adsorption mainly for the treatment of waste liquors, reverse osmosis for concentration of liquors and liquid membranes - both for potential application to treatment of waste streams. Developments in electrowinning will continue to optimize and extend the use of the process.

In the next decade we should see solvent-in-pulp used more frequently (Fig. 4). The industry has accepted resin-in-pulp and carbon-in-pulp and the next stage for a quantum leap in solvent extraction will be the extraction from slurries. SIP with direct pressure reduction to recover a metal powder from the loaded solvent would be an ultimate goal. Solvent extraction reagents with increased selectivity will be required in order to meet the challenge of the fast-developing electronic industry. Also because of the adverse environmental impact of some organic reagents, there will be a requirement for more biodegradable compounds. Improved contactor design and a better understanding of dispersion, coalescence and stable emulsions will help reduce the overall processing costs. Reagent recycle in the scrub and strip circuits will also help to reduce the processing costs. Treatment of specific effluents by membrane technology should develop.

The industry requires ion exchange resins with better selectivity than at present if ion exchange technology is to become more widely accepted. Resins which are of higher density and which will withstand severe abrasion will certainly aid in the exploitation of resin-in-pulp. Such resins would be valuable in

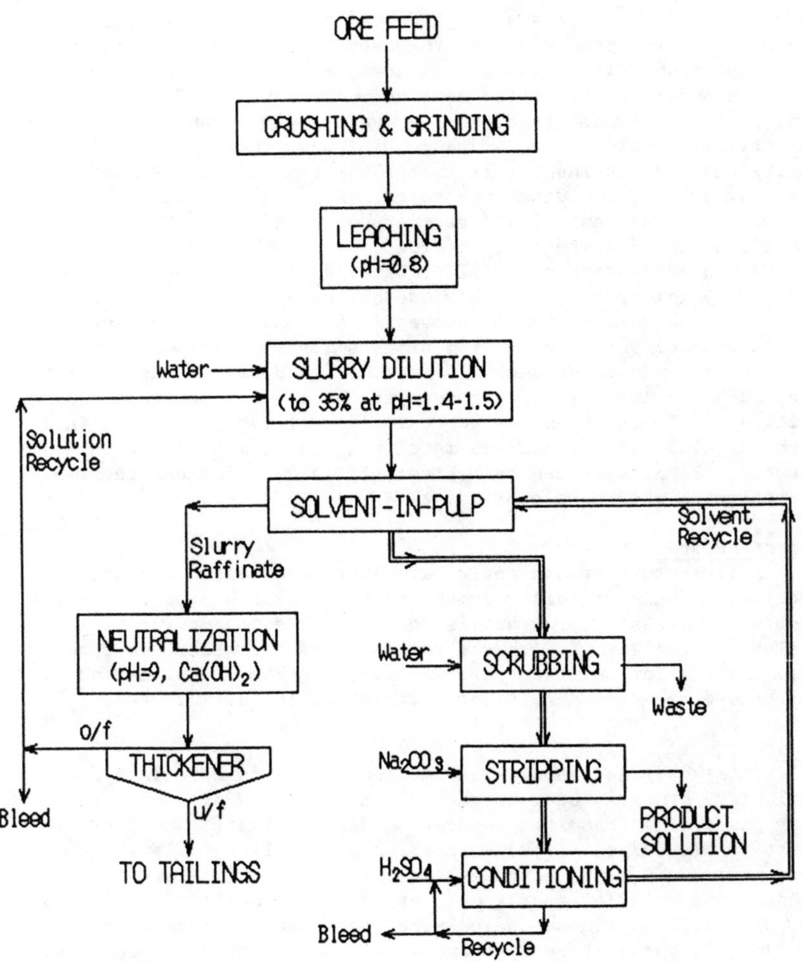

Fig. 4. Schematic of proposed solvent-in-pulp extraction circuit

the recovery of such metals as gold, uranium, precious metals, rare earths, etc. from slurries followed by elution of a concentrated solution which would be extracted in a solvent extraction system in order to attain the desired purification. Electroelution, particularly for gold, will prove to have advantages over conventional elution. Ion exchange will have increased use in the treatment of effluent streams for the removal of noxious constituents prior to discharge to tailings. A good example of this would be the removal and recovery of cyanide and base metals from effluents after recovery of gold.

Carbon adsorption use will probably decline in the application to the gold industry as the use of ion exchange or solvent extraction replaces carbon. However it may have increased use for the removal of organics from process streams.

Electrolytic processes will continue to develop, the challenge being to improve product quality and increase current efficiency. Electrowinning from chloride solutions will continue as the development of chloride metallurgy and its application is accepted. Finally, fused salt electrolysis of solutions such as lead chloride should be operational.

Product Recovery

In the future we can anticipate increased demand for the production of higher purity compounds that would be fed to the electronics industry, the metals industry, the production of catalysts and alloys. With the requirements for selective recovery, after prior solution purification, increased application of pressure reduction and electrowinning can be anticipated.

Chemistry

Little has been said about chemistry and its role in hydrometallurgy, probably because much of the practice of hydrometallurgy has evolved from the chemist and the analyst. However, if we are to effectively improve on the hydrometallurgy flowsheet there must be an improved understanding of various facets of chemistry as relating to the unit operation to achieve separation and purification. These include the speciation of metals in solution; Eh-pH relationships and effects on separation processes; chemistry and thermodynamics of pressure reduction; hydrolytic stripping and atmospheric reduction from organic and aqueous solutions; co-precipitation phenomena; impurity effects (chemistry) on product purity achieved by particular unit processes; and synergistic behavior solvent extraction.

Pilot Plants

Traditionally pilot plants were large but for many years have been decreasing gradually in size. This size reduction will continue, particularly with the increased use of process optimization, process mineralogy and the improved understanding of chemistry, metallurgy and engineering. With the optimization of the process and the resultant decrease in pilot plant size requirements, the development cost to move from bench-scale to pilot plant and plant design will be decreased.

Tailings and the Environment

Environmental concerns have resulted in an increased awareness and concern by the mining companies so that we can expect to see more resources devoted to minimizing the environmental impact of process metallurgy. The attainment of effluent quality and the disposal of solid residues, which remain essentially unaffected during the weathering process in the tailings, will be major objectives. Abatement of acid generation due to sulphide oxidation in the tailings, the possible use of natural cover materials for tailings and the use of indigenous bacteria to aid in tailings disposal will be of importance. Surface chemistry determinations on solids, biogeochemical interactions, speciation of constituents in solution and physical-chemical properties will permit predictive contaminant migration. Improvements of the unit operations within the mill itself will aid in the decrease in the environmental impact due to the tailings discharged. Control of contaminant seepage will be achieved by incorporation of impeding barriers, and certain solid waste may be stabilized by consolidation within synthetic rock materials.

Some Future Flowsheets

Following are given some possible flowsheets that we may see in the future. Many of these have already been closely examined at the bench-scale while others have reached the pilot-scale. Each will be briefly described.

Fig. 5 shows a flowsheet schematic for the heap leaching (with or without bacteria) of an ore followed by metal recovery and purification by solvent extraction, ion exchange or precipitation. The raffinate is returned to the heap. Bacteria requirements are maintained by culture addition prior to recycle. At least one plant has operated in this manner but the success of heap leaching depends upon parameters such as the size of the rock, mineralization, leachant penetration and temperature. We can probably expect to see gold and uranium heap leach operations in the future perhaps even in arctic or sub-arctic conditions.

In Fig. 6 is shown a diagrammatic flowsheet for in-place leaching underground. The ore is crushed and stacked and the lixiviant sprayed on the heap. The leach solution, from a collection sump, is passed through an ion exchange column to recover the metal value(s). Because of the problems of precipitation or elution underground, the resin would be transferred in cannisters to surface to be subjected to the elution and recovery process. The resin would be returned underground in the same cannisters.

Fig. 7 shows a possible flowsheet for the treatment of uranium ore (49). Following pressure leaching with oxygen and $CaCl_2$ the solubilized uranium (and radionuclides) are contacted in a solvent-in-pulp system for uranium recovery. Uranium is recovered by stripping followed by a reduction-precipitation stage to produce a high purity product. The slurry raffinate is sent through a series of wash thickeners, the solution returned to leaching and the solids washed with brine to remove any radionuclides which are recovered by ion exchange. The neutral tailings going to waste thus require no neutralization with lime and should not contain any sulphides or radionuclides. Fig. 8

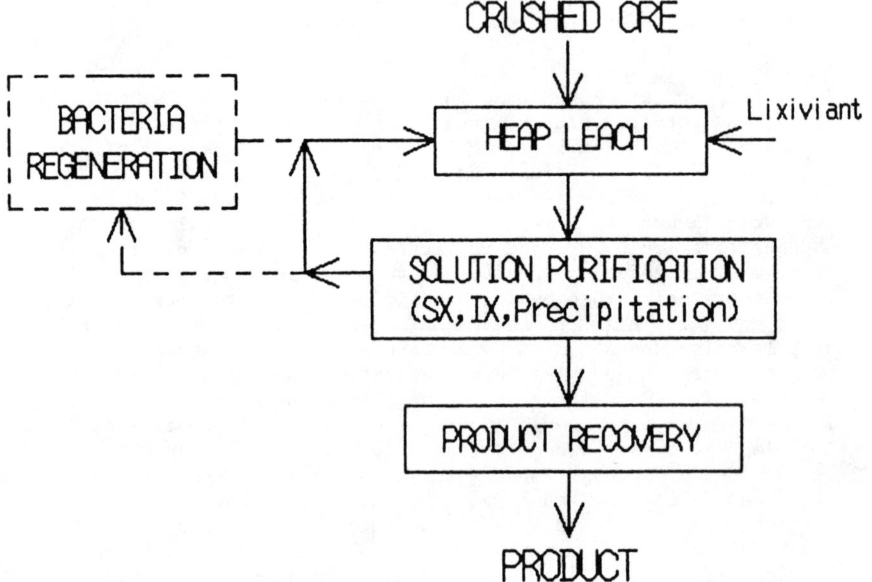

Fig. 5. Heap leaching and metal recovery

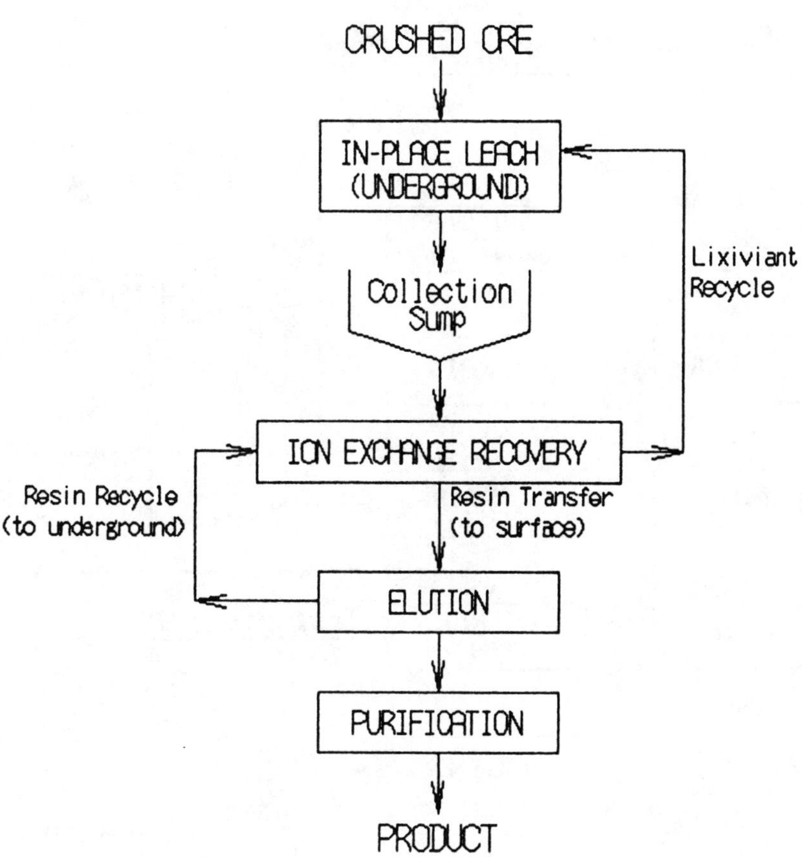

Fig. 6. In-place leaching and metal recovery

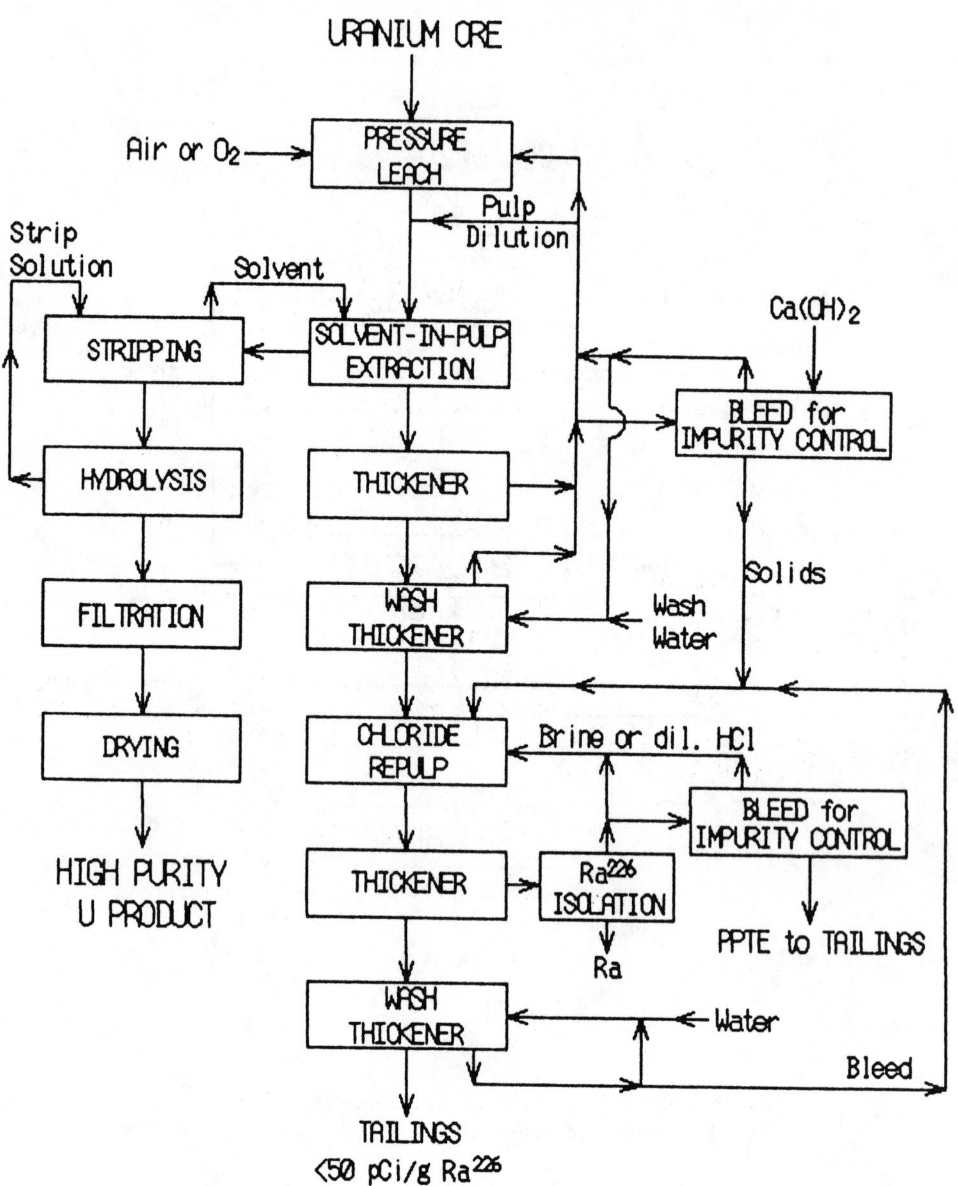

Fig. 7. Future uranium flowsheet incorporating SIP, strip-hydrolysis and solution recycle

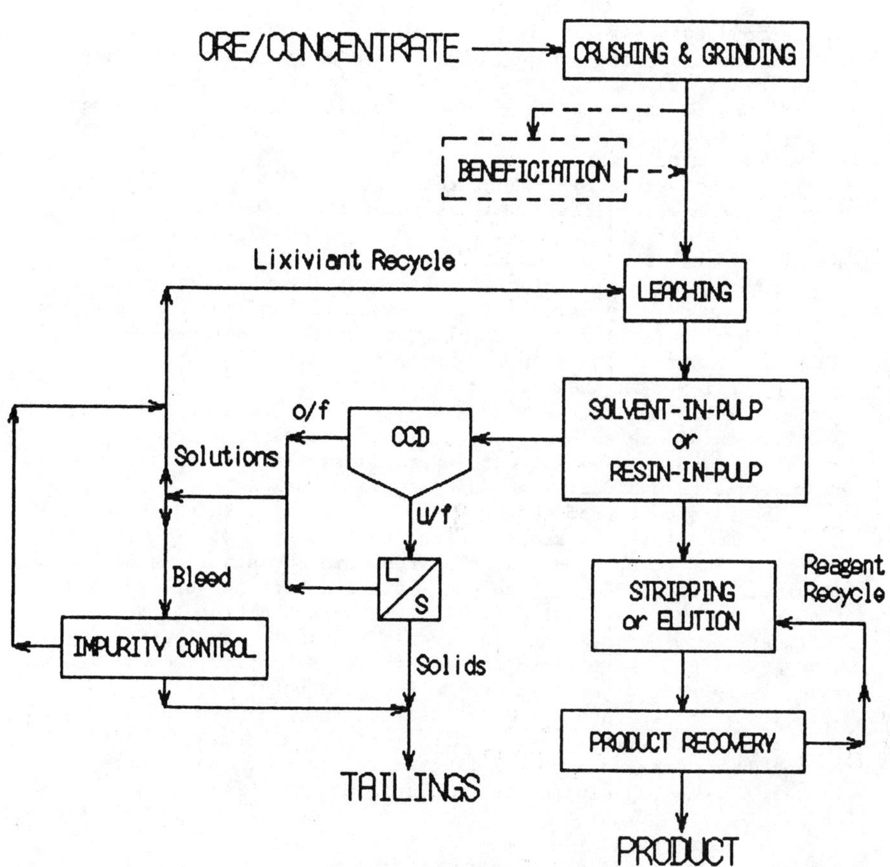

Fig. 8. Generalized flowsheet incorporating SIP and solution recycle

shows a generalized flowsheet incorporating SIP and solution recycle.

During the past 25 years or more there has been considerable research and resources spent by various organizations throughout the world on the attempt to process complex, fine-grained sulphides. These ores, containing Cu/Pb/Zn and other values are particularly found in Canada, Australia, Norway, Spain and other countries. Because of their finely disseminated nature they do not respond successfully to differential flotation to produce separate concentrates without high losses of values to tailings. These losses can amount to perhaps 30% or more of the values.

At CANMET and at other research organizations, chloride metallurgy was selected as the most appropriate to:

- increase metals recovery
- increase possible by-product recovery
- minimize environmental problems associated with disposal of pyrite

and the process must of course be economical.

Although many possible flowsheets have been investigated for various reasons they have failed to result in an operating plant.

In Fig. 9 is shown a possible flowsheet, based on minimizing the labor and reagent-intensive costs of cementation processes usually employed in the various flowsheets (50). Lead is recovered from the ferric chloride leach liquor by cooling and crystallization of the solution after liquid-solids separation. The copper is recovered by solvent extraction. Electrowinning could be achieved from a $CuCl_2$ or $CuSO_4$ strip solution.

The raffinate from copper solvent extraction is subjected to a second extraction stage. Since this is essentially a separation of Zn from the bulk of the iron, minimum stages would be required. Strong HCL or high salt-stripping effectively removes the zinc, leaving all but a few ppm Fe on the solvent. The raffinate from this extraction circuit, which is the $FeCl_2$ ($FeCl_3$) plus Ag and Cd, is returned to leaching after controlled oxidation with Cl_2 to $FeCl_3$. A bleed of the $FeCl_2$ liquor is oxidized with O_2 to precipitate Fe_2O_3 (to control iron build-up during leaching). The filtrate from this iron removal stage, containing Ag and Cd, is passed through activated carbon to recover these values. Following elution the mixed concentrate is sent to a refiner.

The stripped solvent from the second circuit (first Zn) is subjected to a water strip to remove and recover the iron as $FeCl_3$. This can, in part, be recycled to leaching into TBP, stripped with $ZnCl_2$ (dilute). The strip solution is passed through activated carbon to remove organics prior to electrowinning. The chlorine evolved is returned to leaching. The raffinate from extraction is returned to the first Zn extraction circuit as the strip feed solution.

In Fig. 10 is shown a solvent extraction purification circuit followed by pressure reduction stripping to recover a metal powder (51). Fig. 11 shows a flowsheet incorporating SIP and pressure reduction or hydrolytic stripping for recovery of a

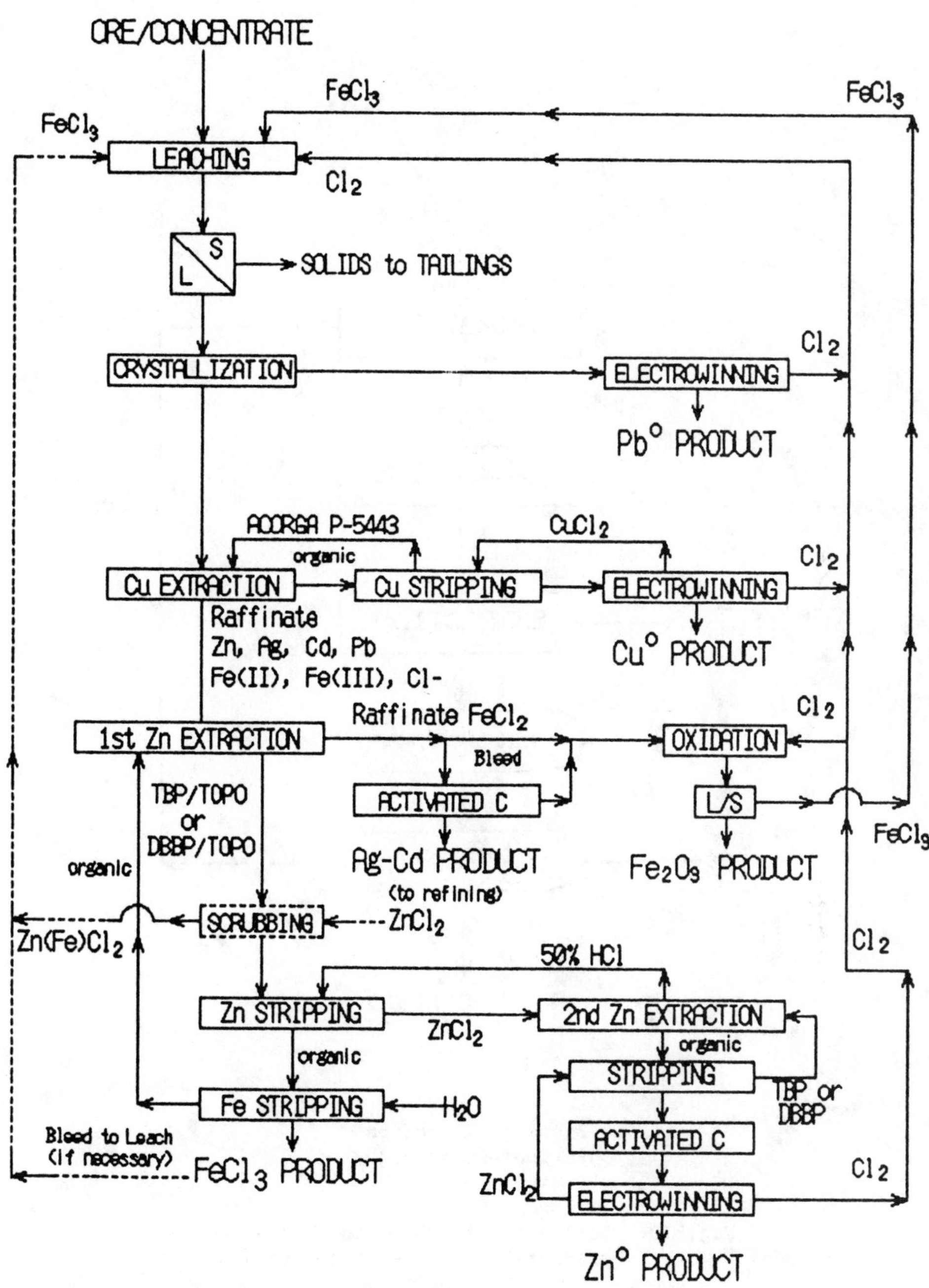

Fig. 9. Conceptual flowsheet for $FeCl_3$ leaching of complex sulphides and recovery of products

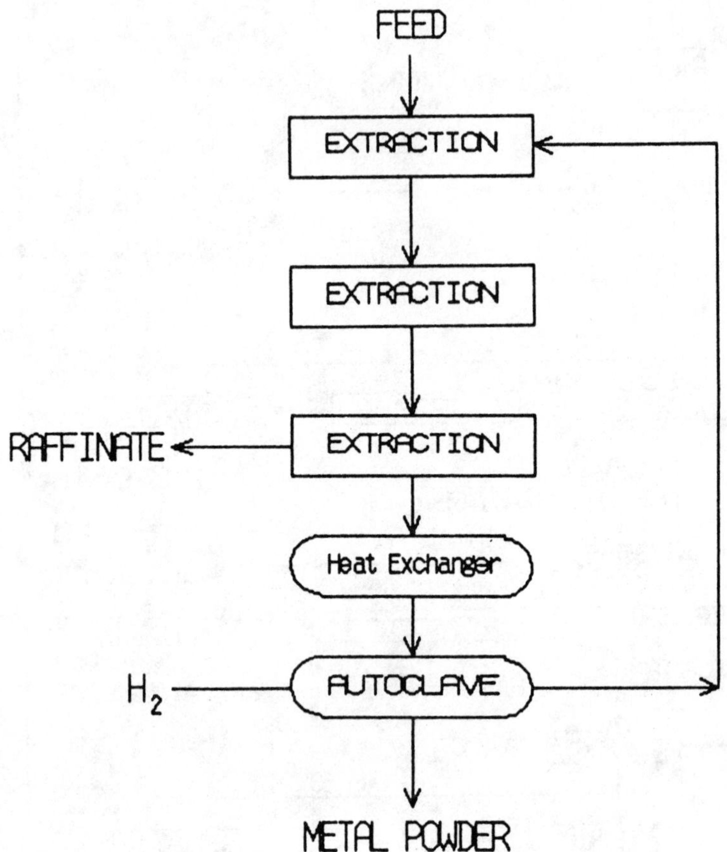

Pressure Reduction
Stripping Circuit

Fig. 10. Schematic flowsheet incorporating pressure stripping-precipitation

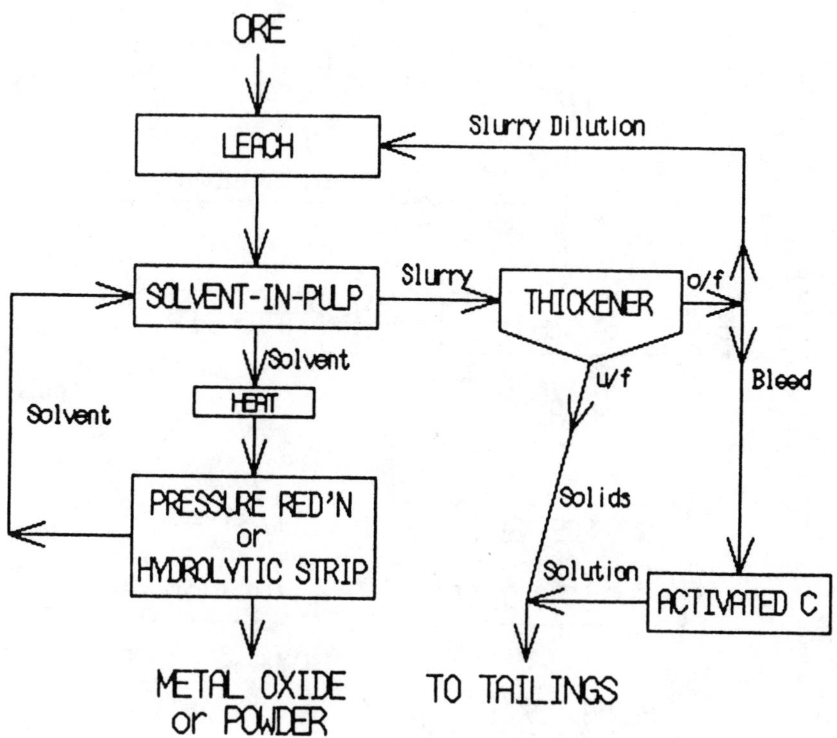

Fig. 11. Future flowsheet incorporating SIP and pressure reduction or hydrolytic stripping and solution recycle

high purity product (52). Fig. 12 (53) shows where solvent leaching followed by hydrolytic stripping could possibly produce a final product.

Finally, Fig. 13 is a schematic for the overall treatment in the mill (54). After the recovery of the metal by solvent extraction, trace constituents of metal and anions are removed and recovered by ion exchange (or solvent extraction). Any solvent is removed by activated carbon to provide for solvent recovery as well as treated water for recycle to the plant or discharge to the environment.

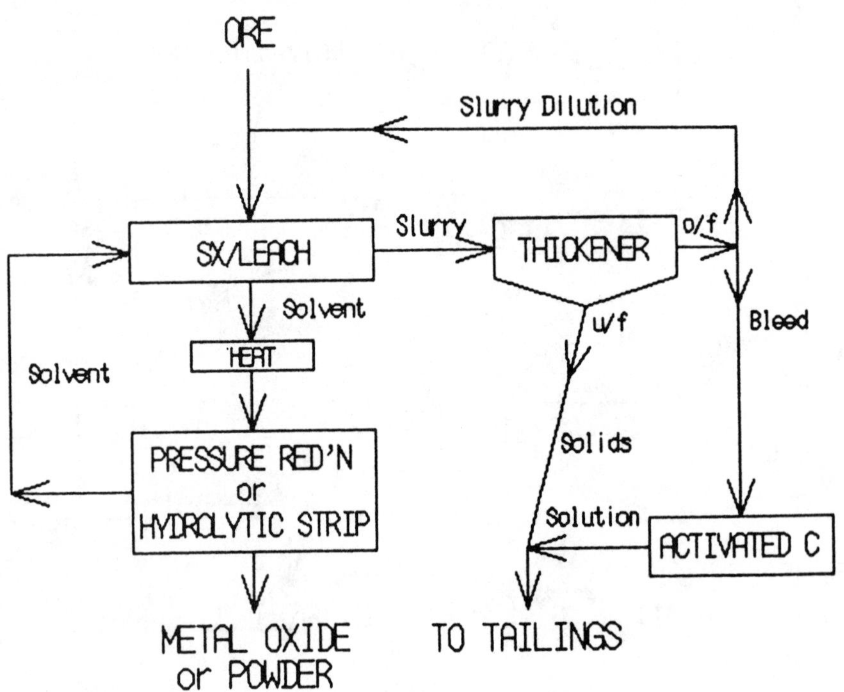

Fig. 12. Future flowsheet incorporating solvent-leaching, pressure reduction and solution recycle

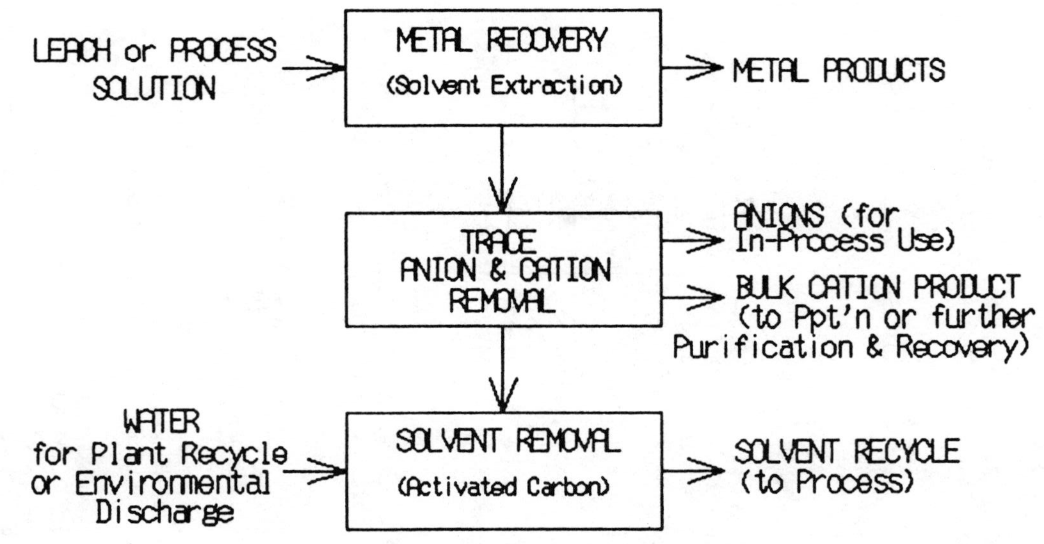

Fig. 13. Schematic outline of purification route

REFERENCES

1. P.W. Thrush and Staff of Bureau of Mines, A Dictionary of Mining, Mineral and Related Terms, U.S. Dept. Interior, 1968.

2. J.R. Boldt and P. Queneau, in The Winning of Nickel, Pub. Longmans Canada Ltd., Toronto, 1967.

3. D.J. Crouse and K.B. Brown, Oak Ridge National Laboratory, Oak Ridge, Tennessee, ORNL-2720, 1959.

4. K.L. Power, in Proceedings of International Solvent Extraction Conference, ISEC '71, The Hague, Pub. Soc. Chem. Ind., London, 1971, pp. 1409-1415.

5. P.G. Thornhill, E. Wigstol and G. Van Weert, Journal of Metals, July 1971, pp. 13-18.

6. M. Ando, M. Takahashi and T. Ogata, "Separation of Cobalt from Nickel in N.M.C. Process", in Hydrometallurgy - Research, Development and Plant Practice", Proceedings of Pub. by AIME, 1983, Ed. K. Osseo-Asare and J.D. Miller.

7. J.M. Demarthe and A. Georgeaux, "Hydrometallurgical Treatment of Complex Sulphides", in Complex Metallurgy '78, ed. M.J. Jones, I.M.M. London, 1978, pp. 113-120.

8. E.D. Nogueira, J.M. Regife and A.M. Arcocha, "Winning Zinc through Solvent Extraction and Electrowinning", Engineering Mining Journal, 180(10):92-94, 1979.

9. J.A. Eisele, L.E. Schultze, D.J. Berinati and D.J. Bauer, "Amine Extraction of Iron from Aluminum Chloride Leach Liquors", U.S. Bureau of Mines, Rep. Invest., RI 8288, 1976.

10. C.F. Acton and W.D. Charles, "Current Gold and Silver Recovery Practice", XIV International Mineral Processing Congress, Toronto, Canada, Oct. 17-23, 1982, pp. II-21.

11. G.M. Ritcey, R. Molnar and G. Pouskouleli, "Recovery of Thorium from Ores and Concentrates and the Production of a Saleable Product", for Publication in Gmelin Handbook Volume on Thorium, 1986.

12. G.M. Ritcey, "Use of TBP to Separate Zirconium from Hafnium", in Science and Technology of Tributyl Phosphate, Vol. 1, Chapter 2, to be published by CRC Press, Boca Raton, Florida, 1986.

13. G.M. Ritcey, "The Process and the Environment - Is Solvent Extraction the Answer?" Paper to be presented at International Solvent Extraction Conference, Munchen, Germany, Sept. 11-16, 1986.

14. J.A. Holmes and J.F.C. Fisher, "Development of a Process for Extraction of Copper from Tailings and Low-Grade Materials at Chingola Mines, Zambia". Presented at the Advances in Extraction Metallurgy and Refining Symposium, I.M.M., London, 1971.

15. G.E. McClelland and S.D. Hill, "Silver and Gold Recovery from Low Grade Resources", Mining Congress Journal, May, 1981, pp. 17-41.

16. R.A. MacGregor, "Bacterial Leaching of Radioactive Conglomerates", Paper presented at the annual meeting, AIME, February, 1969.

17. L.E. Murr, "Theory and Practice of Copper Sulphide Leaching in Dumps and In-Situ", Minerals Sci. Engineering, Vol. 12, No. 3, July, 1980, pp. 121-189.

18. A. Bruynesteyn, "Bioleaching of Refractory Gold/Silver Ores and Concentrates", in <u>First International Symp. on Precious Metals Recovery</u>, Reno, Nevada, June, 1984.

19. R.G. Schulze, "New Aspects in Thiourea Leaching of Precious Metals", J. Metals, June, 1984, pp. 62-65.

20. G.M. Ritcey, B.H. Lucas and R. Molnar, "Pilot Optimization of Flowsheet for the Treatment of Pyrochlore Concentrate for the Recovery of Niobium and Tantalum", CANMET, unpublished data.

21. G.M. Ritcey and A.W. Ashbrook, "Treatment of Cobalt Arsenide Feeds for the Subsequent Recovery of Cobalt", CIM Bulletin, October 1981, pp. 103-112.

22. E.J. Fulton, "Caro's Acid - Its Introduction to Uranium Acid Leaching in Australia", Aus. I.M.M. Conference, Darwin, N.T., Aug. 1984.

23. D.M. Muir, G.M. Ritcey and J.H. Canterford, "Recent Developments in Chloride Hydrometallurgy", Aus. I.M.M., Melbourne Branch Symposium on Extractive Metallurgy, November, 1984.

24. M.C. Campbell and G.M. Ritcey, "Applications of Chloride Metallurgy to Base Metal Sulphide and Uranium Ores at CANMET", in Extraction Metallurgy '81, Ed. M.J. Jones, pp. 76-90, 1981. (I.M.M., London).

25. J.M. Skeaff and J.J. Laliberte, "Continuous High Temperature Chlorination of Uranium Ores", Presented at AIChE Meeting, Portland, Oregon, August 17-20, 1980.

26. K.E. Haque, B.H. Lucas and G.M. Ritcey, "Hydrochloric Acid Leaching of an Elliot Lake Uranium Ore". CIM Bulletin, Vol. 73, No. 819. July, 1980, pp. 141-147.

27. G.P. Demopoulos, G. Pouskouleli and P.J.A. Prud'homme, Can. Pat. Appl. No. 456,044-2 (June 7, 1984); U.S. Pat. Appl. No. 741, 564 (June 5, 1985).

28. W.A. Gow, and G.M. Ritcey, "The Treatment of Canadian Ores - A Review", CIM Bulletin, December 1969.

29. G.M. Ritcey, "Some Economic Considerations in the Recovery of Metals by Solvent Extraction Processing", CIM Bulletin, June, 1975.

30. F.D.L. Cloete, M. Streat and A. Matler, AIChE, I. Chem. E. Symp. Ses. No. 1, 1965, p. 54.

31. A.K. Haines, "The Development of Continuous Fluid Bed Ion Exchange in South Africa and its use in Recovery of Uranium", J. South Africa Inst. of Min. & Met., July, 1979.

32. M.J. Slater, B.H. Lucas and G.M. Ritcey, "Use of Continuous Ion Exchange for Removal of Environmental Contaminants from Waste Streams", CIM Bulletin, August, 1978.

33. A. Himsley, "Production of Pure Yellow Cake by Ion Exchange Processes Using Sulphate Elution", Paper Presented at Int. Atomic Energy Agency Meeting, June, 1979.

34. G.M. Ritcey, "Crud in Solvent Extraction Processing - A Review of Causes and Treatment", Hydrometallurgy, 5 (1980) 97-107.

35. G.M. Ritcey and E.W. Wong, "Influence of Cations on Crud Formation in Uranium Circuits", Hydrometallurgy, 15 (1985) 55-61.

36. C.F. Bonney, G.A. Rowden and D. McKee, "The Combined Mixer-Settler: A New Development in Solvent Extraction"; Paper presented at Annual Meeting AIME, Chicago, Feb. 1981. Paper A81-50.

37. P. Castillon, D. Goodman and G. Cheneau, "Industrial Expansion with the Krebs Solvent Extraction System", Paper presented at Hydrometallurgy Meeting, CIM, August, 1981. Paper 28-1.

38. G.M. Ritcey, "The Process and the Environment-Is Solvent Extraction the Answer?" Paper to be presented at International Solvent Extraction Conference, Munich, Germany, Sept. 1986.

39. G. Thorsen, A.J. Monhemius, "Precipitation of Metal Oxides from Loaded Carboxylic Acid Extractants by Hydrolytic Stripping". Paper presented at 108th AIME meeting, Feb. 1979. Paper A79-12.

40. G.P. Demopoulos, G. Pouskouleli and G.M. Ritcey, "A Novel Solvent Extraction System for the Refining of Precious Metals". Paper to be presented at International Solvent Extraction Conference, ISEC '86, Munich, Germany, Sept. 1986.

41. R.S. Shoemaker, et al., "Carbon-in-Pulp Processing of Gold and Silver Ores - The Experts View the Problems", Mining Engineering, September-October, 1981.

42. J.R. Ross, H.B. Salisbury and G.M. Potter, "Pressure Stripping Gold from Activated Carbon", AIME Annual Meeting, Chicago, February, 1973.

43. D.J. MacKinnon and J.M. Brannen, "Zinc Electrowinning from Chloride Electrolyte", Paper presented at AIME Meeting, Las Vegas, February, 1980.

44. P.K. Everett, "The Dexter Lead Process", in Hydrometallurgy Research, Development and Plant Practice, Ed. K. Osseo-Asare and J.D. Miller, AIME Proceedings 1983.

45. V.I. Lakshmanan and G.M. Ritcey, unpublished data.

46. E.G. Joe and D.E. Pickett, "Water Reuse in Canadian Ore-Concentration Plants - Present Status, Problems and Progress", Paper No. 19, Inst. of Mining and Metallurgy, London, 1974.

47. J.E. Dutrizac, "Jarosite-type Compounds and their Application in the Metallurgical Industry", in Hydrometallurgy Research, Development and Plant Practice, AIME Conference Proceedings, 1982.

48. G.M. Ritcey, "Tailings Disposal, Management and Effluent Control in the Mining Industry", Textbook to be published 1986/87.

49. G.M. Ritcey, "Solvent Extraction - Projection to the Future", Separation Science and Technology, 18 (14 & 15), pp. 1617-1646, 1983.

50. G.M. Ritcey, Unpublished data. CANMET, Dept. Energy, Mines and Resources, Ottawa, Canada, 1986.

51. R.X. Li, G.P. Demopoulos and P.A. Distin, "Nickel Recovery from Loaded Kelex 100 by Precipitation Using Hydrogen", Paper presented at the 21st CIM Annual Conference of Metallurgists, Toronto, Ontario, August 1982.

52. G.M. Ritcey, M.J. Slater and B.H. Lucas, "A Comparison of the Processing and Economics of Uranium Recovery from Leach Slurries by Continuous Ion Exchange or Solvent Extraction", in Proceedings Symposium on Hydrometallurgy. Pub. AIME, pp. 419-473, 1973.

53. G. Thorsen, U.S. Patent 4,200,504, April, 1980.

54. G.M. Ritcey and A.W. Ashbrook, Solvent Extraction - Principles and Applications to Process Metallurgy, Part I., Pub. Elsevier Scientific Publishing Company, 1979.

IRON (III) REMOVAL FROM ALUMINIUM (III) SULPHATE SOLUTIONS BY AMINE PRIMENE 81R

F.J. Alguacil, S. Amer, A. Luis
Centro Nacional de Investigaciones Metalúrgicas
Avda. Gregorio del Amo nº 8, Ciudad Universitaria,
28040 Madrid, Spain

As a part of a research program done in the CENIM (Centro Nacional de Investigaciones Metalúrgicas) to obtain alumina from domestic non-bauxitic ores (clay), a method for extracting iron (III) from aluminium sulphate solutions was investigated. Iron (III) being the main impurity to be eliminated, the extraction system Primene 81R - Iron (III) - Sulphate was studied and extraction curves were determined. The influence of variables such as temperature, aluminium, sulphuric acid and amine concentrations was also studied. Stripping curves with sulphuric acid were determined and the use of basic stripping agents was examined. In countercurrent extraction studies with aluminium sulphate solutions, iron concentration in raffinate can be decreased from 2 g/l to 0.015 g/l after three-five stages. The loaded organic phase was regenerated, after stripping, by washing it with basic solutions.

INTRODUCTION

Alternative methods for alumina production from non bauxitic sources have been a subject of research since World War II. Domestic reserves of non bauxitic ores, especially clay, are large and non expensive. Therefore several processes have been developed in different countries, although no commercial plants have been built, because these methods are not competitive with the known Bayer process for treatment of bauxite. However they still have an interest with regard to the international raw materials situation. Some studies have been done on Spanish non bauxitic ores, as the work on treatment of clays, which lead to a method of leaching these materials via sulphuric acid (1, 2).

The electrometallurgical reduction of alumina according to the known process of Hall-Heroult requires a high purity alumina. Iron, silica, titanium, etc. are among the main impurities that aluminium ores contain. Requirements of purity for alumina are very strict and those impurities must be eliminated to a very low content. In the Bayer process the elimination of iron is accomplished easily because of the insolubility of iron hydroxide in sodium hydroxide.

Iron elimination is a common problem in many industries and not always easy to solve. In certain metallurgical industries, as that of zinc, this is accomplished by the known hydrometallurgical process of jarosite. Iron elimination is also a major problem of purification of leach solutions when an alternative method to produce alumina from clay is developed. In the Spanish case, as the clay tested has a high iron content, a double step for iron elimination is necessary. In the first place most iron is eliminated by means of a precipitation known as jarosite.

To accomplish jarosite precipitation, heating up to 353º-373ºK and stirring for a certain time (4 to 7 hours) have to be done. Some manganese bioxide is added during this operation to oxidize iron (II) to iron (III), as iron (II) does not precipitate as jarosite. Most of iron (III) will be precipitated in the form of jarosite, and a simultaneous and complete elimination of potassium and most of sodium is produced, provided that there will be an excess of iron (III). The yield of leaching recovery for aluminium is as high as 90 - 94% of aluminium present in clay. As the precipitation of jarosite produces free sulphuric acid, the addition of activated clay will neutralize sulphuric acid, letting the precipitation continue. Only about 35 - 45% of activated clay is now dissolved. However some iron (III) will remain in solution as shown in table 1.

Jarosite precipitation can be performed at a temperature between 343ºK and 373ºK. Potassium precipitates as jarosite preferentially to sodium, and only when there is a concentration of 0.1 g/l to 0.6 g/l of potassium, sodium will start to precipitate. Small amounts of aluminium also precipitate, and this can be done by an isomorphic substitution of iron by aluminum (3). The concentrations of both metals in solution are related by the equation

$$\log \frac{[Al^{3+}]_0}{[Al^{3+}]} = \lambda \log \frac{[Fe^{3+}]_0}{[Fe^{3+}]}$$

being λ a coefficient which changes with temperature and the nature of monovalent ion precipitating into the jarosite. For potassium and for the temperature interval of 343º - 373ºK λ changes from 0.01 to 0.0336. Kinetics of jarosite precipitation are faster for potassium than for sodium and ammonium and increase with temperature. A careful control of temperature during the precipitation of jarosite can lower the iron (III) concentration in solution to 1 g/l or even less.

Some impurification of leach solution is introduced by the manganese (∿1 g/l) used for oxidation of iron (II) to iron (III). Manganese (II) can be eliminated and recovered as bioxide by means of an oxidation with ozone, which takes place with precipitation of an hydrated manganese bioxide (4). This operation can be used as a second step for iron (III) elimination, because iron (III) is coprecipitated by this hydrated manganese bioxide, being 333ºK the most convenient temperature. According to literature, a compound of formula $2Fe(MnO_3H)_3$ is precipitated and some free sulphuric acid is liberated, which makes a slight change of pH value of aluminium sulphate solution. This manganese (IV) precipitate can be recycled back to the leach step and used as oxidant for iron (II).

Some experiments performed by varying the ratio manganese/iron in solution and by oxidizing manganese (II) with ozone will form precipitates with variable and different compositions to that mentioned above (3). This shows that a coprecipitation phenomenon takes place instead of the formation of a compound of defined stoichiometry. The yield of oxidation of manganese by ozone is 75 - 80%, when the operation is performed in a volume of 10 - 12 liters of solution. If the mixture of air plus ozone is bubbled through two vessels, one after another, the overall yield of oxidation is of 90%. The content of iron in the aluminium sulphate solution can be lowered to ∿0.1 g/l by this method, which is an acceptable value.

As the research done on treatment of Spanish clays leads to a method somehow different to other published methods, a brief outline of it can be of interest. The steps of the process are:

- Activation of clay at 923ºK in a fluid bed furnace.
- Mixing of activated clay with the stoichiometric amount of concentrated sulphuric acid (80 - 96%) and heating up to 553-573ºK, which is selfsustained, because there is an exothermic reaction, and a solid product is formed.
- Leaching of this product with water and simultaneous precipitation of iron, potassium and sodium as jarosite. Potassium elimination is complete, but some iron and sodium will remain in solution.
- Purification of aluminium sulphate solution for iron elimination.
- High temperature hydrolysis (∿423ºK) of aluminium to produce a compound of formula $Al_7(SO_4)_5Na_2(OH)_{13} \cdot 3H_2O$ (5).
- A reductive and thermal decomposition of the mentioned compound to alumina, sodium sulphate and sulphurous gas, which can be recycled to make sulphuric acid. By washing out this solid product with water, sodium sulphate can be dissolved and pure alumina will be obtained.

We have discussed the steps for iron (III) elimination by hydrometallurgical precipitation. As an alternative way for the second step of purification, that is, the coprecipitation of iron by hydrated manganese bioxide, a solvent extraction method can be used. Primary amines should be used in a sulphate medium. The most known of them Primene JMT has been used with the same purpose of purifying leach aluminium sulphate solutions from clay (6). There is very little information about the uses of Primene 81R as extractant (7). In this paper the results of a research of a solvent extraction method for eliminating iron (III) in leach aluminium sulphate solutions from clay by means of primary amine Primene 81R are presented.

EXPERIMENTAL

Single stage experiments were carried out by contacting equal-volume phases unless otherwise stated, which were placed in separatory funnels thermostated at the required temperature and provided with mechanical shaking. Kinetic studies showed that five minutes time were usually required to complete extraction. For this reason, to ensure that equilibrium state has been reached, ten minutes time were used in all experiments made. On completion of extraction, the phases were separated and the correspondent metallic ion analysis was carried out for both phases.

To perform multistage studies, a 9 - 10 cell mixer-settler unit, also thermostated at the required temperature, was used. The capacity of this unit had a maximum value of 100 ml/min flow for each phase. Mixing and settling volumes were comprised between 35 - - 200 and 140 - 700 ml respectively.

Amine Primene 81R is manufactured by Rohm and Haas Co. The general formula is t-$C_{12}H_{25}NH_2$ to t-$C_{14}H_{29}NH_2$ and its average experimental molecular weight is 200, calculated from titration of dissolved amine in ethanol with standard sulphuric and hydrochloric acids by using a pH meter Beckman 3560. The kerosene diluent was supplied by Campsa. All other chemicals were reagent grade. Synthetic solutions were prepared by dissolving in water chemicals as iron (III) and aluminium (III) sulphates. Leach solutions were prepared by the method outlined previously. Compositions of activated clay and leach solutions after jarosite precipitation are shown in table 1.

Table 1. Composition of calcined clay and leach solutions.*

Analysis of calcined clay

Al	14,9%	Mg	0.46%
SiO_2	51.6%	Na	0.31%
Fe	9.0%	Ca	0.39%
K	2.45%	Ti	0.65%

	Leach Soln. I	Leach Soln. II	Leach Soln. III
Al	44.7	45.0	41.0
Fe	2.32	1.92	2.0
Na	0.5	0.5	0.6
Mn	1.28	1.0	
SO_4^{2-}	245.6	234.5	234.5
H_2SO_4	no	no	0.093M
pH	2.5	2.4	2

* Concentrations in g/l.

RESULTS AND DISCUSSION

Extraction

The behaviour of the amine Primene 81R with regard to the extraction of iron (III) will depend on temperature and composition of aqueous solution, which influence the nature of extracted species (8, 9). A salt of formula $n(RNH_3)_2SO_4 \cdot (FeOHSO_4)_2$ is extracted into the organic phase, being n = 3, 4, when iron is the only metal ion present in aqueous solution. At 323ºK n = 3 and at 281ºK n = 4, a mixture of both species is extracted at temperatures between 281ºK and 323ºK and its concentrations are linearly dependent on temperature. When in addition to iron there is a concentrated aluminium sulphate solution, n = 2 in the above mentioned compound for the organic phase. When the free amine is present in the organic phase and there is no sulphuric acid in the aqueous phase, a compound of formula $4RNH_2 \cdot Fe_2(SO_4)_3$ also can be extracted into the organic phase. In this case the extracted species are the same whether aluminium sulphate will be or not present in the aqueous phase.

In the presence of aluminium sulphate the extraction will be enhanced and the larger the concentration of this salt is, the larger the extraction of iron (III) will be. This must be due to a salting out effect, also observed for this system in the presence of other metallic ion sulphates (10). In table 2 this effect is shown. As leach solution will provide 40 - 50 g/l (1.49 - 1.85M) of aluminium the extraction efficiency should be near the maximum. Iron (II) is not extracted by this amine, although, as has been explained before, an oxidation to iron (III) has been previously performed.

Table 2. Influence of aluminium (III) concentration on iron (III) extraction and aluminium coextraction by 0.203M amine Primene 81R, 5% n-decanol in kerosene*.

Al(III) (Mol/l) aqueous phase	Fe(III) org 298ºK	Fe(III) org 323ºK	Al(III) org 298ºK	Al(III) org 323ºK
0.0	6.40	7.56	-	-
0.1	6.60	7.60	0.0018	0.0022
0.2	6.81	7.65	0.0069	0.0081
0.35	7.00	7.78	0.019	0.028
0.5	7.19	8.00	0.050	0.067
0.75	7.53	8.26	0.143	0.17
1.0	7.80	8.54	0.29	0.32
1.25	8.14	8.77	0.41	0.41
1.50	8.62	8.99	0.48	0.42
1.67	9.38	9.58	0.48	0.42

* Concentrations in g/l

The influence of amine Primene 81R concentration on iron (III) extraction into the organic phase (at saturation) from a concentrated aluminium sulphate solution (1.67M) at 298ºK and 323ºK is shown in figure 1. As sufficient sulphuric acid is present in aqueous solution, amine Primene 81R sulphate, which is the real

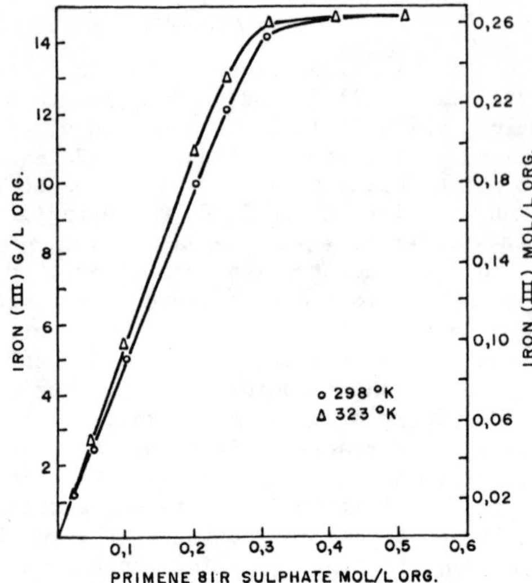

Figure 1. Influence of Primene 81R sulphate concentration on iron (III) concentration in organic phase.

extractant, will be formed in the organic phase. A concentration of 0.1 to 0.2M amine sulphate showed to extract iron (III) satisfactorily for the iron concentration range in leach solutions. Most of the work has been done with a 10% v/v amine Primene 81R (0.203M) solution in kerosene. The addition of half of this concentration of a modifier, such as 5% n-decanol, should be convenient. In order to know if aromatic / aliphatic ratio in diluent composition had any influence on iron (III) extraction some diluents were tested. No significant differences were observed when using Solvesso 150, Escaid 100, Escaid 110 and kerosene in an organic solution containing 0.203M Primene 81R and 5% v/v n-decanol at a temperature of 323ºK to extract iron (III) from an aqueous solution 1.67M (45 g/l) of aluminium sulphate, 0.2M sulphuric acid. The maximum difference of concentration of extracted iron (III) in organic phase among these diluents was of 2.5%, being Solvesso 150 and kerosene the best. As phase separation was better with kerosene, this one was selected for all the work.

Temperature also has a certain effect as shown in table 3. By increasing temperature, better extraction will be reached up to 323ºK, keeping this value from 323º-343ºK, and decreasing at higher temperature. However the effect of temperature on phase separation is more important, where optimum conditions are achieved at 323º - 333ºK. This is related to viscosity of aqueous concentrated aluminium sulphate solutions. The most convenient temperature to operate and to efficiently extract iron (III) will be 318º - 323ºK, when kerosene is used as diluent. Higher temperatures should be avoided because of kerosene flash point.

Table 3. Influence of temperature on iron (III) and aluminium (III) extraction by 0.203M amine Primene 81R, 5% n-decanol in kerosene. Aqueous phase: 0.269M iron (III), 1.67M aluminium (III), 0.2M sulphuric acid.

Temperature ºK	D_{Fe}	$D_{Al} \cdot 10^{-3}$
283	1.66	10.8
293	1.89	8.29
303	2.16	5.81
313	2.43	3.79
318	2.59	2.67
323	2.68	2.67
328	2.68	2.67
333	2.68	2.67
343	2.68	2.67

The concentration of sulphuric acid has an important effect on extraction equilibrium for this system. The iron (III) load in the organic phase has a maximum value at an acid concentration of 0.1M and then decreases as acid concentration increases. In table 4 the influence of sulphuric acid concentration on iron (III) extraction is shown. Since in the leaching and iron (III) removal as jarosite step, acidity was controled by the addition of activated clay, and a sulphuric acid concentration about 0.1M is present, extraction should be near the maximum efficiency.

A point which should be convenient to know is the degree of coextraction of aluminium (III). Aluminium (III) is also extracted by Primene 81R and for this reason some experiments were carried out to determine the degree of aluminium (III) extraction. These results were compared with those of iron (III) extraction in the

Table 4. Influence of sulphuric acid concentration on iron (III) and aluminium (III) extraction by 0.203M Primene 81R, 5% n-decanol in kerosene. Aqueous phase: 0.269M iron (III), 1.67M aluminium (III).

H_2SO_4	D_{Fe}		$D_{Al} \cdot 10^{-3}$	
Mol/l	298ºK	323ºK	298ºK	323ºK
0.0	4.0	2.96	2.45	2.00
0.025	4.75	4.32	2.90	2.00
0.05	6.25	5.82	3.34	2.00
0.075	8.43	8.68	3.57	2.00
0.101	13.71	13.85	3.57	2.23
0.150	5.82	6.58	5.36	2.45
0.203	2.01	2.68	7.16	2.67
0.305	1.29	1.25	6.04	2.00
0.406	1.05	1.25	5.59	2.00
0.5	0.96	1.25	4.69	2.00
1.0	0.81	1.25	4.69	2.00
1.5	0.81	1.25	4.69	2.00

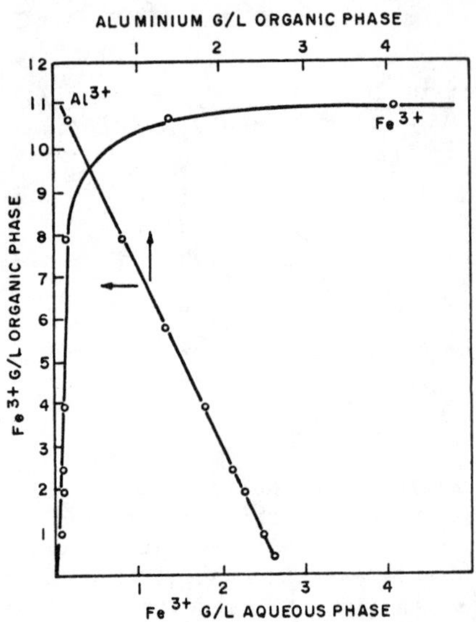

Figure 2. Extraction of iron (III) and aluminium (III) by 0.203M amine Primene 81R, 5% n-decanol in kerosene solution. Aqueous phase: various iron (III) concentrations, 1.67M aluminium (III) and 0.2M sulphuric acid.

presence of aluminium (III) sulphate. Figure 2 shows the extraction of iron (III) and that in the presence of iron (III) the concentration of aluminium (III) decreases with the increasing of iron (III) load in the organic phase, aluminium (III) is extracted only if insufficient iron (III) is present, but it is displaced by iron (III) as the organic phase becomes loaded. Table 2 also shows the coextraction of aluminium (III).

Stripping

Stripping can be performed by the use of sulphuric acid or by means of basic agents as sodium hydroxide or sodium carbonate. Experiments were carried out with sulphuric acid solutions of different concentrations and an organic phase of 0.203M Primene 81R and 5% v/v n-decanol loaded with iron (III) at 298ºK and 323ºK of temperature. At 323ºK a larger amount of iron (III) is stripped than at 298ºK. There is not a great difference on iron (III) stripped with sulphuric acid concentrations comprised between 2M and 5M although phase separation improves in the interval from 3 to 6M. The equilibrium stripping curves are represented in figure 3, which shows that for those concentrations of acid, equilibrium favors the iron (III) stripped into the aqueous phase. The use of these acid concentration allow to use a high ratio of organic to aqueous phases (O/A) in the stripping stage and yield high iron (III) concentrations in the aqueous solution.

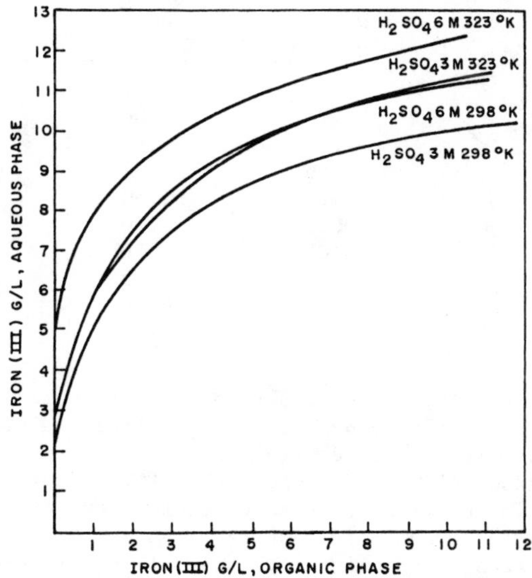

Figure 3. Iron (III) stripping curves by sulphuric acid in the presence of aluminium (III). Organic phase: 0.203M amine Primene 81R, 5% n-decanol in kerosene, fully loaded with iron (III).

Sulphuric acid can be extracted by the amine Primene 81R and when a stripping of iron (III) operation is performed, the sulphuric acid is extracted by the amine. This extraction is quantitative until all the amine is converted to the amine sulphate form. At 298ºK the conversion to bisulphate is complete if sufficient excess of sulphuric acid is present, but at 323ºK only a part of amine will be converted to amine bisulphate, which corresponds to a ratio of Primene 81R sulphate to Primene 81R bisulphate of 1.66. This ratio keeps constant when sulphuric acid concentration in aqueous phase increases. This can be due to the formation of a mixed salt of formula $5(RNH_3)_2SO_4 \cdot 3(RNH_3HSO_4)$ in organic phase. However the bisulphate of amine Primene 81R does not extract iron (III). According to this when an organic phase loaded with iron (III) is stripped with sulphuric acid and the cycle iron (III) extraction - - stripping is repeated consecutively several times with the same sample, a loss of capacity for iron (III) extraction is observed. As can be seen from Table 5, there is a continuous decreasing of extraction. The lower the acid concentration is, the lower iron (III) load in organic phase will be, because of amine bisulphate formation.

This is a limitation for the use of this extractant. Consequently a study for the regeneration of amine was carried out by means of the use of basic agents as sodium hydroxide, sodium

Table 5. Effect of successive extractions and decreasing of iron (III) load in an organic phase 0.203M Primene 81R, 5% n-decanol in kerosene. Aqueous phase: 0.0359M iron (III), 1.67M aluminium (III), 0.101M sulphuric acid. Temperature 323ºK.

Nº successive extractions	D_{Fe} 1.5M H_2SO_4	D_{Fe} 3M H_2SO_4	D_{Fe} 6M H_2SO_4
1	27.57	27.57	27.57
2	5.20	12.00	14.10
3	1.42	4.41	6.51
4	0.56	2.42	4.48
5	0.20	1.50	2.65
6	0.09	0.90	1.70
7	0.06	0.56	1.16
8	0.048	0.35	0.70
9	0.042	0.24	0.52
10	0.038	0.13	0.32

carbonate, ammonium hydroxide and ammonium carbonate. The experiments were performed by contacting, after the stripping of iron (III), the organic phase with a solution of the regeneration agent, at a phase ratio of 1:1 and 323ºK of temperature. For a full regeneration of organic phase concentrations of 2M ammonium hydroxide, 2.5M of ammonium carbonate or sodium hydroxide will be necessary, whereas with sodium carbonate it is not possible to obtain a complete regeneration. A series of experiments performed in this way for consecutive iron (III) extraction - stripping operations gave no loss of capacity after 10 cycles. This is shown in figure 4.

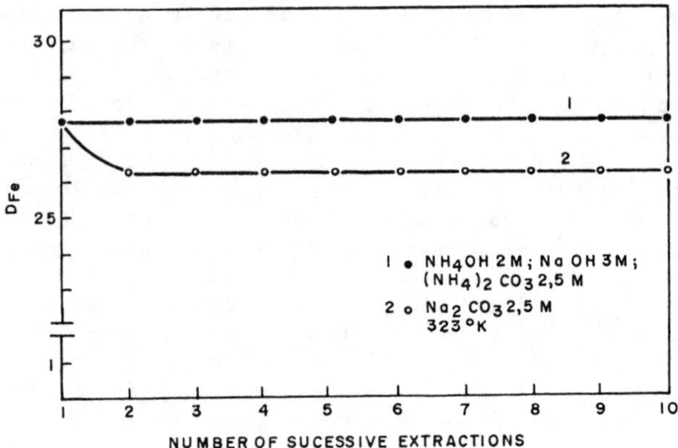

Figure 4. Effect of regeneration agent on successive cycles of iron (III) extraction by 0.203M amine Primene 81R, 5% n-decanol in kerosene. Aqueous phase: 0.0359M iron (III), 1.67M aluminium (III), 0.101M sulphuric acid.

The use of basic stripping agents, namely sodium hydroxide and sodium carbonate, does not present those limitations, as can be seen from table 6. The use of this type of chemical has the advantage of a simultaneous operation of stripping of iron (III) and of regeneration of the extractant as free amine. However a precipitation of iron (III) hydroxide takes place at the same time. For this reason the use of conventional mixer-settlers is not possible and a step of elimination of solids by filtration or centrifugation will be required.

Table 6. Stripping of organic phase with sodium hydroxide. Organic phase: 0.203M Primene 81R as free amine, 5% n-decanol in kerosene. Aqueous phase: 0.0448M iron (III), 1.74M aluminium (III) sulphates. Stripping solution 1.5M sodium hydroxide. Phase ratio: extraction O/A = 1:2, stripping O/A = 1:1. Volume of organic 50 l. Temperature 288ºK. Separation of solids and two liquid phases by means of a centrifugal separator Westfalia SA 1-01-175.

Extraction number	D_{Fe}	$D_{Al} \cdot 10^{-2}$
1	29.3	4.2
2	27.5	3.9
3	29.3	3.9
4	29.2	3.9

Continuous countercurrent operation

The application of the Mc Cabe Thiele diagram to the equilibrium extraction of this system shows that the use of three stages is required, when iron (III) sulphate is extracted by a 0.203M Primene 81R, 5% v/v n-decanol solution in kerosene at 323ºK of temperature from a 1.67M aluminium sulphate solution at an O/A phase ratio of 1:2 to 1:1. Two stages are needed for the stripping with 3 - 6M sulphuric acid at a O/A phase ratio of 2:1. Besides two more steps are required for amine regeneration at O/A ratio phase of 2:1, using a 2.5M ammonium carbonate solution.

Synthetic solutions were first used in countercurrent studies until the system showed that a satisfactory operation was reached. To evaluate the efficiency of the system with leach solutions, several tests of iron (III) extraction from two different leach solutions by 0.203M amine were made. These values were compared with the equilibrium curve from synthetic solutions and fitted it very well. Compositions of those leach solutions are shown in table 1.

The close agreement between calculated values from Mc Cabe Thiele diagram and experimental data shows that the system works very well, as can be seen from table 7. Continuous countercurrent experiments were performed in three circuits, in which organic solution run directly from the extraction circuit to the stripping circuit and from this one to the regeneration of amine circuit and the organic recycled back to the extraction circuit. Two experiments were carried out at 323ºK operating continuously for 30 hours with a synthethic solution and for 20 hours with leach solution III. The raffinate of aluminium sulphate contained 0.015 g/l of iron (III)

Table 7. Comparison of iron (III) content from settler results with calculated values.

	A) For extraction of iron			B) For stripping of iron	
	Settler	Calculated	Experimental	Calculated	Experimental
		Aqueous phase*		Aqueous phase*	
1)	1	0.09	0.09	3.94	3.94
	2	0.02	0.017	0.04	0.06
	3	0.015	0.015		
				Organic phase*	
2)	1	0.13	0.12	0.02	0.03
	2	0.037	0.035	0.0	0.0
	3	0.015	0.016		

Organic phase: 0.203M Primene 81R, 5% v/v n-decanol in kerosene.

Extraction feed:
2g/l Fe(III), 45 g/l Al(III) 1. H_2SO_4 0.05M, flow ratio O/A : 1/1
2. H_2SO_4 0.2M, flow ratio O/A : 1/2

Stripping: 1.97 g/l Fe(III) in organic phase, flow ratio O/A : 2/1, 6M H_2SO_4 aqueous phase.

Temperature: 323ºK.

* Concentrations in g/l

and the recycled stripped organic solution did not contain iron. In each series of experiments each phase of each settler was analyzed for iron at the end of the tests. The losses of organic phase were very small. Aluminium losses into the acid waste stream are small and they were due to the fact that the organic phase was not fully loaded with iron.

When the leaching solution does not contain free sulphuric acid, it requires the use of four or five extraction stages. In this case, as amine Primene 81R is in the form of free amine, aluminium losses are quite small, because free amine does not extract aluminium. Finally a sample of purified aluminium sulphate, after being crystallized, was ignited at 1173ºK to produce alumina and the iron content in this sample was 0.03%, which was about the requirements.

CONCLUSIONS

Iron (III) is extracted from an aqueous sulphate solution by amine Primene 81R. The presence of aluminium (III) in the aqueous phase enhances the iron (III) extraction, because of a salting out effect. There is a preferential iron (III) to aluminium (III) extraction.

Sulphuric acid is also extracted by amine Primene 81R. When there is an excess of sulphuric acid in solution amine bisulphate is formed, which does not extract iron (III).

Organic phase loaded with iron (III) can be stripped by using a sulphuric acid solution, but a loss of capacity to extract iron (III) is shown for subsequent operations. A step of regeneration of amine by washing organic phase with a basic solution should be done. No loss of capacity is shown when stripping is done by means of an alkaline solution. In this case the formation of a precipitate of iron (III) hydroxide takes place, which should be removed by filtration or centrifugation.

Amine Primene 81R can be used effectively to remove iron (III) in a continuous countercurrent system to a content as low as 0.015 g/l.

REFERENCES

1. A. Cuadra, J.L. Limpo, A. Luis, M. Fernández, "Procedure to obtain alumina from solutions containing aluminium sulphate", Spanish Patent nº 479.148 (1979).
2. J.L. Limpo, A. Cuadra, A. Luis, M. Fernández, V Asamblea General del CENIM, Vol. IV, Papers T-79 y T-80, Madrid (1981).
3. J.L. Limpo, A. Luis, A. Hernández Rev. Metalurgia (CENIM), 20, 26 (1984).
4. J.H. Walthall, P. Miller, M.M. Striplin, Trans. Am. Inst. Chem. Engrs., 41, 55 (1945).
5. J.L. Limpo, A. Luis, Rev. Metalurgia (CENIM), 18, 261 (1982).
6. L.E. Schultze, J.A. Eisele et al., Report of U.S. Bureau of Mines RI 8353 (1979).
7. G.M. Ritcey, A.W. Ashbrook, Solvent Extraction, Part II, pp 258, Elsevier Scientific Publishing Co. (1979).
8. F.J. Alguacil, S. Amer, A. Luis, XX Bienal de la Real Sociedad Española de Química, Paper 29-25, Castellón, Spain (1984).
9. F.J. Alguacil, S. Amer, Hydrometallurgy, 15, 337 (1986).
10. F.J. Alguacil, Ph.D. Thesis, F.C. Químicas, Universidad Complutense de Madrid, Madrid, Spain (1985).

Application of a Liquid Membrane for Copper Recycling

P. A. O'Hara
M. P. Bohrer

AT&T Bell Laboratories
Murray Hill, New Jersey 07974

A supported liquid membrane composed of a commercial ion exchange reagent and a microporous teflon film was prepared to selectively transport copper ions from one aqueous solution to another. This type of membrane would have several interesting applications, one of which is the recycling of copper from etching baths to plating baths used in the manufacturing of printed wiring boards. Under conditions typical of such an application, the steady state flux of copper ions across the membrane was measured to be 0.20 g/m^2-min. No degradation in the flux was noted over the entire time period of the experiment ($\sim$14 days).

Experimental results reported here indicate that, at low concentrations of the organic complexing reagent (LIX-54®), the flux is limited by either the diffusion of the complex across the membrane or by the interfacial reaction rate on the extraction side of the membrane. At higher concentrations of the carrier, the flux apparently becomes limited by diffusion in the aqueous solution on the stripping side of the membrane. Experiments are currently underway to independently measure the interfacial reaction rate constants and the diffusion coefficients of the carrier-copper complex in order to formulate a more complete model of the mass transport in this liquid membrane system.

INTRODUCTION

The use of supported liquid membranes for the selective transport of metal ions was first proposed more than 20 years ago (1). Since then, several researchers have examined a wide range of applications for these membranes in which an organic solution of solvent extraction reagents is immobilized in a thin, microporous support (2-8). Excellent reviews of this subject have recently been presented by Danesi (9) and Way and coworkers (10).

In a typical supported liquid membrane, a metal cation, M^+, is selectively transported from one aqueous solution (feed) to another (strip) with the concomitant transport of H^+ in the opposite direction. The organic complexing reagent and its solvent are held in the microporous support by surface tension forces. By complexing with the organic reagent, RH, the metal cation is able to enter the organic phase in which it is otherwise insoluble. Concentration gradients within the membrane cause the complex to diffuse to the opposite side where the metal is released and a proton is picked up. By controlling the environment on each side of the membrane (in this case the pH), the formation of the carrier-metal complex can be controlled.

Supported liquid membranes offer many advantages over other separation techniques including high selectivity, the ability to "pump" against a concentration gradient, high fluxes relative to solid membranes, low capital and operating costs, and the ability to use expensive extracting agents. For these reasons, we have performed some preliminary experiments to evaluate a supported liquid membrane for copper transport. One application for such a membrane is the regeneration of copper etching and plating solutions used in the manufacture of printed wiring boards. Standard solvent extraction techniques have been used to recover the copper from spent etchant solutions. An alternative means for recovering this copper and simultaneously replenishing the plating solution is to use a supported liquid membrane as diagrammed in Figure 1.

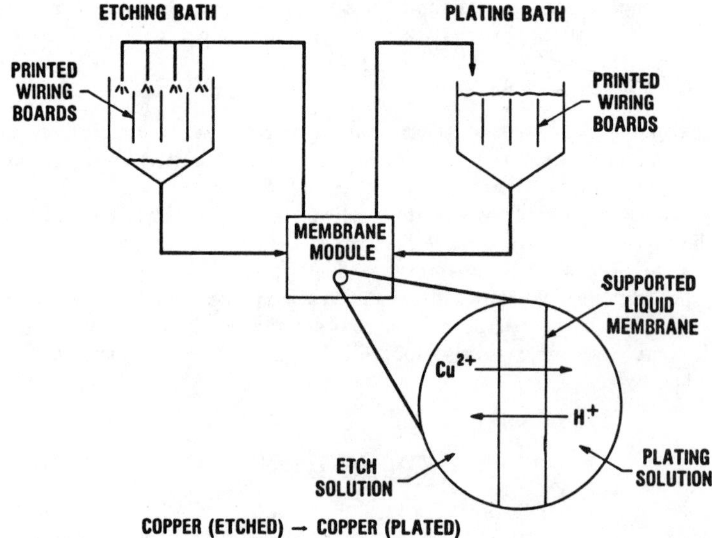

Figure 1. Proposed liquid membrane process to recycle copper from an etching solution to a plating solution used in the manufacturing of printed wiring boards.

For the application reported here, the chemical reactions involving copper are summarized below.

Etching

$$Cu + 4NH_4OH + \frac{1}{2}O_2 \rightarrow Cu(NH_3)_4^{2+} + 2OH^- + 3H_2O$$

Extraction

$$Cu(NH_3)_4^{2+} + 2RH + 2OH^- + 2H_2O \rightarrow R_2Cu + 4NH_4OH$$

Stripping

$$R_2Cu + 2H^+ \rightarrow Cu^{2+} + 2RH$$

Plating

$$H_2O + Cu^{2+} \rightarrow 2H^+ + \frac{1}{2}O_2 + Cu$$

The net reaction is

$$Cu\ (etched) \rightarrow Cu\ (plated)$$

One class of organic compounds which have been shown to selectively complex copper ions at high pH are β-diketones. A commercially available solvent extractant, LIX-54® (Henkel Corp.), is a phenyl alkyl β-diketone dissolved in kerosene. The general structure of this reagent is shown in Figure 2, where the R" is -H or -CN and R and R' can be various phenyl, alkyl substituted phenyl, or alkyl groups.

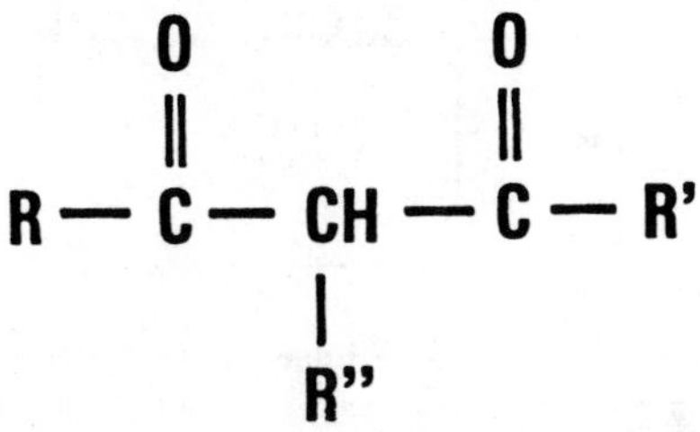

```
      O        O
      ||       ||
R  —  C  — CH — C — R'
           |
           R"
```

Figure 2. General structure of the β-diketones in LIX-54®. R is phenyl or alkyl substituted phenyl, R' is alkyl, alkyl substituted phenyl or chloro substituted phenyl and R" is H or --CN with the provisos that; (1) when R is phenyl, R' is a branched chain alkyl group of at least seven carbon atoms and (2) when R is alkyl substituted phenyl, the number of carbon atoms in the alkyl substituent or substituents is at least 7 and at least one such alkyl substituent is a branched chain.

EXPERIMENTAL PROCEDURES

The supported liquid membranes were prepared by submersing microporous teflon membranes (Millipore Corp.) in a 40 vol% solution of β-diketones in kerosene (LIX-54®) for several days. The teflon membranes had a nominal pore size of 5 μm and an average thickness of 125 μm.

Membrane transport experiments were performed with the diffusion cell shown in Figure 3. The exposed membrane area is 4.34 cm^2. After fixing the membrane in place, 25 ml of etching solution were placed on top of the membrane and the unit was lowered into a bath of dilute sulfuric acid (volume = 500 ml). The etch solution was agitated with a glass stirring rod connected to a variable speed motor and stirring speed was determined with a stroboscope. The acid side was stirred by placing the bath on a magnetic stirrer.

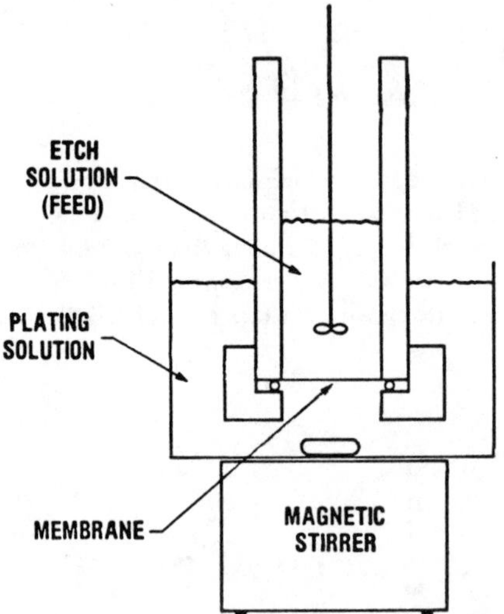

Figure 3. Schematic of diffusion cell used to measure copper ion fluxes. The cell was made of teflon and viton o-rings were used to seal the membrane in place. The volume of etch solution was 25 ml and the volume of the plating solution was 500 ml.

The etching solution used in these experiments was MacDermid Ultra-Etch 50 with an initial copper concentration of 157 g/l. The acid solution was prepared by diluting reagent grade sulfuric acid to the desired concentration. For most experiments, the acid concentration was 2.12 M, a value typical of plating baths. In a few experiments, a lower acid concentration was used to determine the effects of this variable on the measured copper flux.

Copper transport across the membrane was determined by measuring the copper concentration in the acid solution as a function of time. Atomic absorption spectroscopy was used to measure copper concentrations in the acid samples and a standard EDTA titration (PAN indicator) was employed for the etch samples.

EXPERIMENTAL RESULTS AND DISCUSSION

The results of the membrane diffusion experiments are plotted in Figures 4, 5, and 6 as copper flux (in $\mu g/sec\text{-}cm^2$) versus feed solution stirring rate, acid concentration, and complexing agent concentration (respectively). The average flux of copper ions across the supported liquid membrane was found to be 0.20 $\pm$ 0.03 g/m^2–min for a carrier concentration of 40 vol%, etching solution stirring rate of 375 rpm, and 2.12 M sulfuric acid plating solution. The flux was found to be stable for up to 14 days.

The data in Figure 4 are for several feed solution stirring rates between 0 and 550 rpm. They show no dependence of the copper flux on the stirring rate, suggesting that a diffusional boundary layer in the aqueous solution is not transport rate determining.

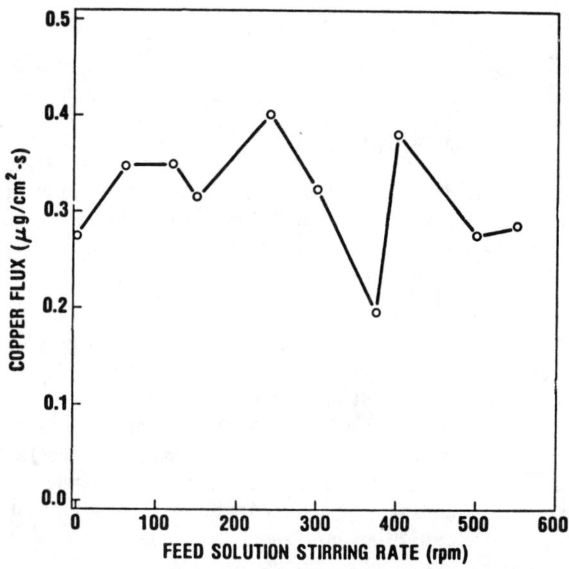

Figure 4. Copper flux data for different stirring speeds in the etch solution. The initial copper concentration in the etch solution was 157 g/l and the sulfuric acid concentration in the plating solution was 2.12 M. The acid solution was stirred at a moderate rate with a magnetic stirrer.

Figure 5 shows the results of experiments in which the concentration of H^+ was varied on the acid side of the membrane. For H^+ concentrations of 0.84 M or greater, the flux of copper remains unchanged. Only when the concentration of H^+ is lowered to 0.08 M does the flux drop off, indicating that, at this low concentration, there are insufficient numbers of protons at the membrane interface to effectively strip the copper from the membrane complex. This result is not surprising since the equilibrium for the stripping reaction has been shown to be shifted strongly to the right.

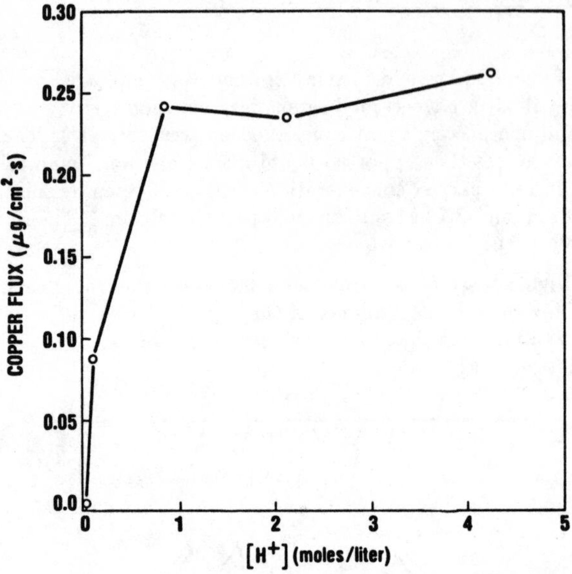

Figure 5. Copper flux data for various acid concentrations in the plating solution. The etch solution was stirred at 375 rpm and the plating solution was stirred at a moderate rate with a magnetic stirrer. The initial copper concentration in the etch solution was 157 g/l.

The effect of carrier concentration on copper flux is shown graphically in Figure 6. There appear to be two separate regimes. At low carrier concentrations, the flux is strongly dependent on carrier concentration, suggesting that mass transport is limited either by diffusion of the copper-carrier complex through the membrane or by the rate of interfacial extraction reaction. At high concentrations of carrier, the flux appears to be less dependent on the carrier concentration, indicating that copper flux is becoming limited by external mass transfer resistances. Since increasing the stirring rate on the etching side of the membrane has no effect on the copper flux, it is postulated that under these conditions, the mass transfer resistance is due primarily to the aqueous diffusion boundary layer on the stripping side.

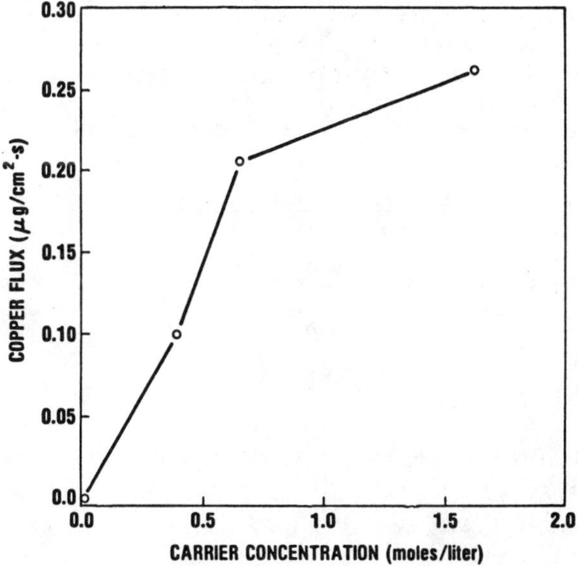

Figure 6. Copper flux data for four different diketone concentrations in the liquid membrane. The etch solution was stirred at 375 rpm and the plating solution was stirred at a moderate rate with a magnetic stirrer.

Previous work on supported liquid membranes for copper transport (11-13) have used various β-hydroxy oximes as extracting agents. The mechanism of the transport in these membrane experiments has been examined. For instance, Lee et al. (11) found in their experiments that the transport of copper was limited by diffusion of the copper complex through the liquid membrane. On the other hand, Komasawa et al. (12) found that the interfacial reaction was the controlling step in their experiments. Using the data of Baker et al. (13), Danesi and coworkers (14) used a model of the transport process to demonstrate that control of the copper flux could be due to aqueous film diffusion, interfacial reaction, or membrane diffusion, depending upon the experimental conditions. Further work will be required to determine the exact nature of the rate controlling step under the conditions examined in the present study.

SUMMARY

A supported liquid membrane for the transport of copper ions from an ammoniacal etch solution to an acid plating solution was prepared and tested. Fluxes of 0.20 g/m^2—min were measured and found to be steady for the entire duration of the experiments (7 - 14 days). With sufficient agitation of the aqueous solutions on either side of the membrane, the flux of copper ions appears to be limited by either the diffusion of the copper complex through the membrane or the interfacial reaction rate. Future experiments are planned to study the transport mechanism in greater detail.

In terms of practical applications, experiments are planned to examine the long term (>1 month) stability of the supported liquid membrane and alternate membrane geometries such as hollow fibers. The selectivity of this membrane for copper ions over other ions present in solution will also be measured.

ACKNOWLEDGEMENTS

The authors are indebted to H. H. Law and V. Tierney for many helpful suggestions.

REFERENCES

1. R. Bloch, in Membrane Science and Technology, J. E. Flin, ed., Plenum, New York, 1970, pp. 171-187.
2. W. C. Babcock, R. W. Baker, E. D. LaChapelle and K. L. Smith, J. Membrane Sci., 7, 89, (1980).
3. E. L. Cussler and D. F. Evans, J. Membrane Sci., 6, 113, (1980).
4. N. N. Li, J. Membrane Sci., 3, 265, (1978).
5. P. R. Danesi, E. P. Horwitz and P. Rickert, Separation Sci. Tech., 17, 1183, (1982).
6. P. R. Danesi, R. Chiarizia and A. Castagnola, J. Membrane Sci., 14, 161, (1983).
7. J. D. Lamb, J. J. Christensen, S. R. Izatt, K. Bedke, M. S. Astin and R. M. Izatt, J. Amer. Chem. Soc., 102, 3399, (1980).
8. J. D. Lamb, J. J. Christensen, J. L. Osearson, B. L. Nielsen, B. W. Asay, and R. M. Izatt, J. Amer. Chem. Soc., 102, 6820, (1980).
9. P. R. Danesi, Separation Sci. Tech., 19, 857, (1985).
10. J. D. Way, R. D. Noble, T. M. Flynn and E. D. Sloan, J. Membrane Sci., 12, 239, (1982).
11. K. H. Lee, D. F. Evans and E. L. Cussler, AIChE J., 24, 860, (1978).
12. I. Komasawa, T. Otake and T. Yamashita, Ind. Eng. Chem. Fundam., 22, 127, (1983).
13. R. W. Baker, M. E. Tuttle, D. J. Kelley and H. K. Lonsdale, J. Membrane Sci., 2, 213, (1977).
14. P. R. Danesi, E. P. Horwitz and G. F. Vandegrift, Separation Sci. Tech., 16, 201, (1981).

SEPARATION OF NOBLE METALS BY DONNAN DIALYSIS

K.Brajter and K.Slonawska
Department of Chemistry
University of Warsaw
02-093 Warsaw, Poland

Donnan dialysis has been demonstrated to be a useful technique not only for enrichment of metal ions but also for their separation. Our studies indicate that method may be applied to separation of gold from some platinum metals. Different receiver electrolytes as well as ligands were investigated to differentiate the transport rate across the ion-exchange membranes.

INTRODUCTION

The possibilities of the application of the Donnan dialysis of metal ions for their preconcentration, speciation and separation is the reason for the great attention paid lately to this method (1-7). In our previous paper we demonstrated that the separation of platinum metal ions on the basis of Donnan dialysis across anion-exchange membranes is possible (7). The most interesting result obtained by Donnan dialysis of platinum metals is their distribution among three phases: sample, anion-exchange membrane and receiver electrolyte. The separation of platinum metal ions by that technique is influenced greatly by the receiver electrolyte composition (7). We confirmed that there is some similarity between the affinities of chloride complexes of platinum metal ions towards anion-exchange membrane and towards anion-exchange resin (7).

The aim of this paper was to test the possibilities of the separation and preconcentration of gold from some platinum group metal ions by Donnan dialysis. We have investigated the influence of the receiver electrolyte and sample composition on Donnan dialysis of gold. We used the complexing agents added to the sample - sodium sulphite and ethylenediamine - and demonstrated that both ligands strongly influence the result of Donnan dialysis.

EXPERIMENTAL

The work was performed with P-1010 cation-exchange membranes and P-1035 anion-exchange membranes (RAI Research Corp., New York). They were pretreated in 1 M HCl prior to initial use. Before each dialysis they were soaked in a solution of the same composition as the receiver electrolyte for 1 h.

The dialysis cell was constructed from a Teflon tube. The membrane (2.5 cn diameter) was used to close one end and was held by a threaded Teflon sleeve. The dialysis cell, containing 5 mL of receiver solution, was placed in 50 mL of sample (magnetically stirred) to initiate the experiment. The dialysis time was 1 h. After dialysis, the receiver chamber contents were transfered to a 10-mL volumetric flask, and the residual sample to a 100-mL volumetric flask. Both were diluted to volume with water. The concentration of the metals was determined by atomic absorption spectrometry (Beckman Model 1272 with a graphite furnace atomizer Perkin-Elmer HGA-74).

All chemicals used were Reagent Grade. Doubly distilled water was used for preparing all solutions. The solutions of Rh, Pd, Pt and Au were prepared from the chlorides of these metals (Koch Light Laboratories). The metals were all in the form of anionic chloro complexes. The pH was adjusted with dilute HCl or NaOH prior to dialysis.

The metal ions were recovered from the cation exchange membranes after dialysis by soaking them in 1 M HNO_3 for 1 h. From the anion-exchange membranes the metal ions were recovered by soaking the membranes as following: for Pd(II) in 1 M HCl for 1 h, for Pt(IV) in 0.5 M NaCl for 3 h and for Au(III) in 0.5 M HNO_3 and 90% v/v acetone for 30 min and then in 1 M HNO_3 for 1 h.

The data are reported in terms of the enrichment factor EF (defined as the ratio of metal concentration in the receiver electrolyte after dialysis to the initial sample concentration) and percent of the metal retained in the sample or on the membrane phase after dialysis. The last one was determined by difference between the initial concentration and the final solution-phase concentration.

RESULTS AND DISCUSSION

The preliminary experiments were performed to establish the influence of the receiver electrolyte composition on the Donnan dialysis of chloroauric complexes. The following electrolytes were used: chlorides (NH_4Cl, NaCl), nitrate (NH_4NO_3) and sulphates (Na_2SO_4, $(NH_4)_2SO_4$) (Table 1). No marked differences were observed when the receiver solution was changed. The anionic species of the receiver electrolyte influenced to some degree the retention of chloroauric complexes on the membrane phase.

Table 1. Effect of the receiver electrolyte on the Donnan dialysis of Au(III) [a].

Receiver [b]	EF	% of Au retained in sample	% of Au retained on membrane
NH_4Cl	0.01	62	37
NaCl	0	51	49
NH_4NO_3	0.01	53	46
Na_2SO_4	0	37	63
$(NH_4)_2SO_4$	0	49	51

[a] Anion-exchange membrane, dialysis time, 1 h, the sample concentration of $AuCl_4^-$ 5.9×10^{-5} M,
[b] All receivers 0.5 M.

There is some agreement between the affinity of anionic species of the receiver electrolyte towards fixed sites of membrane and retention of chloroauric complexes on the membrane phase. Anions characterized by low affinity (sulphates, chlorides) enable high retention. Those characterized by high affinity (nitrates) cause a decrease of retention. This result supports the hypothesis of the influence of the receiver electrolyte composition on the chemistry at the sample-membrane inner face (1). The nitrate favors the transport of chloroauric complexes into the receiver solution probably due to their highest affinity towards anion-exchange membrane and chlorides as the species favoring the chloroauric complexes formation. The similarity of the affinities of chloride and sulphate of sample composition towards strong basic anion exchanger was established. The very important result is the distribution of chloroauric complexes between two rather than three phases as the result of Donnan dialysis (7).

Dialysis of chloroauric complexes with ethylenediamine as a sample matrix

A set of similar experiments, in which anion or cation-exchange membranes were used, was performed. In all experiments gold was taken as the chloroauric complex. In the samples the concentration of ethylenediamine was changed to 0.029 M. The composition of the receiver solution was constant (0.5 M NH_4NO_3). In Donnan dialysis experiments performed with anion-exchange membranes (by increasing ethylenediamine concentration) a drastic decrease of the amount of gold on the membrane was obtained (Fig.1a). Dialysis of chloroauric complexes - without ethylenediamine as sample matrix - resulted in distribution of the gold into 2 phases only: membrane and sample. In the presence of amine it was also the case. It was observed that at some ethylenediamine concentration level transport of gold into the membrane was not permitted. The amount of Au on the membrane and the receiver solution was zero.

When cation-exchange membranes were used (under the same experimental conditions), with increasing ethylenediamine concentration, an increase of the enrichment factor as well as an increase of the retention of gold (on membrane) was obtained (Fig.1b) The results of the above experiments can be explained assuming that ethylenediamine acts as a ligand transforming the chloroauric complexes of gold into other species, probably $(AuenCl_2)^+$, four coordinated positively charged mixed complex. The increase of ethylenediamine concentration favors the formation of the new complex which is excluded from the anion exchange membrane due to its positive charge. There is a very low probability that the amine (weak electrolyte) may influence the rate of transport of the chloroauric complexes (assuming that formation of $(AuenCl_2)^+$ does not occur). Naturally the composition of auricethylenediaminechloro complex is only assumed.

Dialysis of chloroauric complexes with sulphite as a sample matrix

The experiments were performed using anion exchange membranes and gold as the chloroauric complex. Increasing the concentration of sodium sulphite (sample matrix) gave practically quantitative retention of gold in the sample after Donnan dialysis (Table 2).

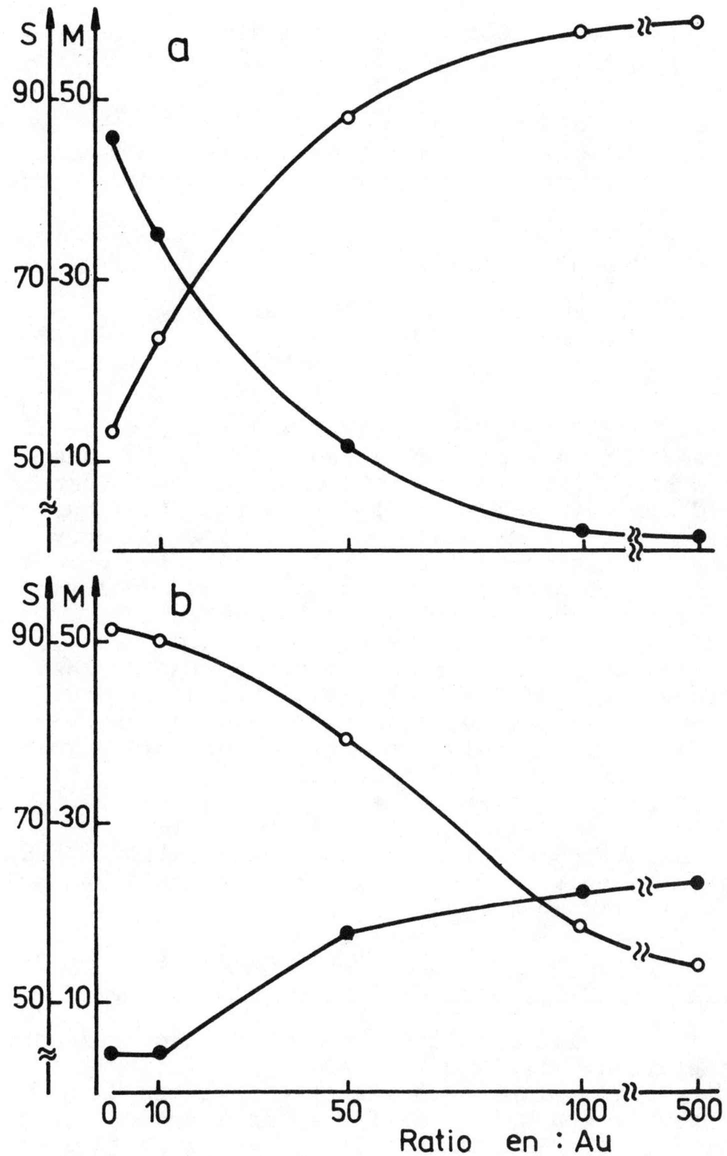

Fig. 1. Effect of ethylenediamine as a sample matrix on the Donnan dialysis of Au complexes, a) anion-exchange membrane, b) cation-exchange o - percent of Au retained in the sample (scale S), • - percent of Au retained on the membrane (scale M), receiver 0.5 M NH_4NO_3, dialysis time, 1.5 h, concentration of Au(III) in the sample 5.9×10^{-5} M.

Table 2. Effect of the presence of Na_2SO_3 in the sample on the Donnan dialysis of Au complexes.

Concentration of Na_2SO_3	% of Au retained in		
	sample	receiver	membrane
0	53	0.8	46
0.06 mM	54	1.4	45
0.3 mM	55	35	10
0.6 mM	74	30	0
6.0 mM	87	22	0
30.0 mM	83	10	7
60.0 mM	96	7	0

It is difficult to postulate the mechanism of the reaction between the chloroauric complex and sodium sulphite, but it is evident that the complexes formed are not dialyzed. It is possible that auric sulphates are formed (although they are not very stable). It is more likely that polymerized aurous sulphite complexes are developed (Au(I)) complexes are usually monomeric though not invariably). In the literature there is no information about these complexes. At a sulphite concentration less than about 0.6 mM the difference of ionic strength between the sample and the receiver solution was great enough to permit the transfer of gold across the anion-exchange membrane. At a higher concentration of Na_2SO_3 the enrichment factor decreases (Figure 2).

In all the experiments no precipitation of gold (Kassius purple), after addition of sulphite as well as amine, was observed.

Separation of mixtures by Donnan dialysis in the presence of ethylenediamine and sodium sulphite

The results obtained when mixtures of gold and some platinum metals were dialyzed in the presence of ethylenediamine are summarized in Table 3. The results are not - as far as gold is concerned - similar to those obtained with single gold samples. But as far as the separation of mixtures of Au - platinum group metals is concerned, the results are very promising, although the separation factors are not the best. Since almost selective preconcentration of single metals on the membrane was achieved, it gives support for development of a new nonconventional method of separation based on the distribu-

Table 3. Separations of mixtures by Donnan dialysis [a)] in the presence of ethylenediamine [b)].

	Sample component	at pH	EF	% retained on the membrane
1.	Au	4.9	0.6	51
	Rh		0.3	1.5
	Au	6.4	0.4	42
	Rh		1.4	0
	Au	7.3	1.3	38
	Rh		0.5	0
2.	Au	7.3	0.4	36
	Rh		0.5	5
3.	Au	6.4	0.7	16
	Pt		0.8	0
	Au	7.3	0.7	33
	Pt		0.8	5
	Au	8.3	0.6	25
	Pt		1.0	0.4
	Au	9.1	0.6	23
	Pt		0.8	0
4.	Au	9.1	0.5	60
	Pt		0	4

a) Cation-exchange membrane, receiver 0.5 M NH_4NO_3, dialysis time, 1 h.

b) In all cases 100-fold excess of ethylenediamine to gold was used.
Concentrations in the samples: 1 - Au(III) 5,9 x 10^{-6} and Rh(III) 9.5 x 10^{-5}, 2 - Au(III) 5.9 x 10^{5} and Rh(III) 9.5 x 10^{-5} M, 3 - Au(III) 5.9 x 10^{-5} M and Pt(IV) 5.9 x 10^{-5} M, 4 - Au(III) 5.9 x 10^{-6} M and Pt(IV) 5.9 x 10^{-5} M.

tion among three phases: membrane, receiver solution and sample.

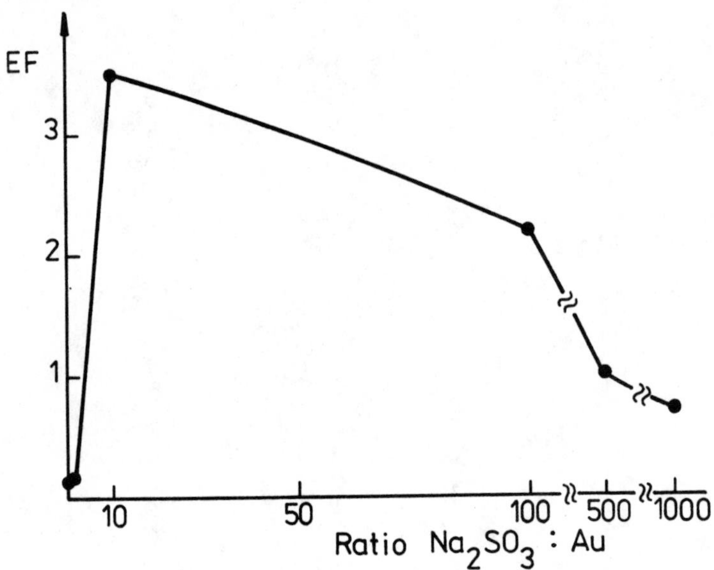

Figure 2. Effect of sodium sulphite as a sample matrix on the Donnan dialysis of Au complexes. Anion-exchange membrane, receiver 0.5 M NH_4NO_3, dialysis time, 1 h, concentration of Au(III) in the sample 5.9 x 10^{-5} M.

When the mixture of Au-Rh (at pH 7.3) was dialyzed, selective isolation of gold on the membrane phase was obtained. It should be pointed out that when there was a 10-fold excess of rhodium, preconcentration of gold was excellent. Rhodium remained almost quantitatively in the sample.

After dialysis of Au-Pt mixture (at pH 9.1) in the presence of ethylenediamine, platinum is almost quantitatively left in the sample when gold is preconcentrated on the membrane phase.

Sulphite does not influences the dialysis of platinum group metals (Table 4). But the presence of some platinum metal ions seems to influence the dialysis of gold in the presence of sodium sulphite.

Table 4. Effect of the presence of Na_2SO_3 in the sample on the Donnan dialysis of Rh, Pt and Pd.

Sample component	at pH	EF	% retained on the membrane
Rh	4.1	0.7	0.1
Pt	5.1	0.4	24
Pd	4.9	0.4	4

At comparable concentrations of sodium sulphite, gold was retained on the membrane phase to a great extent after the dialysis of Au-Rh mixture. The data in Table 5 suggest that certain separations can be performed on the basis of distribution of metal ions into the three phases. Gold can be separated from rhodium; Rh(III) at pH 8.2 is neither dialyzed nor retained on the membrane, when gold is retained on the membrane to a great extent. If the weight of both metals is considered, an excellent preconcentration factor for gold was obtained.

In the case of the Au-Pd mixture (at pH 8.3) in the receiver solution, only gold is present and palladium is preconcentrated on the membrane phase. When the mixture of Au-Pt was dialyzed, preconcentration of platinum on the membrane was observed, and for gold a high EF was obtained.

Metals can be recovered from the membrane phase after the dialysis by soaking the membranes in suitable solutions (see Experimental).

All data obtained demonstrate the possibilities of the separation of mixtures of gold and some platinum metal ions by Donnan dialysis in the presence of non-typical complexing agents due to the partitioning of metal ions among three phases.

Table 5. Separation of mixtures by Donnan dialysis [a] in the presence of sodium sulphite [b].

	Sample component	at pH	EF	% retained on the membrane
1.	Au	5.0	2.1	26
	Rh		0.7	5
	Au	6.1	0.2	23
	Rh		0.5	4
	Au	7.2	1.6	24
	Rh		0.7	0.2
	Au	8.2	1.4	24
	Rh		0.01	1
2.	Au	5.2	1.9	16
	Pd		0.8	10
	Au	8.3	1.0	2.5
	Pd		0	16
3.	Au	5.2	2.7	24
	Pt		0.3	18
	Au	8.3	6.4	4
	Pt		3.0	43

a) Anion-exchange membrane, receiver 0.5 M NH_4NO_3, dialysis time, 1 h.

b) Concentration of sodium sulphite - 3 x 10^{-4} M. The sample concentrations were:
1 - Au(III) 5.9 x 10^{-6} M and Rh(III) 9.5 x 10^{-5} M,
2 - Au(III) 5.9 x 10^{-6} M and Pd(II) 1.1 x 10^{-4} M,
3 - Au(III) 5.9 x 10^{-6} M and Pt(IV) 5.9 x 10^{-5} M.

REFERENCES

1. J. A. Cox and J. E. DiNunzio, Anal. Chem., 49, 1272 (1977).

2. J. A. Cox and Z. Twardowski, Anal. Chem., 52, 1503 (1980)

3. J. A. Cox, E. Olbrzch, and K. Brajter, Anal. Chem., 53, 1308 (1981)

4. R. L. Wilson and J. E. DiNunzio, Anal. Chem. 53, 692 (1981)

5. J. E. DiNunzio, R. L. Wilson, and F. P. Gatchell, Talanta, 30, 57 (1983)

6. J. A. Cox, K. Slonawska, F. P. Gatchell, and A. G. Hiebert, Anal. Chem., 56, 650 (1984)

7. K. Brajter, K. Slonawska, and J. A. Cox, Anal. Chem., 57, 2403 (1985)

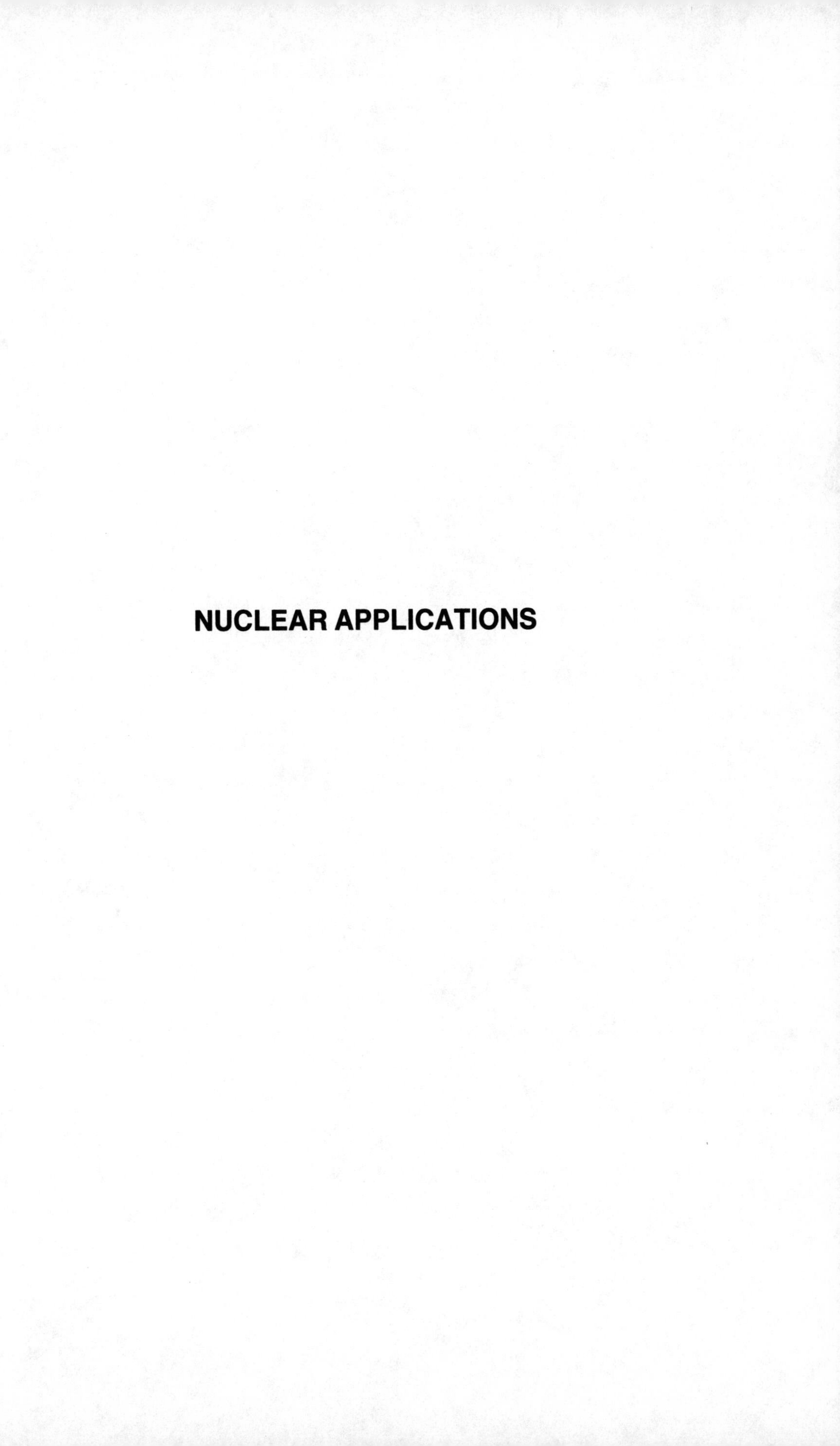

NUCLEAR APPLICATIONS

FUTURE DIRECTIONS FOR SEPARATION SCIENCE IN NUCLEAR AND RADIOCHEMISTRY

David J. Pruett
Chemical Technology Division
Oak Ridge National Laboratory*
Oak Ridge, Tennessee 37831

Solvent extraction and ion exchange have been the most widely used separation techniques in nuclear and radiochemistry since their development in the 1940s. Many successful separations processes based on these techniques have been used for decades in research laboratories, analytical laboratories, and industrial plants. Thus, it easy to conclude that most of the fundamental and applied research that is needed in these areas has been done, and that further work in these "mature" fields is unlikely to be fruitful. A more careful review, however, reveals that significant problems remain to be solved, and that there is a demand for the development of new reagents, methods, and systems to solve the increasingly complex separations problems in the nuclear field. Specifically, new separation techniques based on developments in membrane technology and biotechnology that have occurred over the last 20 years should find extensive applications in radiochemical separations. Considerable research is also needed in such areas as interfacial chemistry, the design and control of highly selective separations agents, critically evaluated data bases and mathematical models, and the fundamental chemistry of dilute solutions. Nonaqueous separation methods, such as pyrochemical and fluoride volatility processes, have traditionally played a more limited role in nuclear and radiochemistry, but recent developments in the chemistry and engineering of these processes promises to open new areas of research and application in the future.

---

*Operated by Martin Marietta Energy Systems, Inc., for the U.S. Department of Energy under Contract No. DE-AC05-84OR21400.

INTRODUCTION

Historically, the ties between separation science and technology and nuclear and radiochemistry have been very strong. It can be fairly stated that the needs of nuclear and radiochemistry have been one of the major driving forces behind the development of many modern separations techniques and processes. Conversely, the numerous and important applications of nuclear and radiation chemistry would not have developed without the rapid progress taking place in separation science and technology in the twentieth century.

Table 1 illustrates this close relationship. The isolation of polonium and radium from pitchblende, which marked the beginning of radiochemistry, was carried out using the classic separation techniques of precipitation, electrolysis, and distillation. Later progress in nuclear and radiochemistry, such as the discovery of isotopes and noninteger isotopic weights, required the development of new separation methods such as mass spectrometry, gaseous diffusion, and gas centrifugation. Later refinements in separation technology made it possible to produce isotopically pure, macroscopic samples of elements for a variety of uses in research, medicine, energy production, and weapons. Similarly, the need to separate and identify the newly discovered transplutonium elements spurred the development of ion exchange and related chromatographic methods. The modern hydrometallurgy industry was founded on the technology that developed in response to the war-time need for processes to purify multikilogram quantities of plutonium and enriched uranium.

Table 1. Some important separations in the history of nuclear and radiochemistry

Year	Development	Separation Method(s)
1898	Discovery of polonium and radium	Precipitation, electrolysis, distillation.
1913	Discovery of isotopes	Mass spectrometry, gaseous diffusion
1919	Discovery of noninteger isotopic weights	Mass spectrometry
1934	Separation of chlorine isotopes	Gas centrifuge
1940	Discovery of transplutonium elements	Ion exchange
1944	Macro-scale production of plutonium	Solvent extraction
1944	Macro-scale production of enriched U-235	Electromagnetic separation, gaseous diffusion
1965	Laser isotope separations	AVLIS, MLIS

Although Table 1 is not an exhaustive listing, it does illustrate the influence of nuclear and radiation chemistry on the development of separation science and technology and vice versa. New separation methods were invented and old ones improved because of the important scientific, military, and economic needs that evolved from the development of nuclear and radiation chemistry.

Will this symbiosis continue? In all probability it will. The development of laser isotope-separation methods and the need for improved separation methods for purifying radiopharmaceuticals may be cited as ongoing examples of nuclear and radiochemists working with separation scientists and engineers to solve problems. There is, however, a new driving force that will affect the directions of both separation science and nuclear and radiochemistry, and that is the increasing pressure to satisfy the ever more stringent regulations for environmental discharge, product purity, and related health, safety, and environmental considerations. Many of these regulations are designed to control the radioactive and hazardous wastes generated in the nuclear industry. Statutes are already in force that require purities at or below the analytical level of detection (1). Developing the separations and purification methods needed to achieve these limits is a great challenge for separations scientists and engineers. The remainder of this paper will examine some of the separation techniques that are especially important in nuclear and radiochemistry and will explore the research areas likely to be most fruitful.

Solvent Extraction

The premier forum for reporting progress in solvent extraction science and technology is the International Solvent Extraction Conference (ISEC), which is held every three years. The proceedings of the most recent conference in this series (2), accurately reflect the research areas and applications that are of greatest concern to workers in this field. The proceedings of a recent symposium sponsored by the Industrial and Engineering Chemistry (I&EC) Division of the American Chemical Society (ACS) are also a significant source of up-to-date information in this area (3). Bravo et al., have also presented an excellent review of extraction technology (4).

The most important fundamental research areas now under investigation by extraction chemists and chemical engineers include studies of interfacial phenomena (especially the kinetics and mechanisms of interfacial transport), the effects of external electromagnetic fields, the development of new, more selective, solvent extraction reagents, investigation of pharmaceutical and biological extraction processes, mathematical modeling of extraction processes, and improved designs for solvent extraction equipment.

There is almost universal agreement that the major challenge in solvent extraction chemistry is the development of new, highly selective, solvent extraction reagents. Ideally, one reagent (or perhaps a pair of synergistic reagents) would be developed for each component of a mixture. Even though this goal may never be realized, it is the guiding principle behind much of the current solvent extraction research. Advances in our understanding of the kinetics and thermodynamics of coordination chemistry are being applied to the design of new solvent extraction reagents by inorganic chemists, who are building more complex molecules to satisfy the specific coordination demands of a given molecule or ion, and by biochemists, who are working to determine the structure of the active sites in complex biological macromolecules (e.g., enzymes) in order to build simpler molecules that will retain the functionality of the active sites without the excess molecular structure. Advances in coordination chemistry, structural chemistry (solid and solution), biochemistry, and quantum mechanics, all of which are needed, will also contribute to realizing the goal of producing "designer molecules" for solvent extraction. In addition to greater selectivity, these new, custom-designed reagents should have higher resistance to chemical, thermal, and radiolytic degradation, as well as improved solubility and emulsification characteristics.

The chemical and physical properties of interfaces have been largely neglected in the study of solvent extraction processes, while vast amounts of information have reported on the equilibrium thermodynamics of the bulk systems. The kinetics and mechanisms of interfacial transport have been particularly difficult to study and interpret, but progress is now being made (5). The results of excellent, detailed studies on drop-interface coalescence are available (6). These and similar studies are of considerable importance to the solution of some of the long-standing problems in the application of solvent extraction, such as the origin, nature, and cleanup of interfacial emulsions (often referred to colloquially as interfacial "cruds"). As more information and improved theories of solvent extraction kinetics become available, separations based on kinetic effects rather than on equilibrium distributions are being designed. (A few separation methods based on kinetic effects are already in use, but the number is small.)

There is a great need for the development of techniques to study liquid-liquid interfaces that are as powerful as those which have evolved for the study of solid surfaces. Trace amounts of surface-active materials can drastically alter the performance of a solvent extraction system, because their influence is concentrated at the critical interface through which transport must occur. Modern analytical tools, such as nuclear magnetic resonance (NMR), neutron scattering, and laser light-scattering, show some promise for studying the structure and reactions that occur at liquid interfaces and in bulk solutions, but the thinness of the interface presents a considerable challenge to the application of most analytical methods. Nonetheless, such applications are at least theoretically possible, and the potential for progress in our understanding of interfacial chemistry and for improvements in process and equipment design that would result from this knowledge is very high.

The effects of the application of external fields (electromagnetic, gravitational, etc.) on solvent extraction systems have begun to receive considerable attention (7). These studies are certain to lead to new separations processes and are already being used to study interfacial properties.

There is a need for improved mathematical modeling in almost all separations processes, and solvent extraction is no exception. Current models are generally (though not always) based on equilibrium data (since these are often the only data available) and may be based on theoretical considerations, empirical relations, or both. These models will be greatly improved when the information gained from the study of kinetic and interfacial effects is incorporated. Although a great deal of equilibrium data has been obtained for solvent extraction systems, the data base is scattered, contains large amounts of incompletely documented or otherwise unreliable data, and is not easily retrievable. Methods for improving the accessibility and reliability of the solvent extraction data base have been the subject of recent discussion in the solvent extraction community.

Modeling complex solvent extraction systems is difficult and will almost certainly require a sophisticated knowledge of applied mathematics that is beyond that of a typical chemist or chemical engineer. Collaboration between these two disciplines is currently quite rare, but it shows promise for advancing our ability to model separation systems and to optimize their design.

Ion Exchange

Much of what has been said about solvent extraction applies equally to ion exchange. Indeed, it is useful to think of ion exchange as a system in which the solvent extraction reagent has been immobilized on a solid support. This analogy may be misleading in some cases, but the techniques share more commonalities than differences. As for solvent extraction, the applications of ion exchange will be enhanced and broadened by the development of improved, more highly selective reagents. Details of the structure and reactivity of the solid-liquid interface are poorly understood, and new techniques are needed to study them. The kinetics of ion exchange processes and the parameters that govern them are not well understood.

An important area in ion exchange technology that is less clearly analogous to solvent extraction chemistry is the development of inorganic ion exchangers. A review of this field has recently been published (8). It is interesting to note that the early synthetic ion exchange materials were primarily inorganic. These were soon replaced by synthetic organic resins, however, because of the difficulty in manufacturing reproducible inorganic materials and because of the instability of these materials in acid solution. Organic resins were found to have limitations as well, especially in applications that required exposure to high temperatures or radiation fields. Thus, research on potential inorganic ion exchangers continued at a low level until a breakthrough came in the mid-1960s, when it was shown that the group IV phosphates and arsenates could be prepared in crystalline form. These materials show great promise, not only as ion exchangers, but also as membranes and as catalysts or catalyst supports.

Membrane Separations

An excellent assessment of membrane technology and applications has recently been published (9). This assessment, along with the proceedings of a symposium sponsored by the Subdivision of Separations Science and Technology of the I&EC Division of the ACS (10), summarizes the current status in membrane technology. A thorough review of the history of membrane technology has been published by Lonsdale (11), while Bravo et al., have also recently reviewed the subject (12). The following comments rely heavily on these sources.

Membrane separation technology is relatively new, but its use has expanded rapidly since the introduction of the first commercial electrodialysis and microfiltration units in the 1960s. The first high-efficiency reverse osmosis membranes were also introduced in the 1960s, but reverse osmosis and ultrafiltration did not become commercially successful until the 1970s. Today, a number of membrane processes are routinely used on both laboratory- and industrial-scale projects. These processes include electrodialysis, reverse osmosis, microfiltration, ultrafiltration, dialysis, electro-osmosis, membrane extraction, pervaporation, and gas separation. Hwang and Kammermeyer (13) have prepared an excellent textbook describing these and other commonly used membrane separation unit operations. Such technologies, however, have been applied almost exclusively to the processing of aqueous streams. The application of membrane separation processes to nonaqueous solutions, gaseous systems, and gas-liquid systems is promising, but relatively unexplored.

It is important to note the close relationship between membrane separations, solvent extraction, and ion exchange. The most striking example is the case of liquid membranes (LMs) with facilitated transport, where a solvent extraction reagent (often called a "liquid ion exchanger" or a "carrier molecule" in this context, depending on the background of the investigator) is used to effect ion exchange or some other form of mass transport into or through the membrane phase. These LMs may take the form of a discrete membrane layer, an emulsion, or may be supported on a solid. They are among the most interesting classes of membranes currently under investigation, and much of the research is being done by scientists who were previously labeled "solvent extraction chemists" and who are reporting their work at solvent extraction conferences (14).

At first glance, this blurring of the distinctions between solvent extraction, ion exchange, membrane chemistry, and (as will be discussed below) chromatography may seem confusing. Once the interrelationships are recognized, however, it becomes obvious that there are many common needs that "cut across" these technologies. Thus, advancements that occur in one area will often result in improvements in others. As with solvent extraction and ion exchange, membrane scientists recognize the need for a better understanding of interfacial and transport phenomena, new carrier reagents (perhaps called "solvent extraction reagents," "ion exchangers," or "stationary phases" by practitioners of other techniques), better mathematical models, and an expanded (or more readily available) data base. The development of inorganic membrane materials is an area with great potential. The most developed of these is the dynamic membrane, although inorganic membranes of material such as palladium, quartz, and silver have been developed for specific applications.

Biomembranes (membranes made from biological material or synthetic membranes designed to mimic the structure and function of biological membranes) show enormous potential for a variety of applications, most notably in the photochemical generation of electricity. Two excellent reviews that detail these applications have recently been published (15, 16).

The greatest limitations to the application of membrane separation technology are the narrow range of useful operating conditions, the lack of understanding of physicochemical phenomena in membrane systems, and the high costs. Most membrane modules are limited to a relatively narrow range of temperature and pH (~45 to 60°C; pH 4 to 10) and are quickly attacked by organic solvents and corrosive gases. The development of new materials, most likely inorganic materials, is necessary if membrane separations are to achieve their full potential in areas such as food processing (where high temperatures are required) nuclear and radiochemistry (where high radiation fields are present) and pharmaceutical chemistry (where organic solvents are common).

Membrane science and technology are currently based on empirical knowledge, and new data are required at each stage in the development of a new membrane separation process. In the short term, this deficiency could be overcome by the systematic development of a broad data base of membrane properties that would allow the construction of empirical models valid over a useful range of operating conditions. In the long term, a much better understanding of fundamental membrane-solute and membrane-solvent interactions, fouling, polarization, boundary-layer theory, and transport phenomena are needed. The manufacture of membrane materials is also an extremely empirical process, since the chemistry of membrane formation is complex and poorly understood. If a sufficiently broad data base and theoretical framework can be developed, then the manufacture of new membrane materials and their application to new processes will be greatly expedited and the costs correspondingly lowered.

The high costs of membrane separation processes are mainly a reflection of the material limitations and the high degree of empiricism in the manufacture and application of membranes. It has been suggested that the short operating life of many membranes is the most important factor contributing to membrane system costs (9). Another factor is the low flux that can be achieved in membrane modules. Both of these problems could potentially be solved by the development of improved membrane materials.

Chromatography

For the purposes of this paper, the general term "chromatography" is used to include the techniques of gas, liquid, paper, thin layer, and affinity chromatography. A thorough review of progress in the development of each of these techniques is published in even-numbered years in the journal **Analytical Chemistry** (17), and good discussions of the general theory of chromatography may be found in several texts (18). That many significant reviews of chromatographic techniques appear in a journal devoted to analytical chemistry highlights a great deficiency in the field, namely, the difficulty in scaling-up most chromatographic techniques to an industrial size. The advantages of speed, resolution, and ability to handle very small samples have made chromatographic techniques among the most useful in analytical separations. The application of chromatographic techniques to large-scale problems, however, has been frustrated by many of the same deficiencies that exist for the separation techniques already discussed. These include the need for technological advances that would allow the practical, cost-effective scale-up of analytical chromatographic systems and the need for theoretical advances, especially in the areas of transport and interfacial phenomena, that would allow accurate modeling of the behavior of these large-scale systems. Many hundreds of references to the theory of chromatography have been reported in recent years, and the development of new packing materials, supports, column designs, column materials, mobile phases, stationary phases, detectors, and instruments continues at an impressive rate. Still, such developments are more evolutionary than revolutionary, and one of the most exciting prospects in chromatography is for a breakthrough that would allow all of these advances to be applied on an industrial scale.

One form of chromatography that already operates on a scale consistent with the industry to which it is most often applied is affinity chromatography. This is an interesting case in which the separation technique (affinity chromatography) is both an application of and is applied to the same science (biotechnology). Two recent papers review the development and status of this relatively new method (19, 20). Affinity chromatography uses highly selective biological molecules immobilized on a solid support to effect the separation of a specific molecule or class of molecules from a mixture. This separation technique depends on the extreme specificity of enzymes, antibodies, and other biological molecules to approach the ideal of "one solute-one separation agent" discussed previously. It should be noted that immobilized inorganic ligands are used to effect the separation of biomolecules as well. The potential of this technique in the biological sciences is enormous, and it represents an active area of research; the potential in inorganic systems is also large (the affinity of certain inorganic ligands for biomolecules works both ways), but is less developed.

In summary, progress in the general field of chromatography depends on advances in many of the same areas that have been discussed for extraction, ion exchange, and membrane technology. The theory of mass transport and the physical chemistry of interfaces in these systems must be developed. New materials for stationary phases, mobile phases, and supports are needed. The highly specific stationary-phase materials being developed by chromatographers should prove useful in other areas, just as new solvent extraction reagents, ion exchange materials, and membrane materials should find use as stationary and mobile phases in chromatography.

Photon-Enhanced Separations

The development of the laser as a source of pure, monochromatic electromagnetic energy has brought about rapid changes in many fields of science and technology. Generally, exposing a mixture to laser light will not cause a separation, but the laser's photons may be thought of as a tunable, highly selective "reagent" that can "react" with one component of a system and change (ionize, oxidize, or reduce) it to a species which can be easily removed from the mixture by some traditional separation technique.

Perhaps the best known examples of the application of laser technology to separations are the laser isotope separations (LIS) methods (21-23). Table 2 summarizes their development. The most developed LIS methods are Atomic Vapor Laser Isotope Separation (AVLIS) and Molecular Laser Isotope Separation (MLIS). While the isotopes of many elements have been separated by LIS techniques since the technology was developed in the early 1970s, most of the work has concentrated on the separation of the isotopes of uranium for use as fuel in nuclear reactors.

In AVLIS, an atomic vapor is produced and exposed to laser photons that have been tuned to selectively ionize a particular isotope (typically U-235, although the same technology can be applied to virtually any other isotope). The positively charged ions can then be removed from the vapor electromagnetically. In MLIS, a molecular vapor is generated (uranium hexafluoride, for example) and then excited by a laser that has been tuned to cause dissociation of the molecules containing the isotope of interest. The dissociated molecule reacts further to form a different chemical species than the feed material, and the new species, now enriched in one isotope, is removed by an appropriate separation process. In the case of uranium enrichment, this separation can be particularly easy, since any lower fluoride formed by the

Table 2. Progress in laser isotope separation*

Year	Development
1960	Invention of the laser
1965	Discovery of the principles of LIS
1966	Tunable dye laser
1972	LIS becomes a viable process
1974	Various LIS methods adopted; economic analyses
1976	Production of approximately 0.1 g 3% enriched uranium
1982	US Department of Energy selects AVLIS over MLIS
1984	Intensive development work on uranium and plutonium LIS; evaluation of industrial photochemistry
1985	AVLIS selected by the U.S. Department of Energy as replacement technology for gaseous diffusion; development of gaseous centrifuge halted
1987	Military LIS of plutonium
1992	Industrial LIS of uranium

*Adapted from ref. 21, Table I, p. 128.

dissociation/reaction of the hexafluoride will have a lower volatility than the starting material and simply fall out of the gas phase as an enriched solid.

Over a period of about 10 years the U.S. Department of Energy evaluated the potential utility of several MLIS and AVLIS processes for the enrichment of uranium. By 1985, a specific AVLIS process, currently being developed at Lawrence Livermore Laboratory (LLL) and the Oak Ridge National Laboratory, had been chosen over all competing LIS methods as well as gaseous centrifugation. The development of AVLIS as the primary enrichment technology in the United States will continue until it is ready for commercial implementation in the 1990s. Details of the key research needed to complete the development of AVLIS for this purpose are classified.

While a single LIS process has been chosen for uranium enrichment in the United States, other processes of this type are being developed and used for a variety of applications (22). In addition, a substantial amount of what has been facetiously called "laser alchemy" has been reported in the literature (23). Unexpected behavior has been observed in a number of chemical systems when they are exposed to laser radiation, especially when the molecules involved are complex. Considerable research will be required before the details of these reactions can be explained. It is clear, however, that these laser-induced reactions have a large, if still somewhat ill-defined, potential as a key step in new separations processes.

In many cases, the highly monochromatic radiation of a laser may not be required to perturb a chemical system in such a way that a separation is feasible. Toth (24-25) and his co-workers have used radiation from a mercury lamp to selectively adjust the oxidation states of various actinides in solution in order to improve the separation of these elements from each other by solvent extraction.

The use of photons, whether from lasers or from conventional light sources, as a "reagent" in a separation system has a number of advantages, including the ease with which they can be produced, their high degree of selectivity, minimal waste generation, and low cost. The potential of lasers as an integral part of new separation systems has begun to be explored only recently. It is reasonable to expect many new techniques and applications to be developed using these powerful tools.

It should be noted that the category of "electron-enhanced" separations could also be discussed. In these techniques, electrons (rather than photons) are fed into (or removed from) the system in order to control an oxidation state. These methods are also relatively unexplored.

Nuclear Fuel Reprocessing

The Purex process remains the only technology that is in widespread use for recycling spent nuclear fuel. Despite more than 40 years of experience, many problems with the process have not been solved. From an operational point of view, methods have been developed to minimize effects of the most severe of these problems, but the fundamental causes and control of solvent degradation, the formation of interfacial precipitates, and marginal decontamination factors for troublesome fission products such as zirconium, technetium, ruthenium, and neptunium are still under investigation. While research in these areas is important, it will result in incremental, rather than revolutionary, improvements in the process. There are also a number of unresolved questions regarding the best method for handling the wastes and useful byproducts that are generated in the Purex process. Further, the decision to defer indefinitely the construction of any civilian reprocessing facilities creates an opportunity to explore radically new process chemistries for possible use in the twenty-first century.

When construction of the AGNS Reprocessing Plant at Barnwell, South Carolina, was halted, and the Clinch River Breeder Reactor Program was canceled, it became certain that no civilian reprocessing plant would be constructed in the United States for several decades. This means that fuel reprocessing on a significant scale will continue only at the existing plants at Hanford, and Savannah River (and on a small scale at Idaho Falls), at least through the end of this century. It is difficult to conceive of a rational scenario in which a new reprocessing plant, civilian or military, will be built in the United States in less than 30 years. Under these conditions, it is clear that we will not be able to maintain a leadership role in Purex processing, since developments in other countries that have vigorous R&D and construction programs already in place are bound to be more rapid.

Thus, the status of nuclear fuel reprocessing technology in the United States is largely defined by efforts currently underway at existing facilities and projects that might be done in collaboration with other countries. A number of important separations problems have been identified at these facilities, and short-term research to improve separations technology in the reprocessing industry should be aimed at their resolution. Long-term research should be focused on developing an improved understanding of the base technology of the Purex process and closely related modifications, such as the use of alternative extraction reagents [e.g., tri(2-ethylhexyl) phosphate, sulfoxides, carbamoylmethylphosphonates, etc.]. Moreover, the long interim between the present and the time when a new reprocessing facility will be constructed affords an excellent opportunity for developing alternative, radically different, reprocessing schemes to the point where critical evaluations can be made concerning the most desirable technology for deployment in the twenty-first century. Examples of such alternative technologies include several nonaqueous methods, such as pyroprocessing (26-27) and fluoride volatilization using dioxygen difluoride or krypton difluoride (28-31), as well as new aqueous methods employing membrane technology or the alternative extractants already mentioned. These novel separation methods were too undeveloped to consider for use in a facility to be built before the end of the century; however, there is now time for extensive development and evaluation of these technologies before a new reprocessing plant is likely to be built in the United States. While the lack of a vigorous reprocessing program based on Purex technology is regrettable, the opportunity for a head start in the development of twenty-first century reprocessing technology is exciting.

One of the most pressing separations problems of immediate concern at existing U.S. reprocessing facilities is a need for a continuous process for recovering cesium from the acid waste generated in the Purex process. The waste is currently made alkaline before storage, and existing methods for recovering Cs-137 (which is in high demand as a gamma source) are adequate to meet market requirements. When the glass encapsulation methods for waste isolation come on-line, however, the waste will not be made alkaline, and the only processes currently available for recovering cesium from acid waste are inefficient, inconvenient, batch methods. A continuous method (preferably based on solvent extraction technology, to ease integration with existing processes) is sorely needed. Ideally, a complete process for recovering and partitioning both cesium and strontium from the waste could be developed.

Nuclear fuel reprocessing generates large volumes of organic wastes that must be disposed of or destroyed. These wastes include both spent solvents (TBP, carbon tetrachloride, hydrocarbon diluents) and aqueous solutions of organic complexing agents. For example, about 4 million gallons of aqueous waste containing significant concentrations of tetraphenyl boron are currently stored at the Hanford Reservation. These solutions, which have boron and carbon contents that are too high for disposal by conventional methods (e.g., in grout), must remain in tanks until a method of removing or destroying the organic species is found. The removal or destruction of organic compounds from aqueous solutions is a generic problem in waste management, and methods developed in the nuclear industry should find wide applicability.

There is a general need in the nuclear industry for a process to separate the lanthanide elements from the trivalent actinides in a nitrate-based (rather than a chloride-based) system. When chloride must be used, it complicates flowsheets by requiring nitrate-to-chloride and chloride-to-nitrate conversions and by accelerating equipment corrosion.

The dissolution of plutonium oxide (and other refractory metal oxides not necessarily related to the nuclear industry) in scrap, waste, or spent fuel is an old problem with no solution. Why is some material so difficult to dissolve, even when fluoride or other catalysts are added? Practical solutions, as well as fundamental explanations, are needed in this area.

A major research program on the main part of the Purex flowsheet is not warranted. The Europeans and Japanese have active programs in this area, but the research is designed to fill the gaps in the data base and to address some of the remaining trouble spots, such as technetium, ruthenium, neptunium, or palladium chemistry, and valence adjustment. These efforts are worthwhile and should be a part of a continuing base-technology program in the United States, but they do not form the basis for a large research program. The prototypic, high-burnup, mixed-oxide fuel that has recently become available from the Fast Flux Test Reactor should, however, be thoroughly studied and used to test proposed reprocessing schemes of all types, including those that are highly speculative (e.g., fluoride volatility processes based on dioxygen difluoride).

Although research in the application of LIS technology to separations problems in the nuclear industry has been directed almost entirely at the separation of uranium isotopes, other applications are possible. For example, Cs-137, a short-lived isotope that can be disposed of simply by storing until it decays away, has already been cited as an important gamma source. Unfortunately, Cs-137 is almost always contaminated with Cs-135, a long-lived isotope that requires elaborate, long-term, retrievable storage. An effective method for separating the useful, more easily handled isotope from the less valuable, more difficult to handle isotope would be most useful. The use of LIS might be an economical solution.

Waste Management

The problems presented by the discharge of hazardous materials to the environment are enormous; solving these problems will challenge separation science and technology on all fronts.

Pollutants that must be separated from the environment may be solids, liquids, or gases; they may be simple anions or cations, organic or inorganic, natural or man-made, dilute or concentrated, radioactive or stable. They come from the smokestack industries, energy production, automobile exhausts, the health-care industry, and residential waste treatment units. Some of the problems are ongoing, while others are left over from an era when little thought was given to how waste materials were discharged to the biosphere. The news media are diligent at pointing out where hazardous and radioactive wastes are, were, or might one day be stored or discharged. Standards for what constitute an environmentally acceptable level of discharge to the environment are becoming increasingly stringent, and there is every indication that trend will continue.

Such a litany of gloom may make the problem seem intractable, but that is far from the case. Great progress has been made, and the techniques and research needs discussed above point the way for even greater progress. The application of highly selective, more efficient solvent extraction, ion exchange, and membrane separation (chiefly reverse osmosis and ultrafiltration) techniques has successfully reduced the amounts of many of the pollutants that were once common in aqueous industrial effluent streams. Emissions of the oxides of sulfur from coal-fired electrical plants have been reduced substantially by pretreating the coal to remove inorganic sulfur and by scrubbing the flue gas after the coal is burned. Biotechnologists have developed remarkable microbes with the ability to decompose hazardous halogenated compounds (32), nitrates (33), and cellulosic materials (34) in the soil, solid wastes, aqueous waste streams, and waste ponds. Supercritical water has been used to destroy organic material in waste streams (35).

The limit of the success of these processes is the point at which the need for research and development of improved techniques begins. Allowable limits for the discharge of many species, such as arsenic and dioxins, are already at or below the limits of detection. New methods of analysis, analytical separations, and process-scale separations are needed. It is easy to understand the difficulty in devising a separations process that can efficiently remove very small amounts of dissolved or suspended materials from very large volumes of aqueous effluents, especially since the chemical and physical behavior of species in very dilute solutions are so poorly understood and reliable data so very scarce.

The successful application of biotechnology to waste problems has been very encouraging, but existing techniques barely scratch the surface of the possibilities for this technology. The working examples of biotechnological solutions to waste problems are impressive, but few. The microbial degradation of halogenated compounds, for example, is still experimental. In addition to the technical questions concerning how well this method will work on a large scale, there are broader philosophical, social, and regulatory questions that must be answered about the introduction of genetically altered microbes into the biosphere. Nevertheless, the potential of developments in biotechnology for solving many environmental waste problems is very great.

The relationship between separations developed to reduce waste and the problem of recovering strategic and valuable materials should also be noted. To the degree that process effluents currently being discharged to the environment contain precious or strategic materials, improved methods of separating, concentrating, and recovering these species before they are discharged will reduce the demand for additional supplies. In favorable cases, the value of the recovered materials may offset all, or most, of the cost.

The reduction in the emissions of the sulfur oxides from coal burning can be cited as an example of an area needing more work and also as an example of progress in waste separations technology. Although sulfur dioxide emissions have been cut dramatically, they are still much too high, and the efforts to reduce emissions of the nitrogen oxide and other pollutants have been much less successful. An integrated, systems approach will be needed to achieve a completely satisfactory solution to the problem of emissions that occur during the burning of fossil fuels. This observation can be generalized to include many processes that produce hazardous and radioactive wastes. Little research has been done to determine when, or if, two less-effective separation processes, perhaps installed at different points in a larger process, might be combined to produce better overall results than one highly effective process being pushed to the limits of technology. Nor have processes generally been designed with full consideration of how the waste streams generated will be treated. As the costs of waste management continue to increase, it will become increasingly important to consider the separation methods used, to ensure that process effluents meet regulatory requirements as an integral part of the overall process, rather than as an add-on to be retrofitted as needed.

The need for closer cooperation among analytical chemists, toxicologists, environmentalists, separation scientists, and regulatory agencies should also be stressed. It is certain that analytical chemists will continue to develop methods for determining species in wastes at lower and lower concentrations, and separations scientists will continue to develop ways to reduce emissions into the biosphere. Considerable work is needed, however, to determine how low the limits for the emission of a potentially hazardous chemical must really be in order to adequately protect the environment and the health and safety of the public. Regulations governing these limits should be based on sound toxicological and environmental data rather than on such arbitrary guidelines as "the limit of detection" or "as low as reasonably achievable." Limits based on criteria such as these could be ruinously expensive to achieve, as analytical and separation capabilities continue to improve.

REFERENCES

1. R. K. Genung and J. H. Stewart, Jr., "Separations Problems in Dilute Solutions Related to the Development of Environmental Control and Waste Management Technologies," presented at the Fourth Symposium on Separation Science and Technology for Energy Applications, Oct. 24, 1985, Knoxville, Tennessee.

2. Proceedings of the International Solvent Extraction Conference, Denver, Colorado (1983).

3. A. S. Kertes and J. D. Navratil, Organizers, Symposium on New Solvent Extraction Reagents, 189th National Meeting of the American Chemical Society, Miami Beach, Florida, May 3-4, 1985.

4. J. L. Bravo, J. R. Fair, J. L. Humphrey, C. L. Martin, A. F. Siebert, and S. Joshi, Assessment of Potential Energy Savings in Fluid Separation Technologies: Technology Review and Recommended Research Areas, Report DOE/ID/12473-1, Separations Research Program, Center for Energy Studies, University of Texas at Austin, Austin, Texas, December 1984.

5. Ref. 2, pp. 1-5, 14-23, 291-320.

6. B. A. Moyer and W. J. McDowell, "Drop-Interface Coalescence Rate in Tertiary Amine Solvent Extraction," Sep. Sci. Technol., 18, 1535 (1983).

7. Ref. 2, pp. 50-51, 172-173, 180-181.

8. A. Clearfield, Ed., Inorganic Ion Exchange Materials, CRC Press, Boca Raton, Florida (1982).

9. S. A. Leeper, D. H. Stevenson, P. Y.-C. Chiu, S. J. Priebe, H. F. Sanchez, and P. M. Wikoff, Membrane Technology and Applications: An Assessment, Report EGG-2282, EG&G Idaho, Inc., Idaho Falls, Idaho, February 1984.

10. W. W. Schulz, Organizer, ACS Separation Science and Technology Award Symposium, 189th National Meeting of the American Chemical Society, Miami Beach, Florida, May 3-4, 1985.

11. H. K. Lonsdale, "The Growth of Membrane Technology," J. Membrane Sci., 10, 2 and 3, 81 (1982).

12. Ref. 4, pp. 77-149.

13. S.-T. Hwang and K. Kammermeyer, "Membranes in Separations," Techniques in Chemistry, 7, Wiley, New York (1975).

14. Ref. 2, pp. 58-67, 282-290, 373-380, 391-400.

15. H. Ti Tien, "Artificial Planar Bilayer Liquid Membranes," in Membranes and Transport, 1, A. M. Martonosi, Ed., Plenum, New York, 165 (1982).

16. J. H. Fendles, "Membrane Mimetic Chemistry," Chem. and Eng. News, 62, 1, 25 (January 2, 1984).

17. G. H. Morrison, Ed., "Fundamental Reviews," Anal. Chem., 56, pp. 48R-63R, 174R-199R, 300R-349R (1984). See also the previous fundamental review issues published each even-numbered year.

18. A. B. Littlewood, Gas Chromatography. Principles, Techniques, and Applications, Academic Press, New York, 2nd Ed., (1970); R. P. W. Scott, Contemporary Liquid Chromatography, Wiley, New York (1976); A. S. Said, Theory and Mathematics of Chromatography, Huthis, Heidelberg (1981); L. R. Snyder and J. J. Kirkland, Introduction to Modern Liquid Chromatography, Wiley, New York, 2nd Ed. (1979).

19. I. Parikh, P. Cuatrecasas, "Affinity Chromatography," Chem. and Eng. News, 63, 17 (August 26, 1985).

20. R. R. Walters, "Affinity Chromatography," Anal. Chem., 57, 1099A (1985).

21. W. de Ruiter, "Laser Separation of Isotopes," Endeavour, New Series, 8(3), 128 (1984).

22. C. P. Robinson and R. J. Jensen, "Laser Methods of Uranium Isotope Separation," pp. 268-290 in Topics in Applied Physics, 35, S. Villani, Ed., Springer-Verlag, Berlin (1979).

23. C. D. Cantrell, S. M. Freund, and J. L. Lyman, "Laser-Induced Chemical Reactions and Isotope Separation," in Laser Handbook, 3, M. L. Stitch, Ed., North-Holland, New York (1979).

24. L. M. Toth, J. T. Bell, and H. A. Friedman, "Photochemistry of the Actinides," in Actinide Separations, J. D. Navratil and W. W. Schulz, Eds.; ACS Symposium Series, 117, American Chemical Society, Washington, D.C. (1980).

25. L. M. Toth, H. A. Friedman, and J. T. Bell, "Photochemical Separation of the Actinides in the Purex Process," in CONF-770506-1 (1977).

26. L. Burris, "The Proposed Fuel Cycle for the Integral Fast Reactor," Trans. Am. Nucl. Soc., 49, 90 (1985).

27. L. Burris, M. J. Steindler, and W. E. Miller, "A Proposed Pyrometallurgical Process for Rapid Recycle of Discharged Fuel Materials from the Integral Fast Reactor," pp. 2-257 in Proc. Reprocessing and Waste Management Mtg., Jackson, Wyoming, Aug. 26-29, 1984.

28. L. B. Asprey, et al., "Preparation and Utilization of Dioxygen Difluoride," submitted to Inorg. Chem. (1985).

29. L. B. Asprey, S. A. Kinkead, and P. G. Eller, "Low Temperature Conversion of Uranium Oxides to Uranium Oxide Using Dioxygen Bifluoride," submitted to Nucl. Tech. (1985).

30. J. G. Malm, P. G. Eller, and L. B. Asprey, "Low Temperature Synthesis of Plutonium Hexafluoride Using Dioxygen Difluoride," J. Am. Chem. Soc., 106, 2726 (1984).

31. L. B. Asprey, P. G. Eller, S. A. Kinkead, and J. G. Malm, submitted to Inorg. Chem. (1985).

32. D. Ghosal, et al., "Microbial Degradation of Halogenated Compounds," Science, 228, 135 (Apr. 12, 1985).

33. B. D. Patton, et al., Design of Fluidized-bed, Biological Denitrification Systems, Report ORNL/TM-7628, Oak Ridge National Laboratory, Oak Ridge, Tennessee, January 1982.

34. T. L. Donaldson, et al., Anaerobic Digestion of Low-Level Radioactive Cellulosic and Animal Wastes, Report ORNL/TM-8535, Oak Ridge National Laboratory, Oak Ridge, Tennessee, February 1983.

35. H. A. Fremont, Extraction of Coniferous Woods with Fluid Carbon Dioxide and other Supercritical Fluids, U.S. Patent 4,308,200 (December 1981).

DEVELOPMENT OF A CHROMATOGRAPHIC SEPARATION PROCESS FOR KRYPTON AND NITROGEN

F.H. Tezel, D.M. Ruthven and T.S. Sridhar*

Department of Chemical Engineering
University of New Brunswick
Fredericton, New Brunswick E3B 5A3
*Atomic Energy of Canada Ltd.
Whiteshell Nuclear Research Establishment
Pinawa, Manitoba R0E 1L0

The final step in purification of the off-gas from nuclear fuel re-processing facilities requires the removal of radioactive Kr from an atmosphere of N2. The objective of this study was to develop an efficient adsorption process for this separation.

A pilot separation unit based on the chromatographic principle was designed and constructed using a column 120 cm x 5.5 cm i.d. packed with H-mordenite. This adsorbent was shown to offer the best compromise between cost and performance in preliminary screening studies. Performance tests were run for different flow-rates and injection pulse sizes in order to determine the optimum conditions for the desired separation. The pilot unit was also tested for continuous operation to check whether the steady-state condition could be maintained. The unit was shown to be capable of handling a flow-rate of 2.5 L/min of mixture (2000 ppm Kr in N2) on a continuous basis, with almost complete separation of Kr. The demonstration unit consists of a single column with injection over a period of 15 seconds and a total time of 450 seconds per feed pulse. Scale-up by a factor of up to 30 could therefore be achieved under the same operating conditions by adding additional columns in parallel.

INTRODUCTION

During the fission of U-235 fuels, radioactive isotopes of Kr and Xe are formed. These gases are released during dissolution of the fuel rods in fuel reprocessing or in the event of a fuel rod failure in a nuclear generating station. Most of these radioactive isotopes have relatively short half-lives and their concentrations decay rapidly to an insignificant level with the exception of Kr-85 which has a half-life of 10.7 years (1). It would take about 100 years for Kr-85 to decay to a stable isotope and its removal is therefore essential. Since the major component in the off-gases from fuel reprocessing is air, considering the adsorption properties and compositions of the different components, the most difficult aspect is the separation of Kr from N2. Present work has been concentrated on this particular separation. Following completion of the adsorbent screening studies (2,3), a pilot unit has been built to demonstrate the actual separation.

A large (5.5 cm I.D. x 122 cm long) chromatographic column is used for the Kr - N2 separation, but, unlike an analytical chromatograph, the system is operated under overload conditions. Large slugs of the Kr - N2 mixture are injected at regular intervals with each pulse being injected as soon as the tail of the Kr peak from the preceeding pulse has emerged from the column. The Kr and N2 products are separated from the He carrier by pressure swing adsorption with the recovered He being recycled to the inlet of the main column.

DESCRIPTION OF THE PILOT PLANT

The pilot plant consists of two units: the main separation system and He recovery subsystem. Schematic diagrams of these units are given in Figures 1 and 2, and photographs of the unit are shown in Figures 3 and 4.

Main Separation System

Recycled carrier gas from He recovery system joins make-up He and passes through the circulation pump. Since the gas temperature is increased at the outlet of the pump, the gas is passed through a heat exchanger to cool it to room temperature. A surge tank with a volume of 5.75 liters is included in the system after the heat exchanger in order to minimize the pulsations from the pump. Flow-rates of the carrier gas helium and the injected mixture of nitrogen + krypton are controlled by mass flow controllers (1) and (2), respectively. Solenoid valve (2) is used as the injection valve and switches the flow from He to N2 + Kr mixture for a specified period of injection time. All the valves in the system, including this injection valve, are automatically controlled by an IBM PC mini-computer. The injected mixture then goes through the main chromatographic separation column (dimensions are given in Table 1).

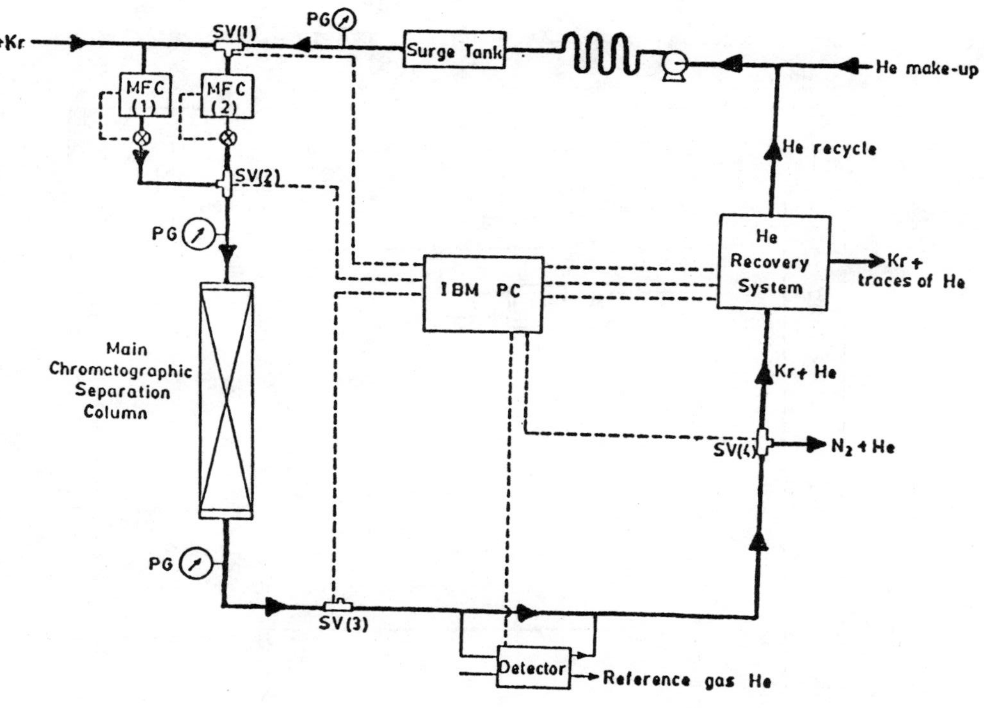

Figure 1. Schematic diagram of the main chromatographic separation system.

PG: Pressure Gauge
SV: Solenoid Valve
MFC: Mass Flow Controller

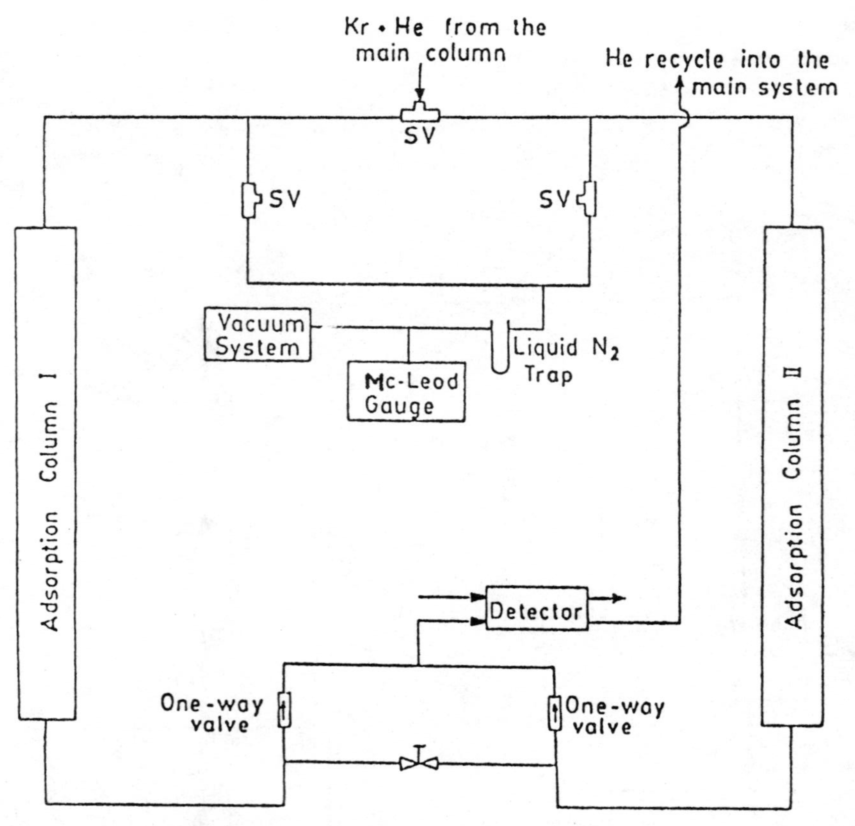

Figure 2. Schematic diagram of the He recovery subsystem.
SV: Solenoid Valve

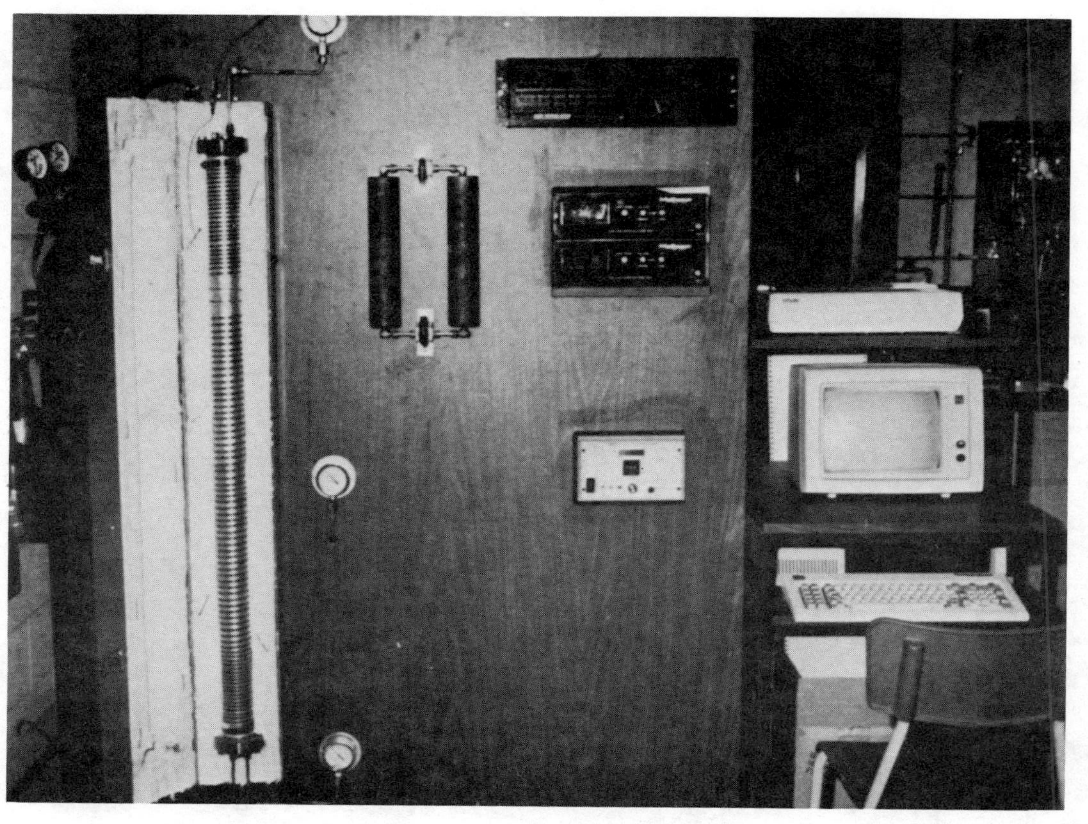

Figure 3. Photograph of the pilot unit (front view).

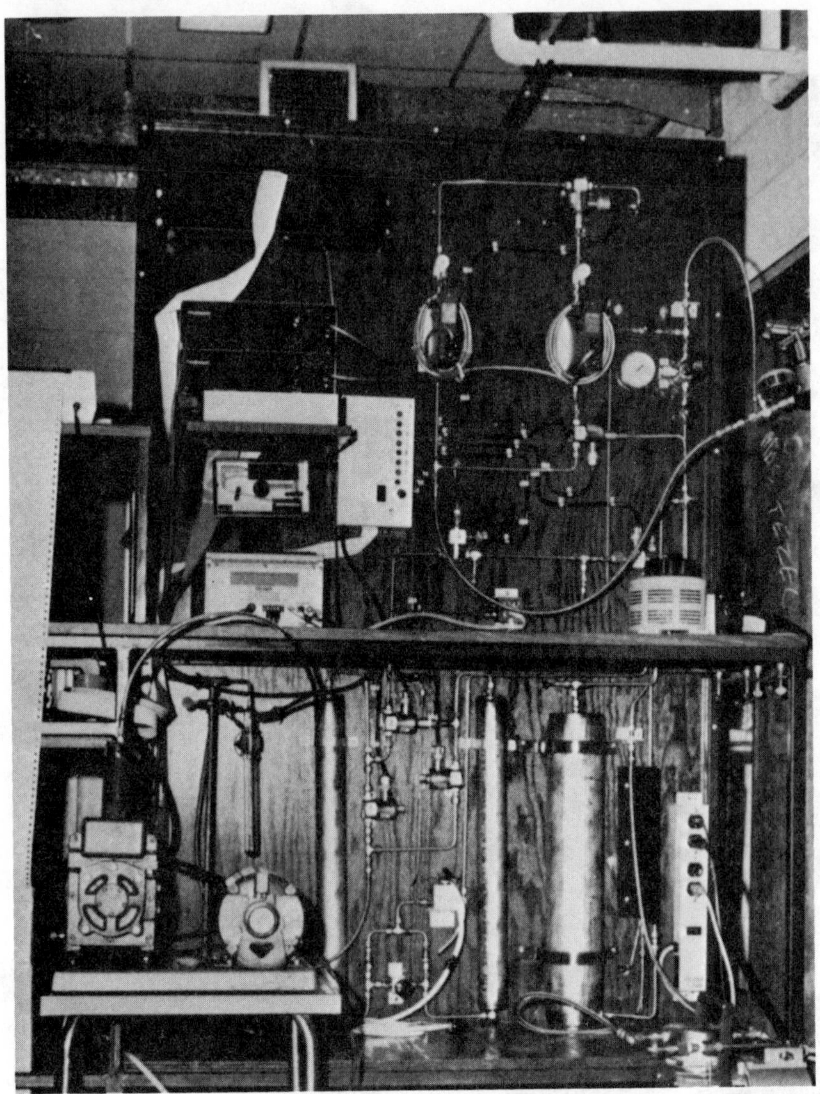

Figure 4. Photograph of the pilot unit from the back of the panel. The He recovery system is shown under the bench.

Table 1. Some dimensions and details for the pilot plant.

Main separation column dimensions:

L = 122 cm
D = 5.5 cm (I.D.)

Adsorbent used in the main column:

12 x 30 mesh H-mordenite from Norton Comp. (Zeolon-900-H)

Columns in He recovery system:

L = 60.5 cm
D = 5.5 cm (I.D.)

Adsorbent used in the recovery columns:

12 x 30 mesh Ca-exchanged Na-mordenite from Norton Comp. (Zeolon-900-Na)

Surge volume dimensions:

L = 60.5 cm
D = 11 cm (I.D.)

Concentration of original N2 + Kr mixture:

0.2% Kr
99.8% N2

The column is surrounded by heating coils and then the insulation material so that the regeneration of the adsorbent can be carried out without the removal of the column from the system (see Figure 3). Temperature is monitored during the regeneration. Pressure gauges at the inlet and at the outlet of the column are used for monitoring the absolute pressure as well as the pressure drop in the column. Four and 3/4 lbs of H-mordenite is used as adsorbent (details of adsorbent are given in Table 1). An air powered vibrator was used to ensure evenness of packing during the loading of the adsorbent into the column.

The composition of the outlet gas from the column is monitored by a thermal conductivity detector. Since the actual flow-rate through the column is greater than 15 liters/minute, a small portion of the total flow is taken as a by-pass through the detector to meet the flow-rate requirements of the detector. Pure He gas has been used in the reference-gas side. Data acquisition is done by an IBM PC mini-computer system which is connected to the detector and output is stored on a diskette. Solenoid valve (4) is used to switch the flow from the detector between vent and He recovery unit. He gas is recovered during three quarters of each pulse cycle during which Kr is eluted, but at present He is not recovered during the remaining part of the cycle while N2 is eluted.

He Recovery System

The He recovery sub-system is essentially a pressure-swing adsorption unit (see Figure 2). It consists of two major columns which are used simultaneously during the recovery. These columns are packed with Ca-mordenite (details of the adsorbent and dimensions of these columns are given in Table 1). While adsorption of Kr is taking place in column I, column II is connected to a vacuum pump and is thus desorbed. When the adsorption cycle is finished, the solenoid valves are activated by the computer so that adsorption takes place in column II while column I is connected to the vacuum regeneration system. In the next cycle, adsorption takes place again in column I, etc. These columns were dehydrated initially at 400°C and then added to the system. Since they are regenerated continuously under vacuum while the system is working, there is no need for any further regeneration. Two non-return valves after each column prevent gas flowing from the main separation system into the vacuum regeneration system. Vacuum is maintained by a rotary pump and monitored by a McLeod gauge. The system is fitted with a liquid nitrogen trap to keep droplets of vacuum-pump oil out of the system.

OPERATING CONDITIONS

Different operating conditions have been tested in order to establish how the quality of separation varies with carrier gas velocity and sample injection time. The best conditions have been found to consist of a carrier flow-rate of 15 liters/minute and an injection time of 15 seconds with an average mixture flow-rate of 1.25 liters/second during the injection period. This corresponds to an injection of about 19 liters of N2 + Kr mixture for each pulse. The change in outlet composition with time for one pulse is shown in Figure 5. Each pulse takes 450 seconds (7.5 minutes) to complete. Therefore the flow-rate of the mixture being separated is 2.5 liters/minute expressed on a continuous basis. A summary of the present operation (expressed

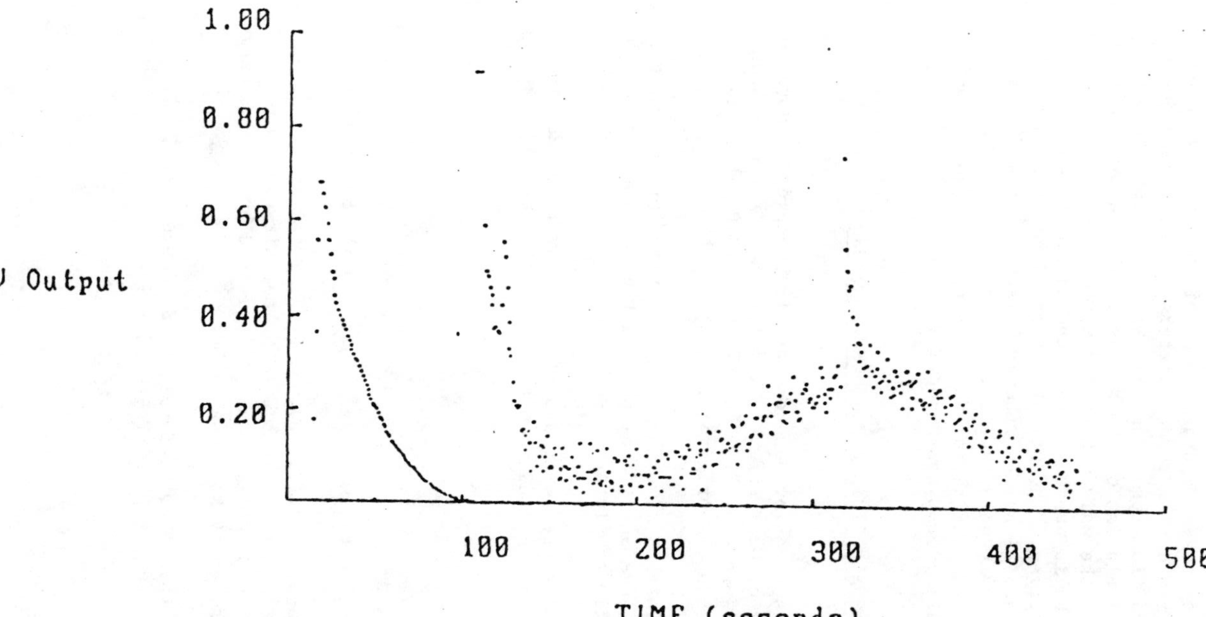

Figure 5. Response of detector at outlet to main column. Time is measured from opening of the injection valve. The initial large nitrogen peak is monitored at a sensitivity of 1. Sensitivity is increased to 500 at 105 seconds in order to monitor the krypton peak and as a result there is a small change in baseline. At 450 seconds elution of the krypton peak is essentially complete.

on a continuous flow basis) is given as a block diagram in Figure 6.

In the present stage of development of this system the He carrier is recovered during three quarters of each pulse cycle while Kr is eluted from the column but it is not recovered during the initial elution of the N2 peak. The recovered He is recycled back into the main separation system. About 2.5 liters/minute of He is needed to make up for the He which is lost with the nitrogen. The He make-up could be reduced by incorporating an additional He recovery system in the N2 stream but this has not yet been attempted. Economic evaluations indicate that if this additional He recovery is added to the system, the operating cost would decrease by about 34%.

CONCLUSIONS AND RECOMMENDATIONS

The pilot plant can provide essentially complete removal of Kr from a mixture of N2 + Kr with 0.2% (volume) Kr composition at an average flow-rate of 2.5 liters/minute (STP) which is somewhat higher than the original design target.

For testing with active Kr and integration into the modular off-gas process, some additional shielding may be needed but the unit is essentially complete.

Additional pressure swing adsorption recovery of He carrier from nitrogen is recommended from the point of view of operating economics.

REFERENCES

1. G. W. Keilholtz, Nuclear Safety 12 (6), 591 (1971).

2. D. M. Ruthven, J. S. Devgun, F. H. Tezel and T. S. Sridhar, Proceedings of the 16th DOE Nuclear Air Cleaning Conference, San Diego, California. Also issued by Atomic Energy of Canada Ltd. as report No. 7004 (1980).

3. D. M. Ruthven, F. H. Tezel, J. S. Devgun and T. S. Sridhar, Can. J. Chem. Eng. 62 (4), 526 (1984).

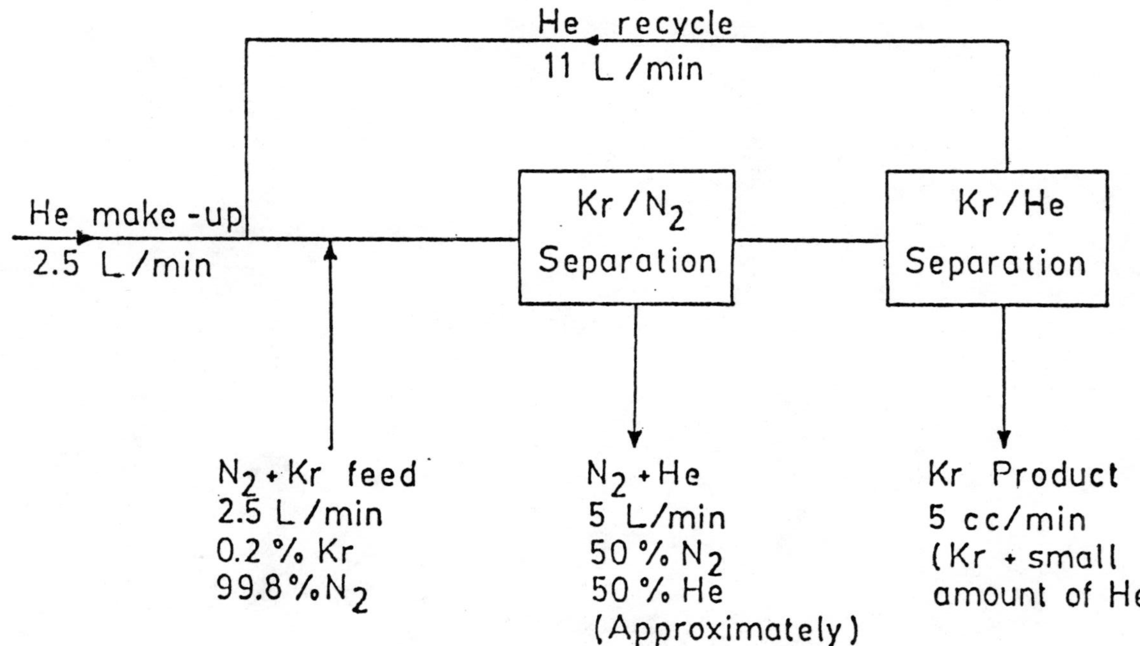

Figure 6. Block diagram of the present operation.

GAS-CHROMATOGRAPHIC INVESTIGATION OF RADIOLYSIS GAS FORMATION IN HIGH-LEVEL REPROCESSING WASTE SOLUTIONS

B.-G. Brodda, K. Hein
Institute of Chemical Technology
Nuclear Research Center Juelich GmbH
POB 1913 - D-5170 Juelich
Federal Republic of Germany

Radiolysis of High Level Liquid Waste solutions could lead to the formation of ignitable gas mixtures reaching explosive limits when such solutions are stored in non-ventilated tanks. An original waste solution was investigated for radiolysis gas formation under different experimental conditions. A considerable build-up of hydrogen, nitrous oxide and carbon dioxide was observed.

INTRODUCTION

The formation of gaseous radiolysis products in the first cycle raffinate stream (High Level Liquid Waste, HLLW) of a PUREX reprocessing flowsheet is of great interest because of the hazard potential connected with ignitable gas mixtures which may reach the explosion limit. This knowledge is especially important in the licensing procedure of transport vessels when assuming that under hypothetic accidental conditions the vessel cannot be emptied or vented for a longer period of time, so that a dangerous build-up of overpressure of a possibly ignitable gas mixture may occur. In Germany, the pressure build-up in a HLLW containing vessel must not exceed 7 bar per year.

EXPERIMENTAL

A waste solution was used as obtained from a reprocessing campaign in the MILLI facility of the Karlsruhe Institute of Hot Chemistry (KfK/IHCh). The reprocessed material was a fast breeder fuel that had been irradiated in the Karlsruhe KNK-II reactor to a burn-up of approximately 76,000 Mwd/t. About 20 liters of the solution were stored in a closed stainless steel container of

30 liters total volume, fitted with a manometer and a gas sampling valve. The experiments were conducted at room temperature after flushing the gas volume in the container with air respectively nitrogen and then sealing the container at atmospheric pressure.

The waste solution was analyzed for its chemical and isotopic composition applying potentiometric titration, x-ray fluorescence analysis, alpha- and gamma-spectrometry and gross-beta measurements. Gas samples were collected in a gas mouse and analyzed on a VARIAN GC model 3700 with TCD and 1/8" packed columns. For hydrogen determination, activated charcoal was used at 50 °C with nitrogen as carrier gas, the other components were determined separately from molecular sieve between -50 and 200 °C with He as carrier gas. The sample loop volume was 2 ml.

RESULTS AND DISCUSSION

The free acidity of the waste solution was potentiometrically titrated as 0.11 molar nitric acid, the density was determined by remote weighing of a volume aliquot as 1.009 g/ml.

The chemical composition of the solution is presented in Table 1, the isotopic composition with regard to the main contributors to the beta- and gamma-activity in Table 2. The value for

Table 1. Chemical composition of the waste solution.

Element	Concentration (g/liter)
Rb	0.05
Sr	0.09
Y	0.01
Zr	<0.05
Mo	0.29
Tc	<0.05
Ru	<0.05
Rh	<0.05
Pd	<0.05
Sb	<0.05
Cs	1.03
Ba	0.31
Ce	<0.05
Pr	<0.05
Nd	<0.05
U	0.28
Am-241	0.02
Cm-242	6.4E-6
Cm-244	39.0E-6

Table 2. Isotopic composition of the waste solution, beta- and gamma-emitters.

Nuclide	Concentration (Bq/liter) Dated to 02/01/84
Ce-144	3.07E11
Cs-134	2.33E10
Cs-137	4.24E11
Eu-155	2.59E09
Rh(Ru)-106	2.18E10
Sb-125	1.48E10
Sr(Y)-90	1.41E12

Am-241 was determined by alpha spectrometry and XRF, those for Cm-242 and Cm-244 were measured by alpha spectrometry.

The measured chemical and radionuclide concentrations were significantly lower than expected, and it could be confirmed that the waste solution had been diluted at the process site to an undefined extent before shipment, and that also a considerable amount of organic material (TBP/kerosene) had been entrained into the solution during the solvent extraction procedure. Both facts explain the analytical data of Tables 1 and 2 as well as the somewhat unexpected results of the radiolysis gas production.

The two containers were available for a time span of about 40 days.

With air as initial atmosphere a continuous gas production was observed leading to a total pressure of 3.5 bar after 35 days (Figure 1). The evaluation of the gas-chromatographic measurements resulted in the data for the composition of the gas atmosphere in the sealed tank as a function of time as presented in Figure 2, and the functions for the radiolytically produced volumes of gases per liter of waste solution versus time after applying a correction for the dissolved fractions as a function of pressure, as given in Figure 3.

The main components formed are carbon dioxide, nitrous oxide and hydrogen with a relative ratio of about 9:3:1 (total of gaseous and dissolved species) and with final concentrations of about 50 %, 15 % and 10 % resp. in the gas compartment of the tank. Oxygen, as contained in the initial air atmosphere, was consumed completely, probably supporting the formation of carbon dioxide and nitrous oxide. Nitrogen did not participate in any chemical reaction, its concentration was lowered just by dilution.

The formation rates of the radiolytically produced gases are constant during the observation time, the changes in

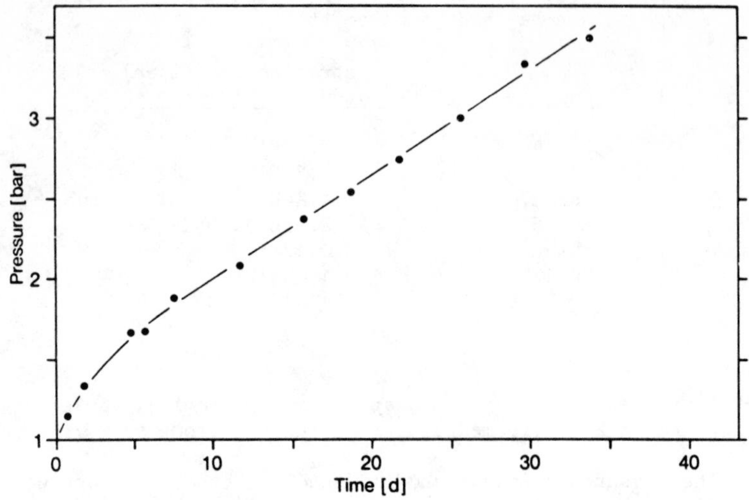

Figure 1. Pressure build-up during radiolysis of HLLW in closed tank. Initial atmosphere: air.

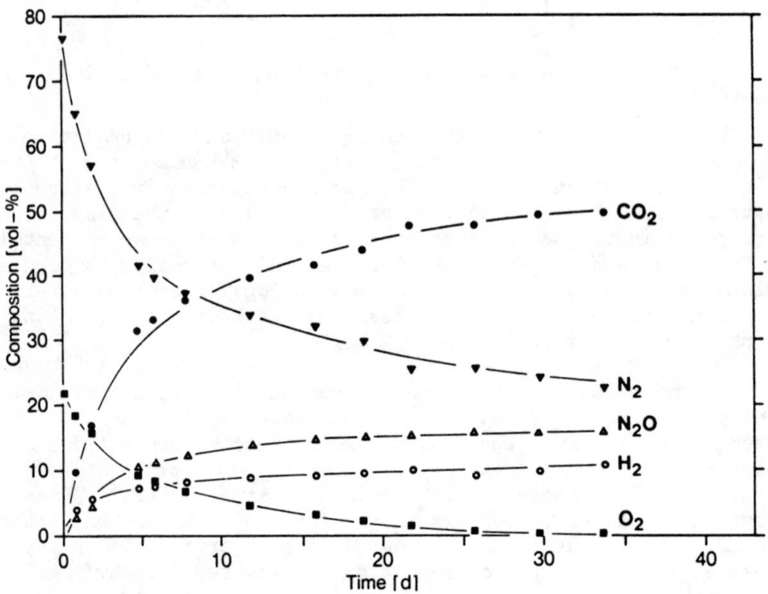

Figure 2. Composition of atmospheric radiolysis gas. Initial atmosphere: air.

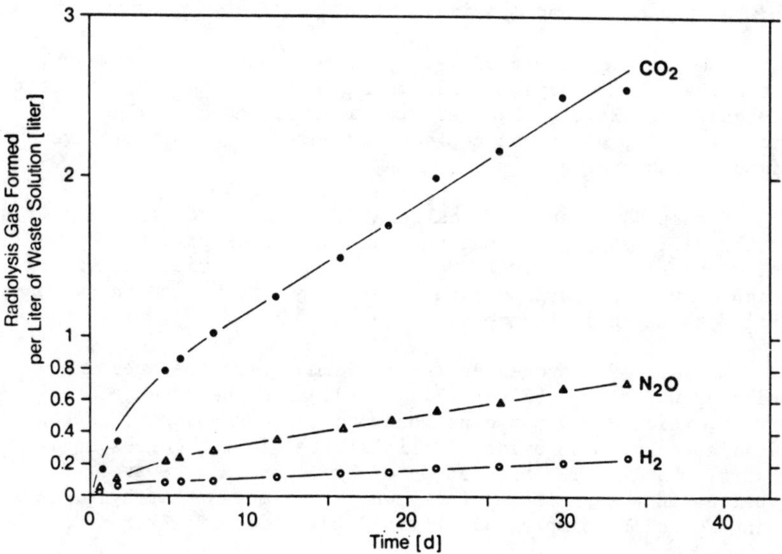

Figure 3. Total radiolytic gas formation (gaseous and dissolved parts). Initial atmosphere: air.

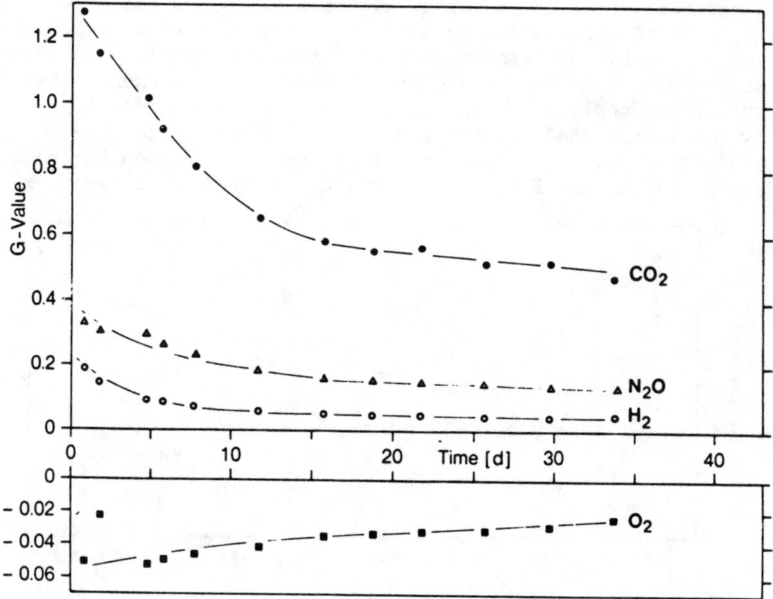

Figure 4. G-Values. Initial atmosphere: air.

concentration flatten out due to solubility effects, especially in the case of carbon dioxide.

The ignition limit of 4 % hydrogen was reached, however, after less than 5 days, when oxygen was present at a sufficient concentration level, together with a simultaneous build-up of nitrous oxide, which can replace oxygen well in an explosive mixture with hydrogen.

Figure 4 presents the calculated G-values for the formation of the radiolysis gases and the consumption of oxygen. The G-values decrease relatively fast from higher initial values, which corresponds approximately to the gas formation characteristics as given in Figure 3.

In case of nitrogen as initial atmosphere the pressure build-up was slower (Figure 5) according to the absence of oxygen, which could not contribute to the formation of carbon dioxide and nitrous oxide, as illustrated in the corresponding Figures 6 and 7. The production of hydrogen was, however, about equal as in the presence of oxygen, which means that hydrogen is being formed by simple radiolytic degradation of water.

Figure 8 shows the calculated G-values in case of nitrogen as initial atmosphere.

The most surprising finding was the production of the rather large quantities of carbon dioxide and nitrous oxide, but no fully verified explanation is available as yet. The carbon source is very probably the beforementioned entrained organic phase, whereas nitrous oxide might have been formed from nitrogen containing reductants applied in solvent extraction or by a possibly radiolytically-enhanced denitration process of the waste solution. The unusual low acidity of the solution, much lower than the value expected from the PUREX flowsheet, may support this hypothesis.

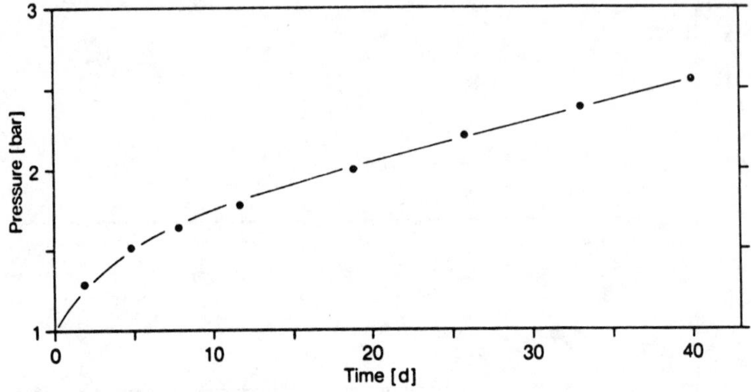

Figure 5. Pressure build-up during radiolysis of HLLW in closed tank. Initial atmosphere: nitrogen.

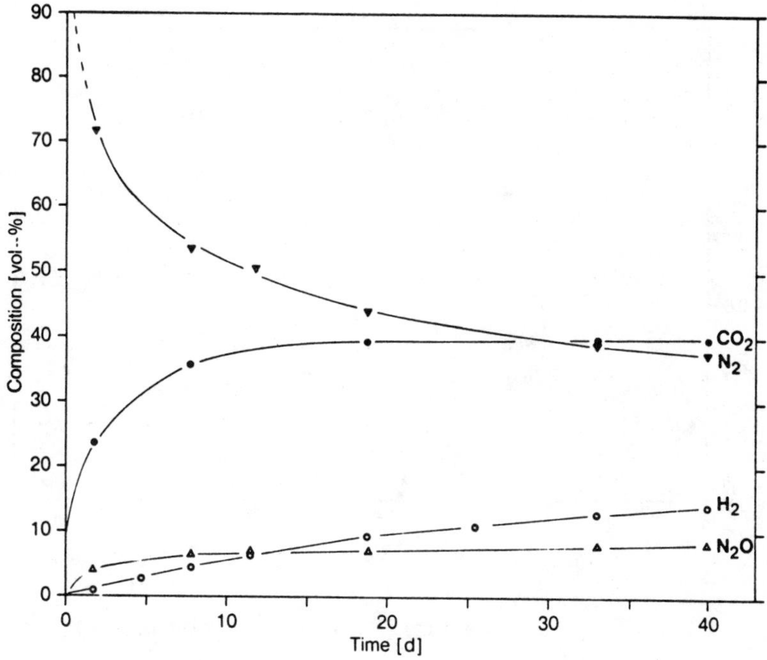

Figure 6. Composition of atmospheric radiolysis gas. Initial atmosphere: nitrogen.

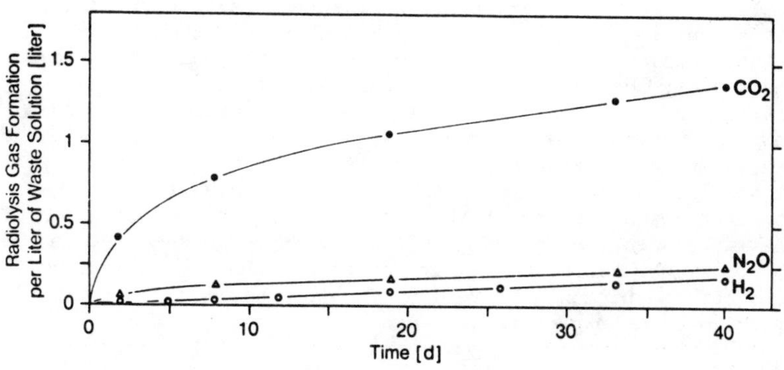

Figure 7. Total radiolytic gas formation (gaseous and dissolved parts). Initial atmosphere: nitrogen.

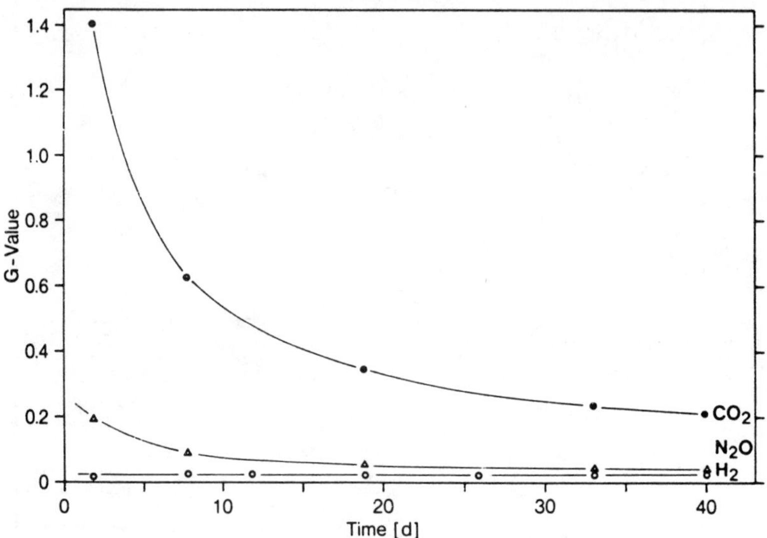

Figure 8. G-Values. Initial atmosphere: nitrogen.

SUMMARY

The investigation of radiolysis gas formation in a HLLW solution from a reprocessing campaign of spent fast breeder fuel resulted in the observation of surprisingly high amounts of carbon dioxide, nitrous oxide and hydrogen, though the solution was more dilute than expected, which should have reduced the radiolytic strength. Due to the fact of an identified entrainment of organic material into the solution before shipment the formation of carbon dioxide becomes explainable. Hydrogen and nitrous oxide may accumulate to explosive mixtures when such solutions are stored in closed tanks. These findings cannot be generalized because of the obviously untypic composition of the waste solution investigated. It seems recommendable, however, to verify the radiolytic behaviour of each homogeneous batch of HLLW before shipment or previous to other chemical or physical operations to be performed.

ACTINIDES-LANTHANIDES GROUP SEPARATION IN MIXER SETTLER

C. Musikas, X. Vitart, J.Y. Pasquiou, P. Hoël
CEA-IRDI-DERDCA-DGR-SEP-SCPR
Centre d'Etudes Nucléaires de Fontenay-Aux-Roses
BP n° 6 - 92265 (France)

Trivalent lanthanides-actinides group separation feasibility using as extractant the mixture tris (2,4,6-(2-pyridyl) 1,3,5-triazine (TPTZ), and dinonylnaphtalene sulfonic acid (HDNNS) in CCl_4 was tested in mixer settler counter current. The aqueous phase was 0.125 N HNO_3.

Encouraging results were obtained but technical problems like slow phase disengagement did not allow observation of the separations factors calculated from batch distribution data.

INTRODUCTION

It has been shown previously (1,2) that lipophilic nitrogen or sulfur donor ligands extract preferentially the 5f trivalent ions rather than the 4f.

Tris (2,4,6-(2-pyridyl)) 1,3,5-triazine (TPTZ) and dinonylnaphtalene acid (HDNNS) mixtures showed promising features for the 5f-4f trivalent ions group separations. This paper deals with the results of separations obtained during three mixer settler runs using different experimental conditions.

EXPERIMENTAL

Reagents : ^{241}Am, ^{152}Eu, ^{141}Ce radioactive tracers were provided by STU (CEN/FAR) and ORIS (CEN/SACLAY).

HDNNS was an "ALFA PRODUCT" delivered as a 50 % solution into heptane. It was purified according to the method described in reference (3). TPTZ was a product MERCK, 99.5 % pure, it has been used without purification.

The other reageants were normapure, PROLABO products, and were used without further purification.

Apparatus : We used a 16 stage mixer-settler battery POLLUX type, commercialized by CERIES (Lyon). The mixer and settler volumes were respectively 10 and 40 ml.

The pumps were syringe type pumps precise at 5 % for flux, in the range 20 to 200 $ml.h^{-1}$.

CHOICE OF THE CHEMICAL CONDITIONS

The distribution ratios of 5f and 4f ions between HDNNS-TPTZ organic solutions and aqueous nitric acid as a function of various parameters were given and discussed in previous papers (2,4,5).

They oriented the choice of the chemical conditions for the mixer-settler separations. However some specific distribution ratios were measured for this work. Distribution ratios of Am(III), Eu(III) and Ce(III) at radioactive trace concentrations between 0.125 N and 0.05 M HDNNS into CCl_4 as a function of TPTZ are plotted in figure 1.

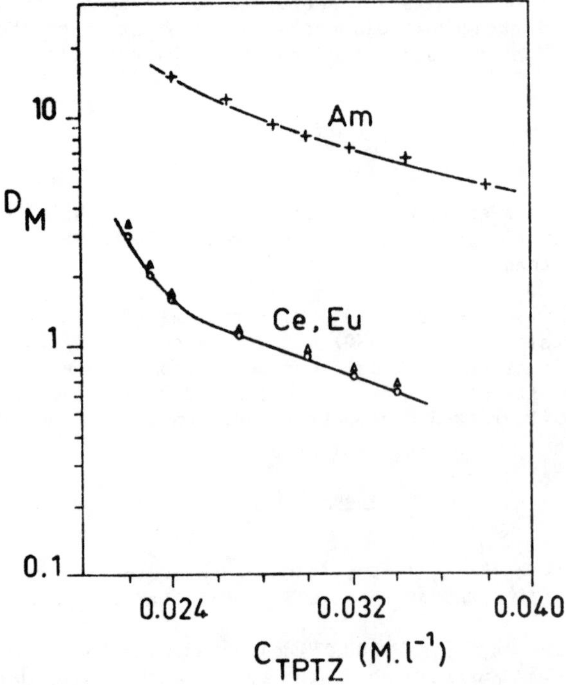

Figure 1 : Distribution ratios of Am(III)+, Ce(III) Δ , Eu(III)o, between HDNNS 0.05 M, TPTZ into CCl_4 and 0.125 N aqueous HNO_3 as a function of the initial TPTZ concentration into the organic phase.

CCl_4 was chosen as a diluent because it led to easier phases disengagement than t-butylbenzene. Distribution ratios of TPTZ as a function of the TPTZ concentration in the organic phase before equilibrium are plotted in figure 2.

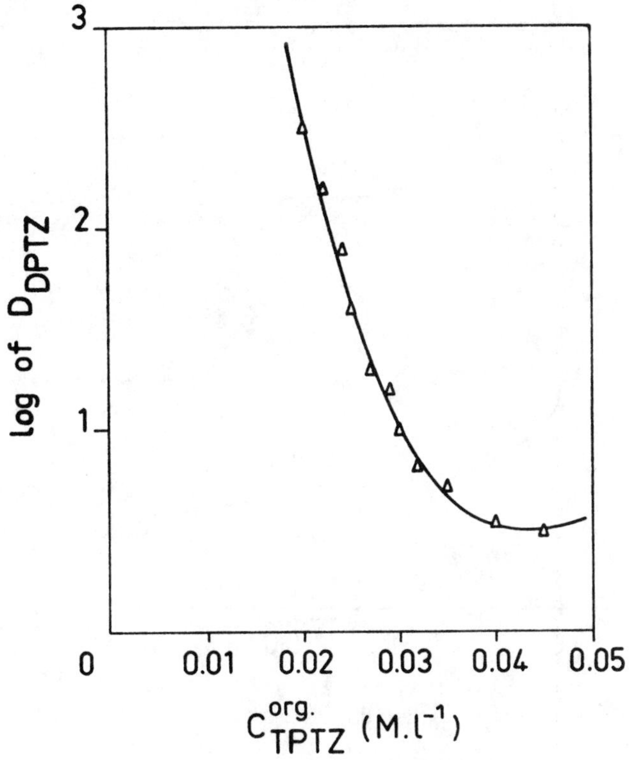

Figure 2 : Distribution ratios of TPTZ between 0.05 M HDNNS into CCl_4 and 0.125 N HNO_3 as a function of TPTZ initial concentration into the organic phase (from reference (6)- with permission).

It can be seen that TPTZ remains mainly in the organic phase only for HDNNS/TPTZ ratios higher than 2. This feature indicates the formation of hydrophobic adducts $(HDNNS)_m(TPTZ)_n$ where $m > 2n$.

The influence of Eu(III) macroconcentrations over traces $^{241}Am(III)$, $^{152}Eu(III)$, $^{141}Ce(III)$ distribution ratios is illustrated by figure 3 plots.

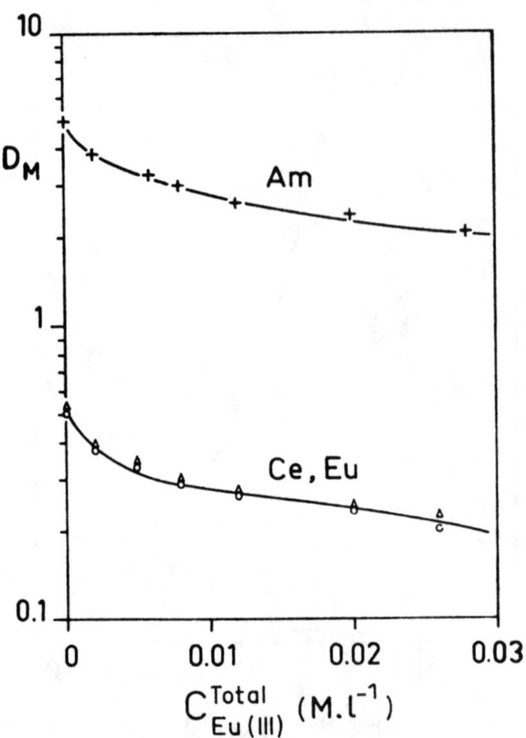

Figure 3 : Influence of Eu(III) concentrations upon the distributions ratios of Am(III)+, Ce(III)Δ, and Eu(III)o, between 0.05 M HDNNS, plus 0.033 M TPTZ into CCl_4 and 0.125 N aqueous HNO_3.

These results were used to calculate the Am(III), Eu(III), Ce(III) distribution into a 16 stage mixer settler battery. The distribution of TPTZ was considered independent from the metal concentration. The computer program used for these calculations took into account the mass balance of each constituent in the different stages of the battery. Distribution coefficients were deduced from figures 1, 2, and 3 plots which were transformed into empirical 3th degree polynomials. Results of calculations are represented schematically in figure 4 for the separation of trace metal concentrations.

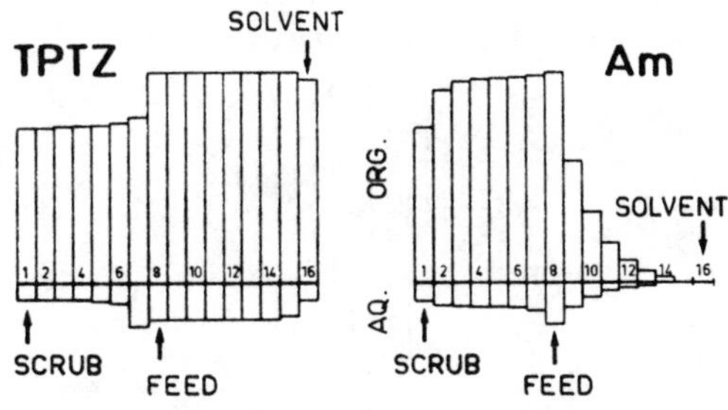

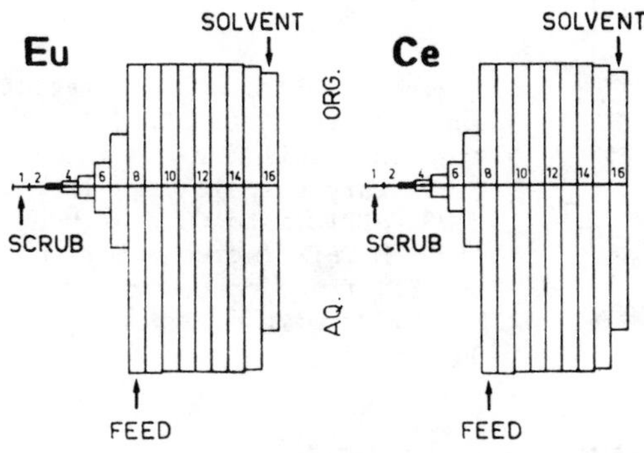

Figure 4 : Schematic representation of TPTZ, Ce(III), Eu(III), Am(III) concentration profil calculated for run I counter current (arbitrary units).
Organic : 0.05 M HDNNS, 0.03 M TPTZ into CCl_4
Aqueous : 0.125 N HNO_3, 0.003 M TPTZ, (scrub and feed) Am(III), Eu(III), Ce(III) trace concentrations in the feed.
Flux : feed 30 $ml.h^{-1}$, scrub 60 $ml.h^{-1}$, organic 30 $ml.h^{-1}$.

RESULTS AND DISCUSSION

Three battery counter current separations were performed. The first one (run I) involved Am(III), Eu(III) and Ce(III) as radioactive traces. The second and the third (run II and run III) contained macro amounts of Ce. After respectively 48 h, 48 h and 24 h of operation samples were taken for quantitative radiometric analysis.

The results and the experimental conditions of these mixer settler separations are given in figures 5, 6 and 7.

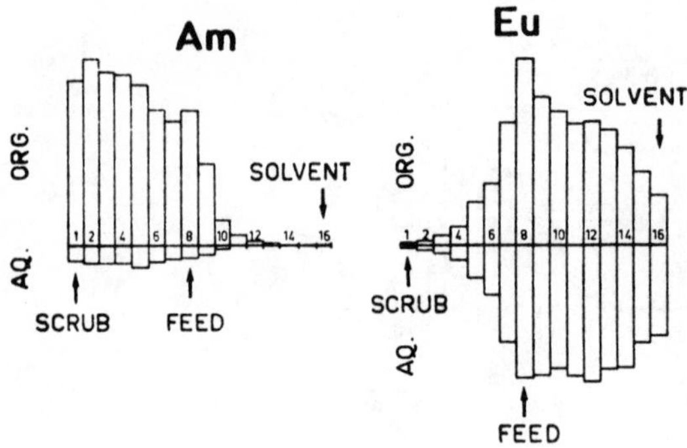

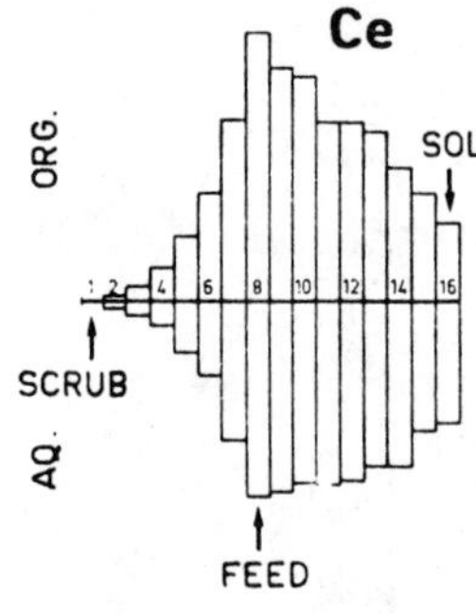

Figure 5 : Schematic representation of the experimental Am(III), Eu(III), Ce(III) concentration profil for run I counter current (arbitrary units). Organic : 0.05 M HDNNS, 0.03 M TPTZ into CCl_4 Aqueous : 0.125 N HNO_3 0.003 M TPTZ (scrub and feed), Am(III), Eu(III), Ce(III) trace concentrations in the feed. Flux : 30 $ml.h^{-1}$, scrub 60 $ml.h^{-1}$, organic : 30 $ml.h^{-1}$.

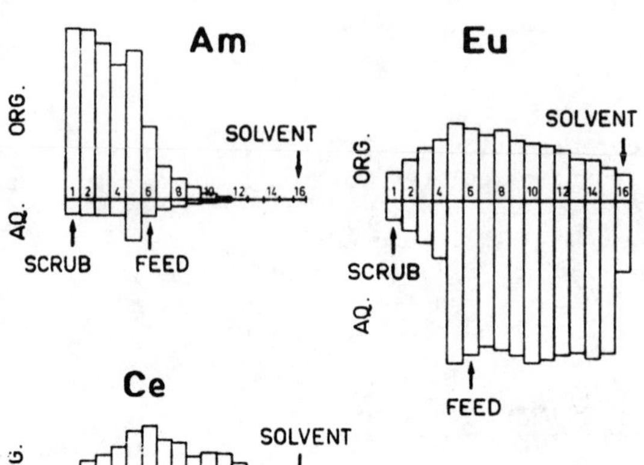

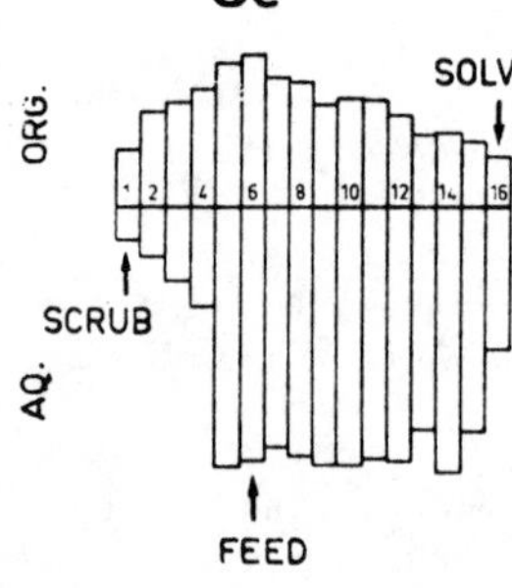

Figure 6 : Schematic representation of the experimental Am(III), Eu(III), Ce(III) concentration profil for run II counter current (arbitrary units). Organic : 0.05 M HDNNS, 0.03 M TPTZ into CCl_4, Aqueous : 0.125 N HNO_3 0.003 M TPTZ 'scrub and feed), Am(III), Eu(III) traces, Ce(III) 0.025 M in the feed. Flux : Feed 30 $ml.h^{-1}$, scrub 20 $ml.h^{-1}$, solvent : 30 $ml.h^{-1}$.

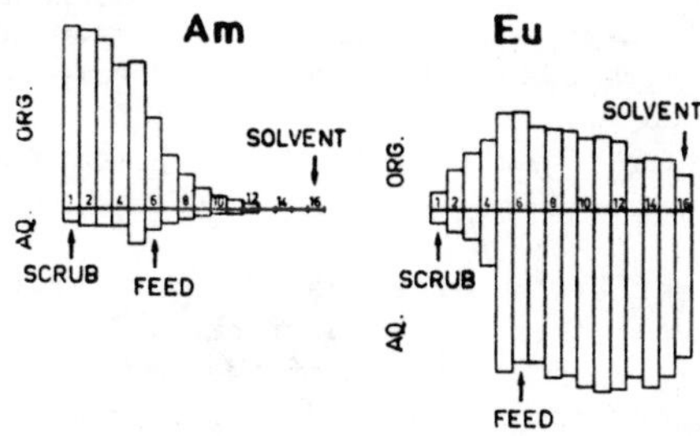

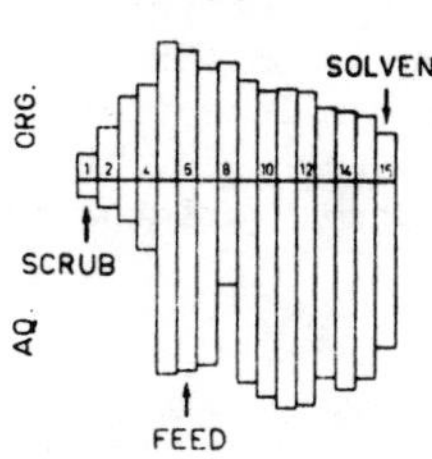

Figure 7 : Schematic representation of the experimental Am(III), Eu(III), Ce(III) concentration profil for run III counter current (arbitrary units). Organic : 0.05 M HDNNS, 0.03 M TPTZ into CCl_4, Aqueous : 0.125 N HNO_3, 0.003 M TPTZ (scrub and feed) Am(III), Eu(III) traces, Ce(III) 0.025 M in the feed, Flux : feed 30 $ml.h^{-1}$, scrub 30 $ml.h^{-1}$, solvent 30 $ml.h^{-1}$

The comparison of the results are summarized in Table 1. It can be seen that a good Am(III) extraction yield was obtained. Small part of the lanthanides followed Am(III) mainly in run II and III involving macro amounts of Ce.

Table 1. Results of mixer settler actinides-lanthanides separation.

Metal $mol.l.^{-1}$	Scrub stages	Extraction stages	Feed $ml.h^{-1}$	Scrub $ml.h^{-1}$	Solvent $ml.h^{-1}$
traces(I)	8	8	30	60	30
0.025 (II)	6	10	30	20	30
0.025 (III)	6	10	30	30	30

	Aqueous effluent (% of feed)			Solvent (% of feed)		
	^{241}Am	^{152}Eu	^{141}Ce	^{241}Am	^{152}Eu	^{141}Ce
(I)	0.1	98	98	99.4	2	2
(II)	0.2	90	89	99.8	10	11
(III)	0.5	95.5	95	99.5	4.5	5

We attributed these losses in performance to two factors. First, the incomplete phase disengagement (worse when macroconcentrations of metals were involved) causes aqueous droplet transportation to the solvent output. Secondly, accidental increases in temperature occurred in the battery and could have caused part of the lanthanides retention. The distribution ratios of Am(III) and Eu(III) increase with the temperature.

However these first mixer settler separations are encouraging because the causes of performance losses can be overcome and better phase disengagements are possible by an increase of the residence time in the settlers. Using bigger batteries could also improve hydrolics, because it will be then possible to work with lower speed of rotation in the mixer. During the present work high speed of rotation was necessary for a good liquid circulation. Discussion dealing with the improvements in the chemical system is out of the scope of this article, but the difficulties encountered during this work suggest to try to replace HDNNS by another strong acid having lower surfactant properties.

REFERENCES

1. C. Musikas, G. Le Marois, R. Fitoussi and C. Cuillerdier
   Actinide Separations, J.D. Navratil, W.W. Schulz ed.
   ACS Symposium Series 117 (1980) p. 131.

2. C. Musikas, P. Vitorge, R. Fitoussi, M. Bonnin and D. Pattée
   Actinide Recovery from Waste and Low Grade Sources
   J.D. Navratil, W.W. Schulz(ed.)
   Hardwood Acad. Publishers (1982) p. 255

3. P.R. Danesi, R. Chiarizia, G. Scibona
   J. Inorg. Nucl. Chem. 3928 (1973).

4. P. Vitorge
   CEA-R 5270 (1984).

5. C. Musikas
   Actinide-Lanthanide separations.
   G.R. Choppin, J.D. Navratil, W.W. Schulz(editors.)
   World Scientific (1985) p. 19.

6. X. Vitart, C. Musikas, J.Y. Pasquiou, P. Hoël
   J. Less. Comm. Metals (to be published).

STUDY OF EXTRACTION RATE FOR URANYL NITRATE-TBP/n-DODECANE IN A SINGLE DROP COLUMN

K.W. Kim, Y.J. Shin, J.H. Yoo, H.S. Park
Korea Advanced Energy Research Institute
Dae-Duk Science Town, Chung-Nam, 300-31, Korea

The rate of mass transfer of uranyl nitrate from the aqueous phase of 2N nitric acid to the organic phase of tributyl phosphate/n-dodecane mixture has been determined in a single drop column without external force, and the motion of the single droplet rising through the aqueous phase has been also observed in preliminary experiments. The mass transfer rate was increased with an increase of uranyl nitrate concentration in the aqueous phase. The effect of the drop size on mass transfer rate was significant at high uranyl nitrate concentration in the aqueous phase but not significant at low uranyl nitrate concentration. The correlations for terminal velocities of droplets with drop size and physical properties of both phases were also presented.

INTRODUCTION

Even if there are many effects on extraction efficiency in the various solvent extraction equipment as mixer-settler, pulsed perforated-plate extraction column, and others, the prime importance is to understand mass transfer rate between a dispersed phase and a continuous phase. It is well known that a study of an extraction rate using a single droplet should provide basic information for understanding the performance of solvent extractors. In accordance with the earlier studies for rate of falling and rising of single liquid droplets, motion of a droplet as a drop deformation, oscillation, and turbulent circulation inside a droplet exists in a single droplet rising through another liquid.(1-6) These phenomena give an effect on mass transfer rate of heavy metal from an aqueous phase to an organic phase, therefore, a hydrodynamic study of a single organic droplet in a stagnant aqueous phase must be carried out as preliminary work. Interfacial turbulence between two phases due to steep concentration gradient near the interface and physical properties of both phases such as viscosity and interfacial tension, etc. is also important on extraction rate.(7)

In the preliminary work of this investigation to understand the motion of the single droplet rising through another stagnant liquid, 2N nitric acid solution and the single droplet of tributyl phosphate (TBP)/n-dodecane were used as a continuous phase and a dispersed phase, respectively. The mass transfer experiment was carried out in uranyl nitrate solution having free acidity of 2N nitric acid and 30% TBP/n-dodecane system. The objectives of the current study are to experimentally observe the motion of the single droplet rising through another liquid, and to determine the extraction rate at various drop size and uranyl nitrate concentration in the continuous phase, using the single drop extraction column without external force.

THEORY

Motion of Droplet through Fluid without External Force

Considering a droplet rising through a fluid without external force, the force balance on the droplet can be represented as follows :

$$\frac{d}{dt}\left(\frac{\pi}{6} D_e^3 \rho_d U_d \right) = \frac{\pi}{6} D_e^3 g (\rho_c - \rho_d) - \pi D_e^2 \rho_c \frac{U_d^2}{2} C_D \quad (1)$$

The droplet quickly reaches terminal velocity, and then equation (1) is changed as follows :

$$C_D = \frac{4}{3} \frac{(\rho_c - \rho_d)\, g\, D_e}{\rho_c U_\infty^2} \quad (2)$$

Otherwise, if the mass transfer continuously occurs between both phases during drop life time, the concept of a terminal velocity will disappear and the equation (1) is changed again as follows :

$$\rho_d \frac{dU_d}{dt} + U_d \frac{d\rho_d}{dt} = g (\rho_c - \rho_d) - \frac{3}{D_e} \rho_c C_D U_d^2 \quad (3)$$

Because the solution of this equation is not simple, approximately establish the empirical equation like equation (4) under the assumption that the rising velocity of the droplet accompanying mass transfer at arbitrary time and location in a single drop column is identical with a terminal velocity of a droplet in no mass transfer system getting the same density difference as the droplet accompanying mass transfer.

$$U_d|_t \simeq U_\infty = f\,(D_e, \Delta\rho\ , \ldots\ldots) \quad (4)$$

Mass Transfer Rate

A mass transfer rate can be represented in a single droplet column as follows :

$$\frac{\pi}{6} D_e^3 \frac{dC_d}{dt} = \pi D_e^2 K_o (C_d^* - C_d) \quad (5)$$

In order to minimize the experimental error for the transfer caused by drop formation and coalescence, the shortest reference column can be used. In this case, the concentration of drop after withdrawal from the reference column approximately equals that after the same length of rising in any longer column. The corresponding time of contact was the difference between the total rising times in the reference column and any longer column. With these boundary conditions, equation (5) is integrated to give the following expression for overall mass transfer coefficient, K_o :

$$K_o = -\frac{D_e}{6t} \ln\left[\frac{C_d^* - C_d}{C_d^* - C_{d,r}}\right] \quad (6)$$

The relation between the overall resistance and individual resistance in two film theory is :

$$\frac{1}{K_o} = \frac{1}{k_d} + \frac{m}{k_c} \quad (7)$$

The individual mass transfer coefficient in the aqueous phase, k_c can be calculated from Higbie penetration theory (8) and West theory (9) as follows :

$$k_c = 2\left[\frac{D_c U_d}{\pi D_e}\right] \quad (8)$$

In equation (8), diffusivity, D_c is also calculated by the following equation used usually in electrolyte.

$$D_c = \frac{2RT}{F^2} \frac{\lambda_+ \lambda_-}{(\lambda_+ + \lambda_-)} \left[1 + \frac{d\ln \gamma_\pm}{d\ln C}\right] \quad (9)$$

EXPERIMENTAL

Physical Properties

To search the effects of physical properties of the aqueous and organic phases on the motion of the single droplet and the extraction rate, densities and viscosities of both phase and interfacial tension have been selected, and those values were determined by changing the volumetric fraction of TBP in the organic phase. The selected physical properties were measured as follows :

density by 'DMA45 Density Meter' (accuracy; 10^{-4} g/ml), interfacial tension by 'Drop Weight Method', and viscosity by 'Falling Ball Viscometer'. All organic phases and the aqueous phase of 2N nitric acid were mutually saturated before the main experiment was carried out. The measured physical properties are shown in Figures 1 and 2. From the current data, the correlations for the interfacial tension and density of organic phase with volumetric fraction of TBP in organic mixture were found as follows.

$$\sigma = \frac{25.4220}{0.3738 + \log 3x} \tag{10}$$

$$\rho_d = 0.7602 + 0.0025\, x \tag{11}$$

where

$$x = \frac{\text{volume of TBP}}{\text{total volume of organic mixture}}$$

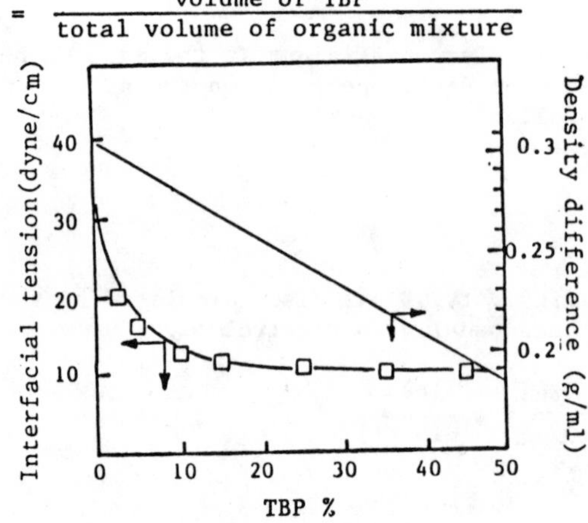

Fig. 1 Interfacial tension and density difference between 2N HNO_3 and organic phase vs. volumetric percentage of TBP in organic phase at 25°C.

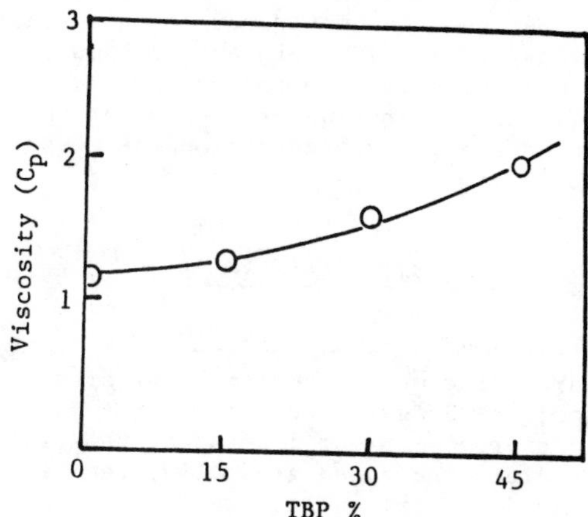

Fig. 2 Effect of volumetric fraction of TBP in organic phase on viscosity of organic phase at 25°C.

Equipment and Procedure

The single drop column of annular type was constructed of two 150cm-long glass tubes with inside diameters of 2.5cm and 4cm. 2.5cm I.D. glass tube was used as extraction column and annular chamber between 2.5cm I.D. glass tube and 4cm I.D. glass tube was used as water jacket to maintain operation temperature of 25°C. Hypodermic needles were used to form the liquid droplets at their tips. The oblique cut on the end of the needle was removed and the tip rounded off with a file to give a sharp-edged circular tip. The inside diameters of the selected needles were 0.278, o.366, 0.431, 0.525, 0.644, 0.811, 1,029, 1.35, and 1.55mm. The size of each needle was measured by light projection method. The drop size was obtained from the number of droplets and the volume of the organic phase passed. The passed organic volume was also measured by burette. To regulate the flowrate of the organic phase and to maintain a steady production rate of single droplet, the pressurized feeding system (35mmHg as Gauge pressure) was adapted in this investigation.

In preliminary work, it was found that droplets attained terminal velocities within an initial height of rising of 30cm from nozzle tip. All terminal velocities of droplets were measured by free rising period of a droplet from 30cm above the nozzle tip. Extraction rate was determined by detection of uranium concentration extracted from the continuous phase to the single droplet and its contact time. The reference column of 30cm-long glass tube was used to eliminate the error caused by mass transfer during formation and breakup of droplet at the top of column. The extracted uranium concentration was measured by 'DMA 45 density meter'.

RESULTS AND DISCUSSION

Motion of Single Droplet without Mass Transfer

The effect of drop size on the terminal velocity of a rising droplet is shown in Figure 3, using 2N nitric acid and TBP/n-dodecane mixture. As drop size was increased, the rising velocities of the droplets increased gradually, reached a maximum, and then slightly decreased again. These effects are in good agreement with the observations reported earlier. It was also found that the droplets rising through the continuous phase were nearly spherical in small sizes, and their deformation increased with an increase of drop size. The terminal velocities of droplets were increased with an increase of TBP content in low volumetric fraction range of TBP(<5%), but those values were decreased with an increase of TBP content in high fraction (>15%). In accordance with the result of physical property change experimental interfacial tension was steeply decreased with an increase of volumetric fraction of TBP in organic mixture until about 10% TBP but almost constant in volumetric fraction range of TBP of 20% above. The difference of density between two phases was monotonously decreased with an increase of volumetric fraction of TBP in organic mixture. From above trend for physical properties, it was suspected that the dominant effect on the terminal velocity was interfacial tension at low volumetric fraction of TBP, and the difference of the density at high volumetric fraction of TBP. When the terminal velocities from observed values were considered with respect to drop size and physical properties as forming two separate regions, region I($D_e \leq D_{e,c}$) where the velocity increased with drop size and region II ($D_e > D_{e,c}$).

Where the velocity decreased with drop size, the following correlations were obtained :

$$0 < D_e \leq D_{e,c} \; : \; U_\infty = 124.07 \, D_e^{0.623} \, \Delta\rho^{0.636} \, \sigma^{-0.225} \quad (12)$$

$$D_e > D_{e,c} \; : \; U_\infty = 14.946 \, D_e^{-0.069} \, \Delta\rho^{0.225} \, \sigma^{0.036} \quad (13)$$

$$D_{e,c} = 0.0054 \, \Delta\rho^{-1.121} \, \sigma^{0.948} \quad (14)$$

The multiple regression coefficients of equation (12) and (13) are 0.88 and 0.94, respectively, and the standard deviations of the data from these equations are 1.73% and 1.09%, respectively.

Motion of Single Droplet Accompanying Mass Transfer

When the mass transfer continuously occurs from the continuous phase to rising droplets of 30% TBP/70% n-dodecane, the concept of terminal velocity cannot be applied to this system. However, under the assumption that the rising velocity of the droplet accompanying mass transfer at arbitrary time and location in a single drop column is identical with a terminal velocity of a droplet in no mass transfer system getting the same density difference with the droplet accompanying mass transfer, the empirical equation of equation (4) could be approximately obtained from the terminal velocities of single droplets using 30% TBP/n-dodecane having various uranyl nitrate concentrations and. 2N nitric acid system. This system almost prevents back-extraction of uranyl nitrate from an organic phase to an aqueous phase because of high distribution coefficient.

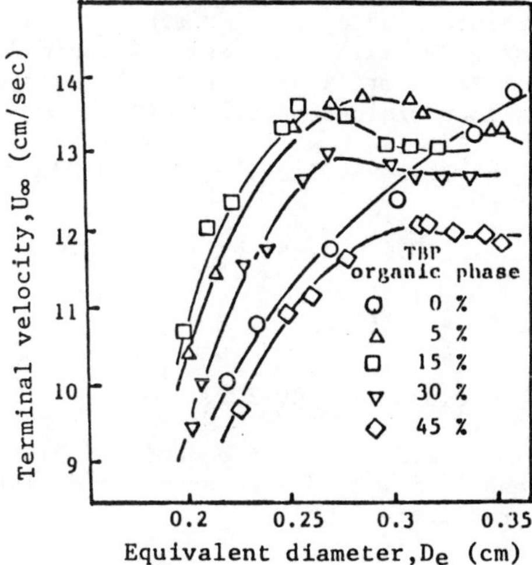

Fig. 3 Terminal velocity VS. drop diameter at 25°C.

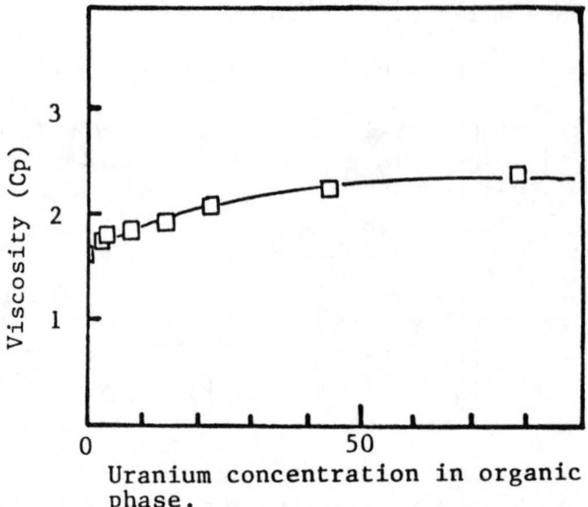

Fig. 4 Effect of uranium concentration in organic phase on viscosity of organic phase at 25°C.

The effect of uranium concentration in the droplet on physical properties was observed and shown in Figures 4 and 5. The effect of drop size on the terminal velocity of rising droplet is also shown in Figure 6, using 2N nitric acid and 30% TBP/n-dodecane having various uranium concentrations. From the current data, the following correlations are obtained.

$$0 < D_e \leq D_{e,c} : U_\infty = -7.924 + 23.295\ D_e + 60.353\Delta\rho \tag{15}$$

$$D_e > D_{e,c} : U_\infty = 7.095 - 5.455\ D_e + 31.279\Delta\rho \tag{16}$$

$$D_{e,c} = 0.583 - 1.667\Delta\rho \tag{17}$$

$$\Delta\rho = \rho_c - 0.8351 - 0.0012\ C_d \tag{18}$$

The multiple regression coefficients of equation (15) and (16) are 0.96 and 0.97, respectively, and the standard deviations are 2.44% and 1.56%, respectively.

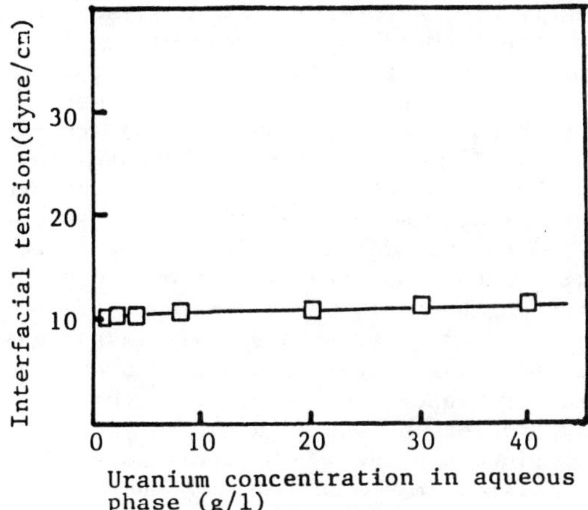

Fig. 5 Effect of uranium concentration in 2N HNO_3 on interfacial tension at 25°C

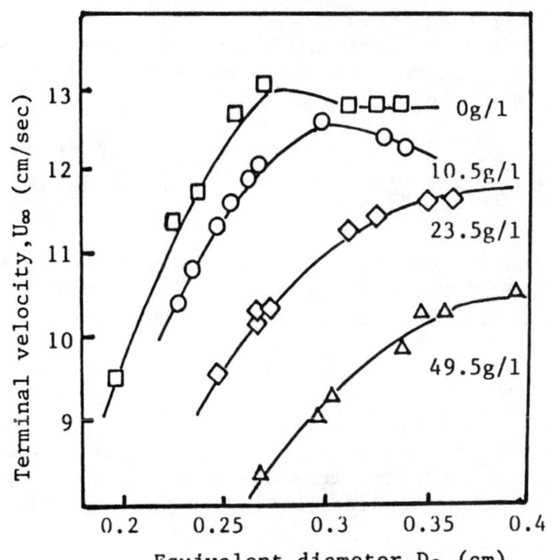

Fig. 6 Effect of drop diameter on terminal velocity of droplet rising through continuous phase of 2N HNO_3 at different uranium concentration of droplet.

Extraction Rate

All experiments for extraction rate were carried out in uranyl nitrate/2N nitric acid -30% TBP/n-dodecane system. There are end-effects of the formation and coalescence of the droplets in the single drop column. These effects can be eliminated by comparing with reference column. The mass transfer flux observed is shown in Figure 7 and the overall mass transfer coefficients calculated by equation (6) are shown in Figure 8. In case of low uranyl nitrate concentration in a continuous phase, the effect of the drop size on the overall mass transfer coefficient was not significant, because the effect of interfacial turbulence was small and the increase of internal circulation with drop size can not overcome the increase of viscosity which occurs in case of mass transfer of uranium to organic droplet. But the overall mass transfer coefficient was increased with the increase of drop size due to the interfacial turbulence and turbulent circulation inside the organic droplet in case of high uranyl nitrate concentration in a continuous phase. The resistance of each phase has been calculated from equations (7) and (8). Here, the values of λ_+, λ_- and $\ln \gamma_\pm / \ln C$ used 349.8cm^2/ohm. mole, 71.4cm^2/ohm.mole, and 0.0886, respectively. The portion of resistance inside the droplet was about 60-80% of total resistance, and this portion was decreased with an increase of uranyl nitrate concentration in a continuous phase, especially at large droplets. (Figure 9).
Newman(10) expressed the individual mass transfer coefficient within rigid drop with no internal circulation and no resistance around the drop as follows.

$$k_{d\ rigid} = -\frac{D_e}{\Delta t} \ln\left[\frac{6}{\pi^2}\sum_{n=1}^{\infty}\frac{1}{n^2}\exp\left(-n^2\pi^2\frac{4\bar{D}_L\Delta t}{D_e^2}\right)\right] \qquad (19)$$

Diffusivity of solute in organic phase $\bar{D}_L$ in above equation was calculated by use of Wilke-Chang equation as 9.186x10^{-6}cm/sec and $k_{d\ rigid}$ was calculated with the error of

$$(k_{d,n}-k_{d,n-1})/k_{d,n-1} < 10^{-6} \text{ by computer.}$$

The ratio of k_d from equations (7)-(8) and $k_{d\ rigid}$ from equation (19) is shown in Figure 10. The ratio was increased with uranyl nitrate concentration in a continuous phase because of the existance of some circulations inside the organic droplet.

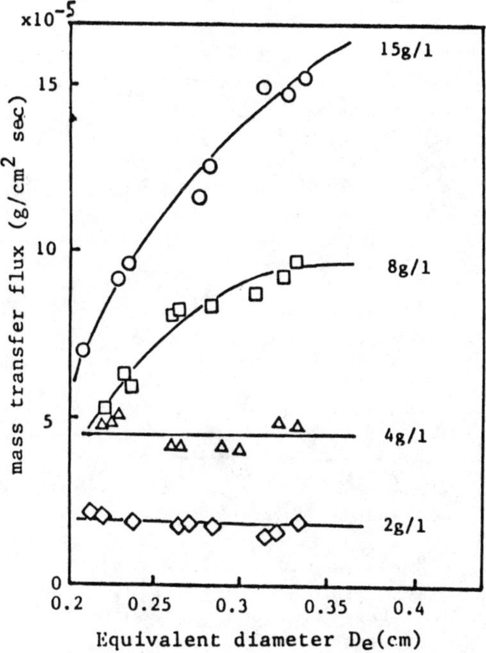

Fig. 7 Effect of drop diameter on mass transfer rate at different uranium concentrations in aqueous phase.

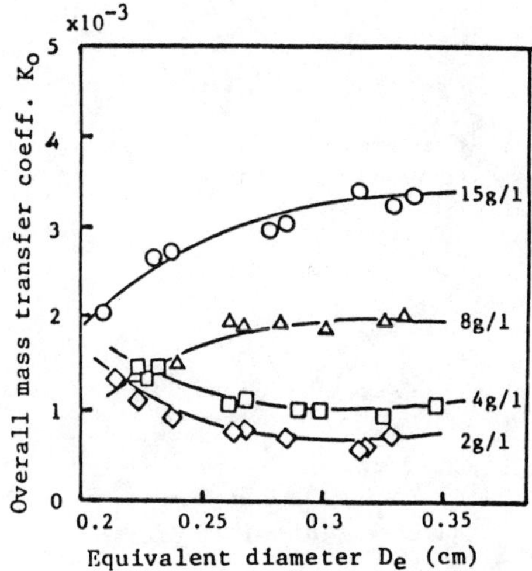

Fig. 8 Overall mass transfer coefficient vs. drop diameter at different uranium concentrations in aqueous phase.

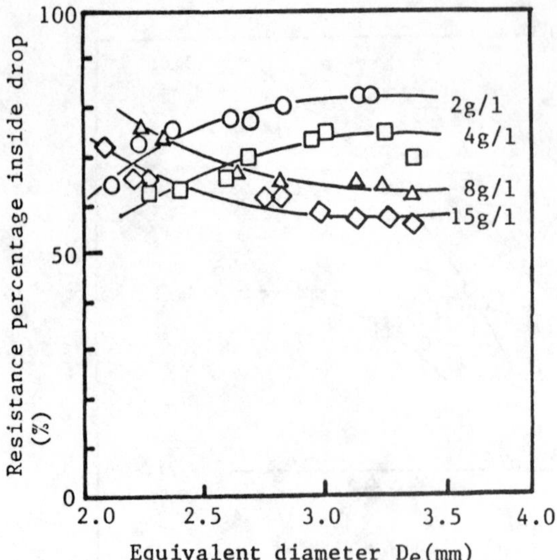

Fig. 9 Resistance percentage inside drop vs. drop diameter at different uranium concentrations in aqueous phase.

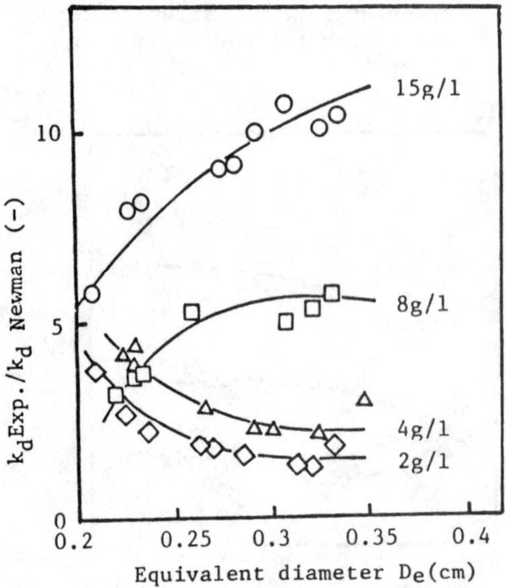

Fig.10 Ratio of kd of experimental to k_d of Newman's model VS. equivalent diameter in different uranium concentration of aqueous phase.

CONCLUSIONS

Spherical shape of a droplet becomes more stable with the increase of uranyl nitrate concentration in a droplet. The mass transfer rate is increased with the increase of uranyl nitrate concentration in the aqueous phase. The effect of drop size on mass transfer rate is significant at high uranyl nitrate concentration in the aqueous phase, and the mass transfer rate is increased with the increase of drop size due to interfacial turbulence and turbulent circulation inside the droplet. In uranyl nitrate-30% TBP/n-dodecane extraction system, the portion of resistance inside a droplet for mass transfer was about 60-80% of total resistance.

NOMENCLATURE

C_D : drag coefficient

C_d : uranium concentration in organic phase, g/l

C_d^* : uranium concentration in organic phase at equilibrium, g/l

D_c : diffusivity of solute in aqueous film, cm^2/sec

$\bar{D}_L$: diffusivity of solute in organic phase, cm^2/sec

D_e : equivalent diameter of droplet, cm

$D_{e,c}$: equivalent diameter of droplet at peak terminal velocity, cm

F : Faraday constant

k_c : individual mass transfer coefficient in aqueous phase, cm/sec

k_d : individual mass transfer coefficient in organic phase, cm/sec

K_o : overall mass transfer coefficient, cm/sec

g : gravitational constant, cm/sec^2

m : distribution coefficient of solute at given free acidity

R : gas constant

t : time, sec

T : temperature, ^oK

U_d : rising velocity of droplet, cm/sec

U_∞ : rising terminal velocity of droplet, cm/sec

x : volumetric fraction of TBP in TBP/n-dodecane mixture

ρ_c : density of continuous phase, g/cm^3

ρ_d : density of droplet, g/cm^3

σ : interfacial tension, dyne/cm

λ : equivalent ion conductivity, cm^2/ohm·mole

REFERENCES

1. Hu, S. and R.C.Kintner, AIChE J.,1(1), 42(1955)

2. Klee, A.J. and R.E.Treybal, AIChE J.,2(4), 444(1956)

3. Lehrer, I.S., AIChE J.,26(1), 170 (1980)

4. Elzinga, E.R. and J.T.Banchero, AIChE J., 7(3), 394 (1961)

5. Handlos, A.E. and T.Baron, AIChE J., 3(1), 127(1957)

6. Inoue, K. and F.Nakashio, Chem. Eng. Sci., 34, 191 (1979)

7. Sterning, C.V. and L.E.Scriven, AIChE J., 5(4), 514 (1959)

8. Higbie, R., Trans. Am. Inst. Chem. Eng., 44, 621 (1952)

9. West, F.B., A.J.Herrman, A.T.Chong, and L.E.K. Thomas, I.&E.C., 44, 621 (1952)

10. Newman, A.B., Trans. Am. Inst. Chem. Engrs., 27, 203 (1931)

RECOVERY OF URANIUM FROM 30 VOL % TRIBUTYL PHOSPHATE SOLVENTS CONTAINING DIBUTYL PHOSPHATE

J. C. MAILEN and O. K. TALLENT
Chemical Technology Division
Oak Ridge National Laboratory
P. O. Box X
Oak Ridge, TN 37831

A number of solid sorbents were tested for the removal of uranium and dibutyl phosphate (DBP) from 30% tributyl phosphate (TBP) solvent. The desired clean uranium product can be obtained either by removing the DBP, leaving the uranium in the solvent for subsequent stripping, or by removing the uranium, leaving the DBP in the solvent for subsequent treatment.

Solid sorbents were tested in shake-outs, and promising candidates were examined in small packed columns. The variables examined included temperature and residence time of the solvent in contact with the solid sorbent. No method for removing DBP while leaving the uranium in the solvent was found. Both cation resins and diethylenetriamine pentaacetic acid (DTPA) deposited on glass beads preferentially removed the uranium, leaving the DBP in the solvent. The DBP could then be readily removed by a number of simple treatments, and the uranium could be recovered by elution with acidified TBP. Both hydroxide-form anion exchange resins and activated alumina (used with dry solvent) removed both uranium and DBP. It is possible that the DBP and uranium could be separately eluted from these sorbents, but it was not successfully tested. These results suggest a number of possible applications in solvent extraction systems.

INTRODUCTION

Solvent cleanup practice in Purex plants is to scrub the contaminated tributyl phosphate (TBP) solvent with a sodium carbonate solution before recycle to the solvent extraction system. This procedure removes the acidic solvent degradation products, including dibutyl phosphoric acid (HDBP) and monobutyl phosphoric acid, and transfers the metallic contaminants to the aqueous phase as carbonate complexes.

Transfer of the metals that are complexed with DBP to the carbonate scrub solution makes their recovery difficult. If the value of the metals is sufficiently high, as for enriched uranium or plutonium, a process for removing the metals prior to such scrubbing may be useful. The desired treatment method would (a) use a solid sorbent to preferentially remove the DBP, leaving the uranium or plutonium in the solvent (this would allow direct recycle of the solvent); or (b) use a procedure which would separately remove both the DBP and uranium or plutonium, yielding a clean product. We prepared a series of solid sorbents which have potential for this separation, performed scouting tests with uranium solutions to determine which sorbents provided separations, and conducted tests in small packed columns using the promising sorbents.

EXPERIMENTAL

Expected Reactions

It is expected that, before reaction with a damp solid, the components of the organic phase will first have to distribute to the aqueous layer on the solid sorbent. Thereafter, the reactions with the sorbent will be the ones expected for the same components in the aqueous phase. In addition, if a component such as uranyl nitrate should react to form an inextractable compound, the uranium would be trapped in the aqueous layer of the sorbent. An example would be reactions that replace the nitrates of uranyl nitrate with other anions such as hydroxide or carbonate; the resulting compounds would likely not be extractable. Sorbents that are active in the dry state (the principal example is activated alumina) may remove components directly from the solvent.

Scouting Tests

The ability of a number of solid sorbents to remove uranium and DBP from 30% TBP in normal paraffin hydrocarbon (NPH) was investigated in smallbatch equilibrations. A ~0.01 M DBP solution was prepared by adding HDBP containing about 3.3 mol % monobutyl phosphoric acid to 30% TBP-NPH. This solution was then contacted with a uranyl nitrate solution containing 5 g of uranium per L (0.02 M uranium) in water (pH = 3) to prepare a uranium-DBP test solution containing ~0.01 M DBP and ~0.01 M uranium. One-half of the uranium-DBP solution was dried by sparging with 1 volume of dry air per min per volume of solution for about 5 h at ~60°C. This procedure has been shown to result in quite dry solvent (removal of ≥97% of the water from water-saturated solvent) (1). It is expected that some solid sorbents, especially activated alumina, will have higher capacities if water is excluded. Most of the solid sorbents were tested by two methods: dried with dry solvent and damp with wet solvent. The dry sorbents were dried at ~100°C, except for activated alumina that was dried at ~200°C. The damp sorbents were stored in a "desiccator" with a layer of water in the bottom for several days before use. Complete saturation of sorbents with a high capacity for water may not have been achieved; later tests with activated alumina exposed for longer periods gave poorer results than those from these scouting tests. Some sorbents required preparation; these procedures were as follows:

Preparation of solid sorbents - Literature information (2) indicates that DBP can be removed from nitric acid solutions containing uranium by extracting the DBP with 2-ethylhexanol and that the extraction occurs by hydrogen bonding of the HDBP to the hydroxyl of the alcohol. The extraction is most effective when the nitric acid concentration is greater than 3.5 M. Sucrose and cellulose, compounds with numerous hydroxyl groups, were tested for their ability to remove DBP. Sucrose was deposited on 12—28 mesh silica gel by dissolving 100 mg of sucrose in 3 mL of water, soaking the silica gel in the solution, draining the solution, and drying the solid. The sucrose-coated silica gel was tested with a solvent of low acidity and with a solvent equilibrated with 1 or 4 M nitric acid. Cellulose (adjustable pipette filters) was tested with a solvent of low acidity and a solvent equilibrated with 4 M nitric acid.

It is known that iron forms a relatively strong DBP complex. A cation exchange resin (BIO-RAD AG MP-50) was loaded with ferric iron, the resin was treated with ammonium hydroxide to precipitate the iron, and the resin was washed with water to remove the excess ammonia.

Diethylenetriamine pentaacetic acid (DTPA) (0.216 g) was dissolved in 10 mL of water by increasing the pH to ~7 to increase the solubility; 1 mL of dried Dowex 21K anion resin in the hydroxyl form was then added, soaked overnight, and then drained. The pH of the solution increased to 7.70 after contact with the resin.

Zirconium oxynitrate (0.5 g) dissolved in 5 mL of water was contacted with 2 g of Norton Z-900 molecular sieve over the weekend; the sieve was drained, and rinsed with 5 mL of water.

The anion exchange resins, which were obtained in the chloride form, were converted to hydroxide form by passing 1 M sodium hydroxide through a packed column of the resin until a negative chloride result was obtained by a silver nitrate test. The resin was then rinsed with water until the effluent water was near neutral. The nitrate form was produced by passing 1 M sodium nitrate through a bed of hydroxide-form resin, followed by rinsing with water. Dowex 21K anion exchange resin in the carbonate/bicarbonate form was prepared by suspending the hydroxide-form resin in water and sparging with carbon dioxide for about 1 h. The pH of the water was decreased from an initial reading of 8 to 9 to about 7 at the end of the sparging.

The cation exchange resins were received in the acid form and were used as-received in most tests.

Scouting test procedure - Either 1 g (most solids) or 1 mL (ion exchange resins) of solid was contacted with 10 mL of the appropriate solvent (dry solvent with dry sorbent, wet solvent with damp sorbent) by shaking for 1 min three times over a 10-min period. The transfer of uranium color from the organic phase was noted. A portion of the organic phase was stripped with water; if the DBP had been removed leaving the uranium in the organic phase, the uranium would strip with water. In those cases where the uranium was quantitatively removed by contact with the solid sorbent, the colorless solvent was contacted with a uranyl nitrate solution in water. The presence of HDBP in the solvent causes extraction of uranium.

Solid sorbents tested - The solid sorbents examined were Dowex 50W-X4 [50—100 mesh, acid form (Bio-Rad Laboratories, Richmond, Calif.)], Duolite CS-100 [20—50 mesh, acid form (Diamond Shamrock Chemical Co., Cleveland, Ohio)], Dowex 21K anion exchange resin [50—100 mesh, hydroxyl form (Bio-Rad Laboratories, Richmond, Calif.)], AG MP-1 anion exchange resin [20—50 mesh hydroxyl form [(Bio-Rad Laboratories, Richmond, Calif.)], Amberlyst A-26 anion exchange resin [14—50 mesh, hydroxyl form (Rohm and Haas Co., Philadelphia, Penn.)], Dowex 1-X4 anion exchange resin [50—100 mesh, chloride form (Bio-Rad Laboratories, Richmond, Calif.)], Amberlyst A-26 anion exchange resin in the chloride form, Dowex 21K anion exchange resin in the chloride form, AG MP-1 anion exchange resin in the nitrate form, Dowex 21K coated with DTPA (prepared by and obtained from D. O. Campbell and S. R. Buxton, Oak Ridge National Laboratory), Norton Z-900 zeolite in the sodium form [0.16-cm-diam. by ~0.6-cm-long cylinders; pore size, ~7 Å (Norton Co., Worcester, Mass.)], zirconium loaded on Norton Z-900 molecular sieve, silica gel [12—28 mesh, grade 408 (Fisher Scientific Co., Fairlawn, N.J.)], sucrose on the same silica gel, Florex attapulgite clay [Florex AA-LVM, 45—60 mesh (Floridin Co., Pittsburgh, Penn.)], hydrous zirconium oxide [100—200 mesh (formerly available from Bio-Rad Laboratories, Richmond, Calif.)], activated alumina [Alcoa F-1 type, 60—120 mesh (Aluminum Co. of America, Pittsburgh, Penn.)], and cellulose from automatic pipette filters.

Results of scouting tests - The only solids that showed promise for the removal of uranium and/or DBP were the Dowex cation exchange resins in the acid form or loaded with iron, anion exchange resins in the hydroxyl or carbonate/bicarbonate forms, and activated alumina. The ion exchange resins must be used damp. The anion exchange resins removed both uranium and DBP, while the cation exchange resin removed only the uranium. The cation resin loaded with iron had less capacity than the as-received resin; it appears that the iron was not effective, but occupied some of the sorption sites. Note that the removal of the DBP after the removal of the uranium and the recovery of the uranium from the cation resin should be relatively easy. The scouting tests indicated that activated alumina could be used either dry or damp, but later column tests showed that the capacity for both uranium and DBP are reduced by the presence of water.

The promising solid sorbents were then tested in small packed columns.

Analysis of Solvents for Uranium and DBP

Uranium in the solvent was analyzed by stripping the uranium (plus any DBP and nitric acid) from the solvent with an equal or greater volume of 0.5 M ammonium carbonate. This stripping produced the uranyl carbonate complex in the aqueous phase, which is determined spectrophotometrically using the absorbance at 450.4 nm. The absorbance was found to be linear, with concentrations up to 0.01 M uranium. The molar concentration in the aqueous phase is given by the absorbance (1-cm path length) divided by 25.1. The molar absorptivity is determined from uranium standards in the same spectrophotometer used for the analyses (Model 200, Hitachi Co., Tokyo, Japan). Under most conditions, uranium forms a 1:1 complex with DBP in TBP solutions (3). The fact that a 0.01 M HDBP solution extracted only about 0.01 M uranium in the current tests is in agreement with this. Extraction of uranium by TBP and DBP-TBP solutions was investigated to develop a simple method for analyzing solvents for DBP. A 0.02 M aqueous uranyl nitrate solution with a pH of ~3 was contacted with an equal volume of clean 30% TBP. The uranium distribution coefficient was 0.0526. Uranium distribution coefficients were then determined for organic phases containing 0.001, 0.0025, 0.005 and 0.01 M DBP. Assuming the uranium-DBP complex was 1:1, the free DBP was calculated for each case. Figure 1 shows a plot of the distribution of uranium to the DBP complex (extracted uranium corrected for extraction by TBP; distribution coefficient of 0.0526) vs the free DBP concentration. The plot is approximately linear, with a slope of 314. An unknown solvent can be contacted with a standard uranyl nitrate solution, and the extracted uranium can be determined; the correlation given in Fig. 1, plus the extraction of uranium by TBP, allows the calculation of the DBP content. When the organic phase itself initially contains uranium, its concentration must be determined to know the total uranium in the system. To determine both the uranium and DBP in a 30% TBP sample, the solution is divided into two parts. The uranium in one portion is determined by stripping with ammonium carbonate solution and determining the absorbance. The other portion is contacted with an equal volume of the standard 0.02 M uranyl nitrate solution. This solution is stripped with ammonium carbonate, and the uranium is determined by absorbance. The DBP concentration is then determined by the following BASIC program:

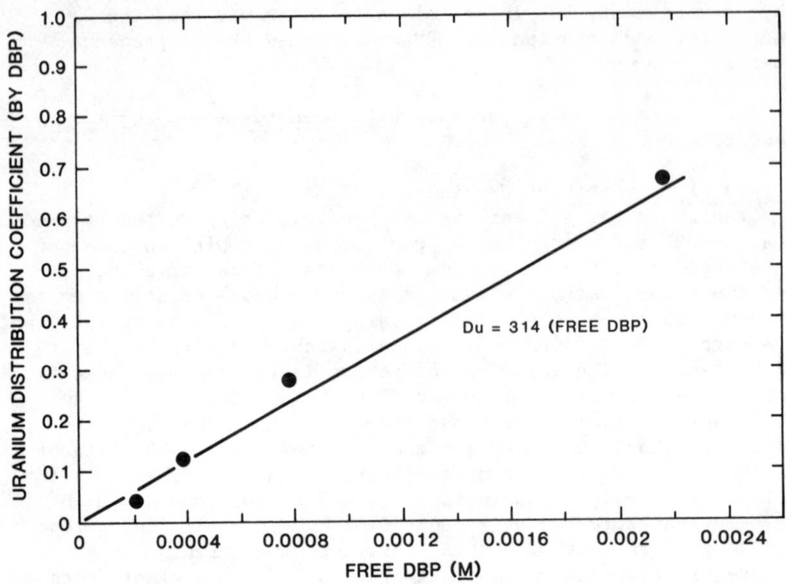

Figure 1. Uranium extraction by HDBP. The quantity of uranium extracted was corrected by substraction of uranium extracted by TBP. Free DBP is DBP not complexed with uranium.

```
10  INPUT "WHAT IS THE ABSORBANCE OF THE FIRST CARBONATE
    STRIP";A1

20  INPUT "WHAT IS THE ABSORBANCE OF THE CARBONATE STRIP
    AFTER URANIUM EQUILIBRATION";A2

30  UT=0.0526*(0.02+A1/25.1)

40  CD=(A2/25.1-UT)/(0.02+A1/25.1-A2/25.1)/314-UT+A2/25.1

50  PRINT "THE DBP CONC. IS ";CD;" MOLAR"

60  PRINT:PRINT:

70  GOTO 10
```

Note that this analytical method may only be used where the nitric acid and nitrates other than uranyl nitrate are near zero.

<u>Small-Column Tests</u>

The column used in the tests is shown in Figure 2. The bed was contained in an 8-mm-OD, 6-mm-ID glass tube with a jack-leg and could be immersed in a beaker of heated water for elevated

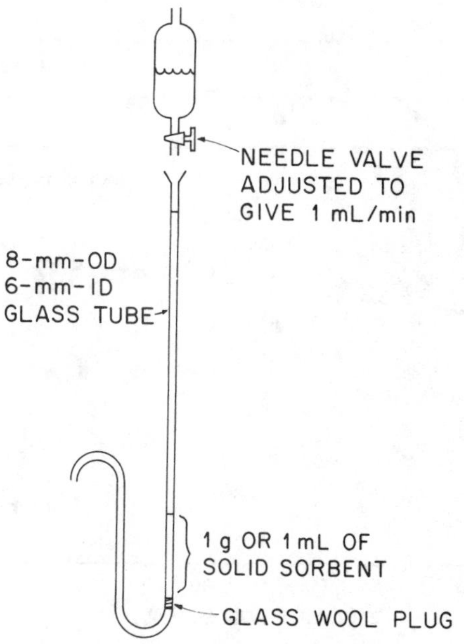

Figure 2. Apparatus used in small-bed tests.

temperature tests. In all cases, the solvent passed through the beds was intended to be ~0.01 M in both uranium and DBP (in one test the DBP concentration was found to be ~0.015 M in the feed) and was prepared as described for the scouting tests.

Loading 20–50 mesh Bio-Rad AG MP-50 - Small-column tests used 1 mL of 20–50 mesh BIO-RAD AG MP-50 macroporous cation resin (acid form) to treat ~1 mL/min of solvent containing ~0.01 M of uranium and DBP. Tests were conducted at 22° (duplicates) and 52°C. Figure 3 shows effluent concentrations of the two loading tests at 22°C. Significant removal of uranium and no detectable removal of DBP were found. Figure 4 shows the uranium and DBP concentrations in the column effluent in the test at 52°C; again, uranium loaded without DBP loading. Uranium removal half-times were ~20 s at 52°C and ~40 s at 22°C. The apparent capacity of the resin was ~0.35 mmol of uranium per mL of resin at 22°C and 0.56 mmol of uranium per mL of resin at 52°C (stated capacity is 1.86 meq/mL). A later test repeated the loading of the cation exchange resin at 50°C (Figure 5). In this case the volume of sorbent was decreased to 0.75 mL, and the flow rate was decreased to 0.5 mL/min. This decrease in volume and flow rate gives an increase of 50% in the residence time of the solvent in the sorbent bed. DBP was not removed. The uranium removal efficiency was significantly improved over that in the earlier tests — the half-time was reduced to ~10 s, and the apparent capacity of the bed

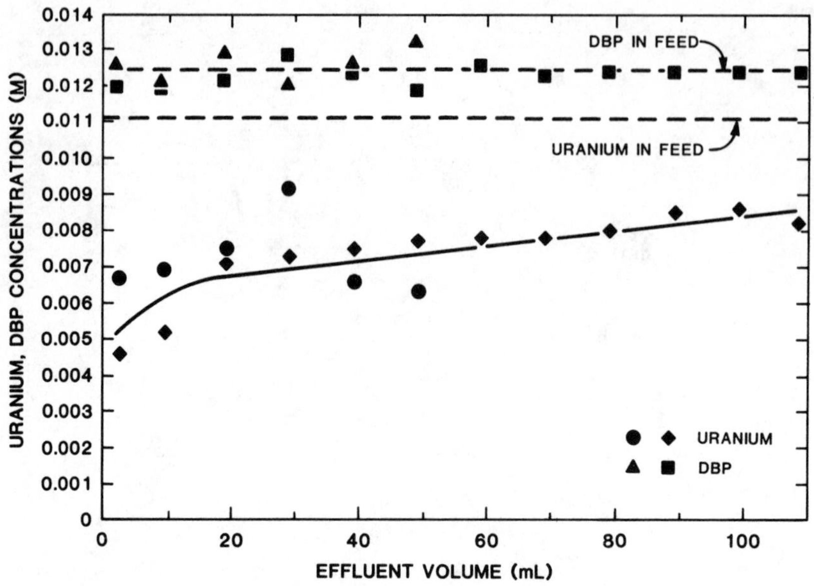

Figure 3. Effluent uranium and DBP concentrations from duplicate test loadings of 1 mL of Dowex 50W-X4 cation exchange resin (acid form) at 22°C. The flow rates of the solvent were 1 mL/min.

was increased to ~1 mmol of uranium per mL of resin; this value is about the theoretical capacity of the resin. In a larger bed where the residence times are larger relative to two half-times for removal, the sensitivity to flow rate may be less.

Eluting 20–50 mesh Bio-Rad AG MP-50 - Elution of the uranium loaded in the 52°C test was approximately 50% complete by the passage of 80 mL of 30% TBP containing 0.06 M nitric acid at a flow rate of 1 mL/min at 22°C (Figure 6). Uranium elution at 22°C from loaded cation resin was tested in duplicate using 30% TBP which had been equilibrated with 2 M nitric acid (0.36 M nitric acid in the 30% TBP) (Figure 7). Essentially all the mobile uranium had been removed with the passage of 80 mL of solvent. The increased acidity of the solvent significantly increased the removal rate; a further increase by operating at elevated temperature should also be possible.

Loading 60–120 mesh Alcoa F-1 activated alumina - Columns containing 1 mL of damp (equilibrated with water-saturated air for about 6 weeks) and 1 mL of dry 60–120 mesh Alcoa F-1 activated alumina were tested at 22°C for the removal of uranium and DBP from damp and dry solvent (dried by sparging with air at 60°C),

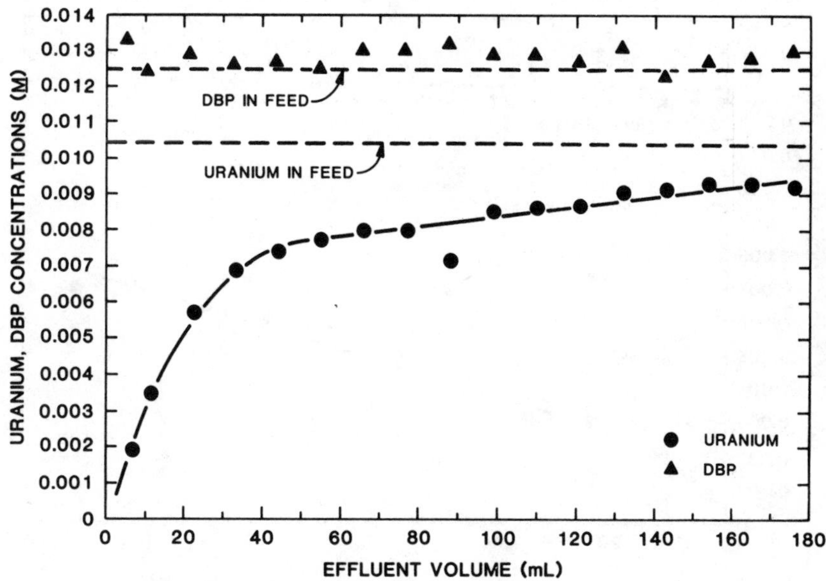

Figure 4. Effluent uranium and DBP concentrations from a test loading of 1 mL of Dowex 50W-X4 cation exchange resin (acid form) at 52°C. The flow rate of the solvent was 1 mL/min.

respectively. The column capacities were more than twice as great with dry alumina and dry solvent as compared with damp solvent and damp alumina. Figure 8 shows the cumulative loadings of uranium and DBP for the dry sorbent and solvent test. Both uranium and DBP were loaded in approximately the same ratio as their ratio in the solvent (solvent ratio, DBP/uranium = 1.17; sorbent ratio = 1.24).

Eluting 60–120 mesh Alcoa F-1 activated alumina - Uranium was eluted from loaded activated alumina using 30% TBP solvent containing 0.36 M nitric acid. These results are shown in Figure 9. The uranium was more easily eluted than from the cation exchange resin, with an initial concentration in the solvent about twice that found when eluting the cation exchange resin. The volume required to elute the resin was also reduced by about one-half.

Loading a column containing DTPA on 3-mm glass beads - A column of 3-mm glass beads coated with ~0.2 g of DTPA (pH adjusted to ~7) was tested for removal of uranium and DBP from the solvent. At about 40 mL of effluent, the temperature of the column was increased from 22° to 50°C. At the lower temperature the loading of uranium and DBP was minimal; at the

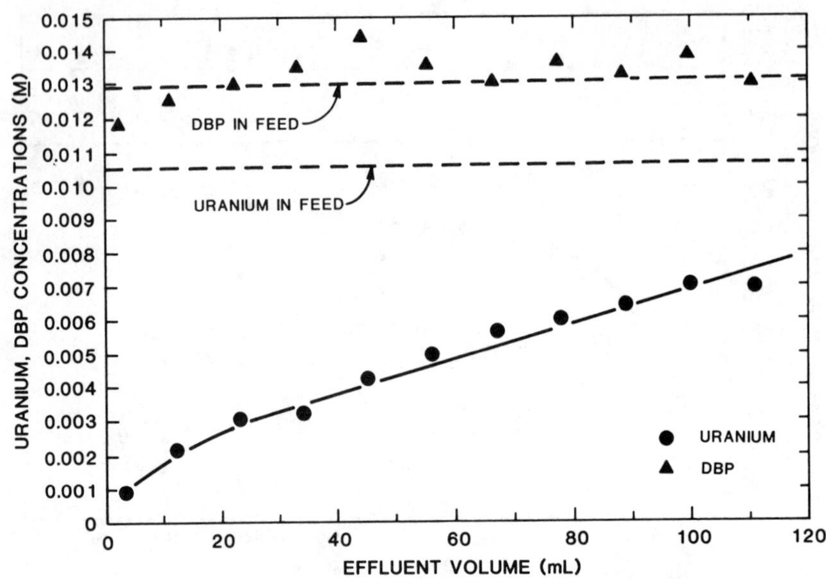

Figure 5. Effluent uranium and DBP concentrations from a test loading of 0.75 mL of Dowex 50W-X4 cation exchange resin (acid form) at 50°C. The flow rate of the solvent was 0.5 mL/min.

higher temperature the loading of uranium was significantly increased. This material appears to be usable for preferential removal of uranium from the uranium-DBP mixture. Removal of the loaded DTPA from the column is very simple since it is water-soluble. Passage of a small amount of water (10–20 column volumes) through the column removed all the coating, with its uranium, from the column.

Loading Dowex 21K (hydroxide form) - A solvent containing ~0.01 M uranium and ~0.015 M DBP was used to load a hydroxide-form anion resin. Both materials were loaded in about the same ratio as their ratio in the solvent.

Separation of uranium and DBP during elution - It may be possible to separately elute uranium and DBP from a sorbent that loads both components such as hydroxide-form anion exchange resins or activated alumina. HDBP can be extracted from acidified aqueous solutions containing both uranium and DBP by 2-ethylhexanol (2). Elution of the DBP by 2-ethylhexanol containing ~0.7 M nitric acid followed by elution of the uranium by 30% TBP containing ~0.6 M nitric acid was attempted. The uranium stayed on the hydroxide-form anion exchange resin and the activated alumina during the treatment with acidified

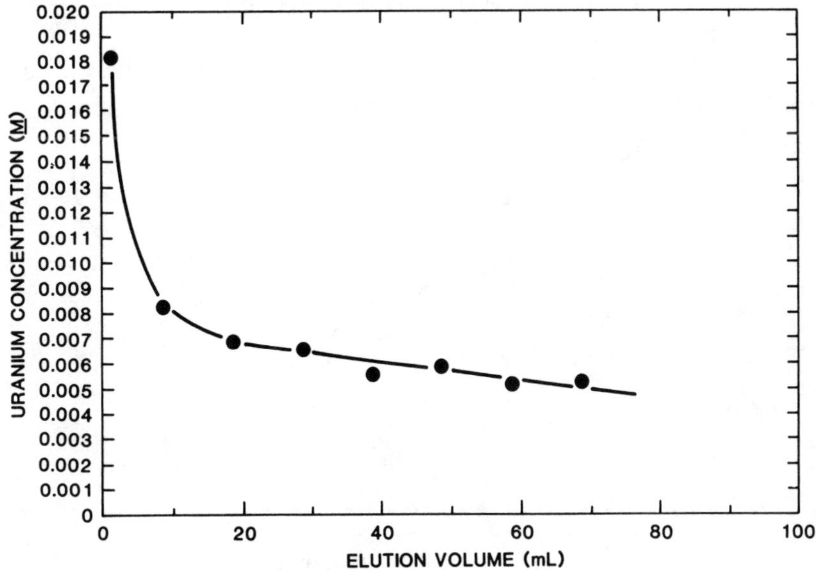

Figure 6. Elution of uranium from 1 mL of Dowex 50W-X4 cation exchange resin (acid form) by 30% TBP solvent containing 0.06 M nitric acid at 22°C. The solvent flow rate was 1 mL/min.

2-ethylhexanol and was removed by the treatment with acidified TBP. Both of these observations were visual; analyses for the components in the eluates were not successful.

POSSIBLE APPLICATIONS

The major potential use of the cation exchange system would be to prevent the transfer of valuable metals (uranium and plutonium) to the solvent cleanup system, where they are more difficult to recover. These metals will be present in the solvent at a low level because of the strong complexes formed with DBP; low-acid stripping does not remove metals below about a 1:1 mole ratio to the DBP. Uranium and plutonium are the major metals present which form strong complexes with DBP; the plutonium complex is much stronger than the uranium complex (4). If the fuel being processed had a high uranium:plutonium ratio, as in light-water reactor fuel, the quantity of plutonium held by the DBP would be quite small; however, in the case of breeder fuel the higher plutonium content would result in more retention of plutonium

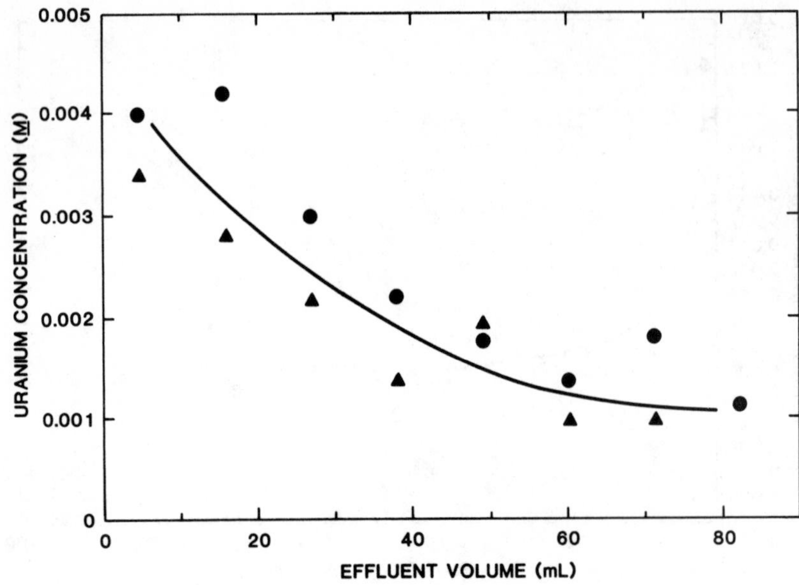

Figure 7. Duplicate tests of elution of uranium from 1 mL of Dowex 50W-X4 cation exchange resin (acid form) at 22°C with 30% TBP solvent which had been equilibrated with an equal volume of 2 M nitric acid. The flow rate of the solvent was 1 mL/min.

than uranium. It is possible that the uranium and plutonium could be removed from the solvent by a cation exchange resin from which they could be easily recovered; the actual behavior of plutonium is not known.

Scrubbing with sodium carbonate solutions has been observed to produce emulsions partly because of the presence of cations which form low solubility complexes with carbonate; these include zirconium and plutonium (5,6). Treatment of the solvent with a cation exchange resin to remove the cations that cause emulsions could significantly improve the operation of the solvent cleanup system. In fact, in the absence of cations, the solvent could be scrubbed with a basic solution without the necessity of a complexing anion such as carbonate. Weak bases such as hydroxylamine hydrate, which can be readily decomposed into innocuous gases with a significant reduction of solid wastes, would be attractive.

The two applications discussed above assume that all cations can be removed from DBP solutions by cation exchange resins. Only the removal of uranium has been demonstrated; additional experiments should be undertaken to examine the behavior of zirconium and plutonium.

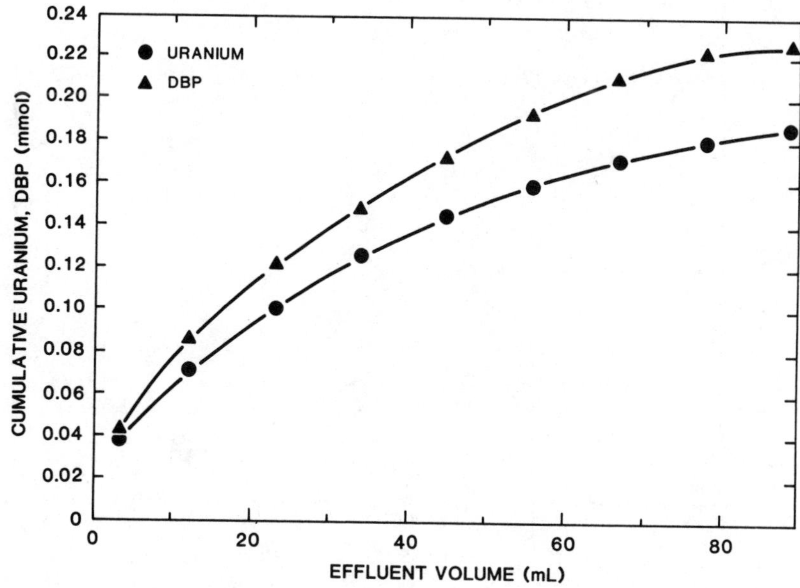

Figure 8. Cumulative loadings of uranium and DBP on 1 mL of dry, activated alumina at 22°C. The dry solvent flow rate was 1 mL/min.

CONCLUSIONS

The tests performed show that it is relatively easy to preferentially remove uranium from solvents containing uranium and DBP, but quite difficult to remove DBP preferentially. The current methods could be used by removing the uranium (as by a cation exchange resin) and then using either an anion exchange resin in the hydroxyl form or a conventional treatment with a basic solution to remove the DBP. Treatment of a solvent with a cation exchange resin could be useful for recovery of valuable metals from solvents containing DBP and might be used to remove cations before scrubbing a solvent with a basic solution to minimize emulsion formation.

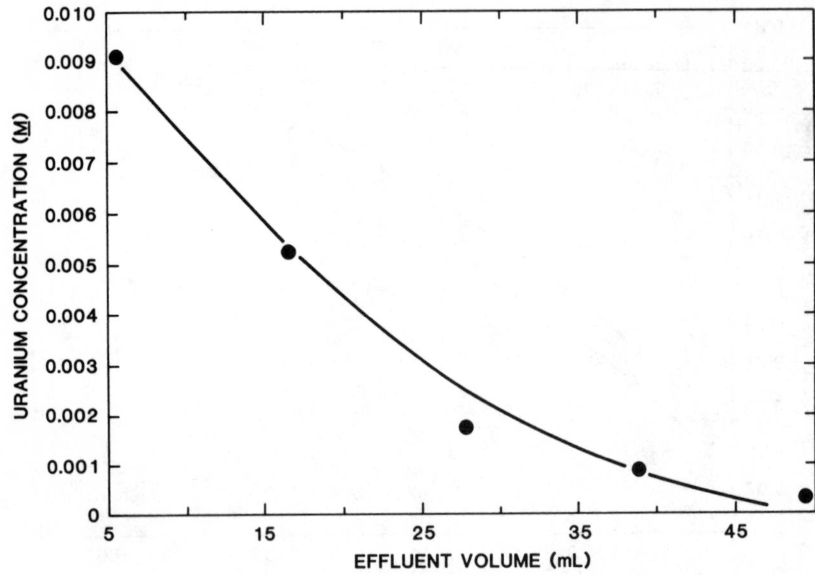

Figure 9. Uranium removal from activated alumina by 30% TBP solvent which was equilibrated with an equal volume of 2 M nitric acid. The solvent flow rate was 1 mL/min.

ACKNOWLEDGMENTS

F. I. Case (Chemistry Division, ORNL) performed all the testing and analyses reported. M. L. Baker (Development Division, Y-12 Plant) participated in planning and performance of several of the tests and in the preparation of solid sorbents. The advice of B. A. Moyer (Chemistry Division, ORNL) in the area of solvent extraction chemistry is gratefully acknowledged.

The Oak Ridge National Laboratory is operated by Martin Marietta Energy Systems, Inc., for the U. S. Department of Energy under Contract No. DE-AC05-84OR21400.

REFERENCES

1. J. C. Mailen and O. K. Tallent, Cleanup of Savannah River Plant Solvent Using Solid Sorbents, Report ORNL/TM-9256, Oak Ridge National Laboratory, Oak Ridge, Tennessee, April 1985.

2. E. P. Horwitz, G. W. Mason, C. A. A. Bloomquist, R. A. Leonard, and G. J. Bernstein, "The Extraction of DBP and MBP from Actinides: Application to the Recovery of Actinides from TBP-Sodium Carbonate Solutions," p. 475–496 in Actinide Separations, J. D. Navratil and W. W. Schulz, Eds., ACS Symp. Ser. 117, American Chemical Society, Washington, D.C. (1980).

3. H. T. Hahn and E. M. Vander Wall, "TBP Decomposition Product Behavior in Post-Extractive Operations," Nucl. Sci. Eng. 17, 613 (1963).

4. L. Stieglitz, W. Ochsenfeld, and H. Schmieder, The Influence of the Radiolysis of Tributyl Phosphate on the Plutonium Yield in the Purex Process for High Plutonium Content, Karlsruhe Nuclear Research Center, KFK 691, Karlsruhe, FRG, November 1968.

5. J. C. Mailen and O. K. Tallent, Assessment of Purex Solvent Cleanup Methods Using a Mixer-Settler System, Report ORNL/TM-9118, Oak Ridge National Laboratory, Oak Ridge, Tennessee, November 1984.

6. J. I. Kim, C. Lierse, and F. Baumgartner, "Complexation of the Plutonium(IV) Ion in Carbonate-Bicarbonate Solutions," p. 318–334 in Plutonium Chemistry, W. T. Carnall and G. R. Choppin, Eds., ACS Symp. Ser. 216, American Chemical Society, Washington, D.C. (1983).

RECOVERY OF URANIUM FROM WET-PROCESS PHOSPHORIC ACID BY EXTRACTION WITH D2EHPA-DBBP

Hsiao-Ming Chen, Haw-Jan Chen, Ying-Ming Tsai,
Te-Wei Lee and Gann Ting
Institute of Nuclear Energy Research
P.O. Box 3-6, Lung-Tan, Taiwan 32500, R.O.C.

A synergistic extractant combination consisting of di(2-ethylhexyl) phosphoric acid (D2EHPA) and dibutyl butyl phosphonate (DBBP) in kerosene is employed in a two-cycle separation process for the recovery of uranium from wet-process phosphoric acid. The addition of the sulfuric acid and water scrubbing steps for the recycled extractant eliminates the contamination and dilution of the phosphoric acid by the ammonium ion and water and also no precipitation occurs in the second cycle extraction step. The amount of ferrous iron used in the stripping unit of the first concentration cycle can be reduced to a great extent when the dissolved oxygen in the organic extractant is removed with inert gas purging. The advantages of this process are lower chemical cost, higher product purity and better phase separation in comparison with the previous process.

INTRODUCTION

Although the amount of uranium in phosphate deposits is very low, most of the uranium is dissolved and accompanies phosphoric acid during the process of producing phosphoric acid. It represents a significant potential source of uranium in case of the large amount of phosphate rock processed. Considerable research on the feasibility of recovering uranium as a by-product from wet-process phosphoric acid has been undertaken since the early 50's and several solvent extraction processes have been developed. Of the various extractant systems studied and reported in the literature, only three are of commercial interest. They are the octyl pyrophosphoric acid system (OPPA) which extracts tetravalent uranium (1,2,3,4,5), the synergistic di(2-ethylhexyl) phosphoric acid-trioctyl phosphine oxide system (D2EHPA-TOPO) which extracts hexavalent uranium(6,7,8,9,10,11) and the octylphenylphosphoric acid system (OPAP) in which tetravalent uranium is extracted (12,13,14,15,16). D2EHPA-TOPO processes based on the laboratory development at the Oak Ridge National Laboratory by Hurst et al. in the early 70's are the most promissing ones, which are further developed and operating at many commercial plants. Advantages claimed for the D2EHPA-TOPO process are that the extractant combination is stable toward high acid concentration as well as the carbonate solutions used for stripping and a high-grade product is recovered. However, the TOPO synergistic agent is expensive and hence total chemicals cost is high. Another drawback is the formation of a precipitate of some kind of phosphatic compounds in the extraction stage of the second purification cycle when the organic extractant recycles directly from the ammonium carbonate stripping stage. Owing to the conversion of di(2-ethylhexyl) phosphoric acid to a highly hydrated ammonium salt, the extracted water and NH_4^+ transfer to the aqueous phase, dilute and contaminate the wet-process phosphoric acid, which is unacceptable to the phosphoric acid plant to produce feed stocks and various phosphate chemicals. These in turn cause unclear phase separtion and loss in overall efficiency of operation.

The synergistic enhancement of uranium extraction by addition of neutral organophosphorus compounds to D2EHPA has been studied by Murthy et al.(17) and Blake et al.(18). It was shown that TOPO gave higher synergistic effect than dibutyl butyl phosphonate (DBBP) did, but in view of the higher cost and unavailability of TOPO in commercial quantities DBBP seems to be a good substitute for TOPO. An improved two-cycle separation process using the extractant combination D2EHPA-DBBP has been developed in our research group(19). Demonstration of a complete process flow sheet showed that the drawbacks mentioned above could be solved by subjecting the stripped organic extractant to sulfuric acid treatment, followed by a water scrubbing operation. Ting et al. (19) claimed that the reductive strip solution used for concentrating the uranium in the first cycle contains about 25~50 g/ℓ of ferrous iron, which is apparently far beyond the expected stoichiometric amount. The presence of dissolved oxygen in the organic extractant was found to result in excessive oxidation and consumption of ferrous iron when the extractant was fed into the stripping operation and this thus accounts for the excessive consumption of ferrous iron(20).

In the present work, factors affecting the uranium recovery from wet-process phosphoric acid by extraction with D2EHPA-DBBP in kerosene were extensively examined first, then a demonstration test for the two-cycle separation process in which the dissolved oxygen in the organic extractant was removed with inert gas purging was further made to verify the improvement of the process efficiency.

EXPERIMENTAL

Phosphoric Acid Samples

Different batches of wet-process phosphoric acid samples (green acid) were obtained from one commercial phosphate plant located at the south of Taiwan for testing. The acid samples were several weeks old when received. Typical values of some of the constituents were shown in Table 1. These samples had phosphate concentration ranging from 4 to 5 M and contained 0.06~0.07 grams of uranium per liter. For maximum efficiency in extraction of uranium with D2EHPA-DBBP, the uranium should be in the hexavalent state. If necessary, oxidation of uranium to this form was accomplished by adding suitable amounts of hydrogen peroxide to the acid liquor.

Table 1. Major compositions of wet-process phosphoric acid.

Composition		Concentration	
	Sample 1	Sample 2	Sample 3
H_3PO_4 (M)	4.8	4.3	4.8
F (M)	1.24	2.24	1.24
U (g/ℓ)	0.067	0.062	0.072
Fe (g/ℓ)	1.22	1.98	1.01
Ti (g/ℓ)	0.035	0.040	0.032
Ca (g/ℓ)	4.56	2.08	3.96
Mg (g/ℓ)	1.24	0.76	1.98
Al (g/ℓ)	0.76	1.82	0.89
Si (g/ℓ)	5.65	5.60	3.85
Mo (g/ℓ)	0.008	0.017	0.013
V (g/ℓ)	0.065	0.070	0.073
SO_4 (g/ℓ)	5.4	6.0	6.4
Cl (g/ℓ)	0.41	0.33	0.47
R.E. (g/ℓ)	0.31	0.21	0.26

Process Flow Sheet

The proposed process consists of two cycles, which are the same as described previously(19) except that the recycled extractant in the first cycle was sparged with nitrogen to remove dissolved oxygen before entering the stripping section.

Extraction of Uranium

Of the various diluents tested, kerosene, being cheaper and commercially available, was selected as the only diluent for D2EHPA-DBBP extractant in all of the subsequent work. It was established in batch shakeouts that equilibrium for extraction of uranium was attained in 5 min (Figure 1) and phase separation time was about 3 to 5 min. As reported previously by Murthy et al.(17) and Blake et al.(18), addition of DBBP to the D2EHPA extractant enhances its uranium extraction coefficient. The data of Figure 2 and Figure 3 indicate that maximum uranium extraction occurs when the D2EHPA : DBBP mole ratio is approximately 5 : 1. The uranium extraction coefficients decrease with increasing phosphoric acid concentration. Extraction in 4M phosphoric acid at 25°C with 0.5M D2EHPA-0.1M DBBP in kerosene, about 90% of the uranium was recovered. However, uranium recoveries were only 60% and 40%, respectively, as shown in Figure 4 when the acid liquor was concentrated up to 6M and 7M.

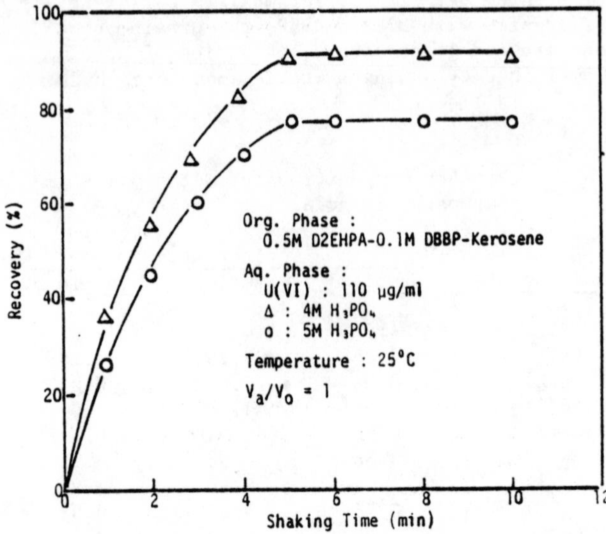

Fig. 1. Uranium recovery as a function of shaking time.

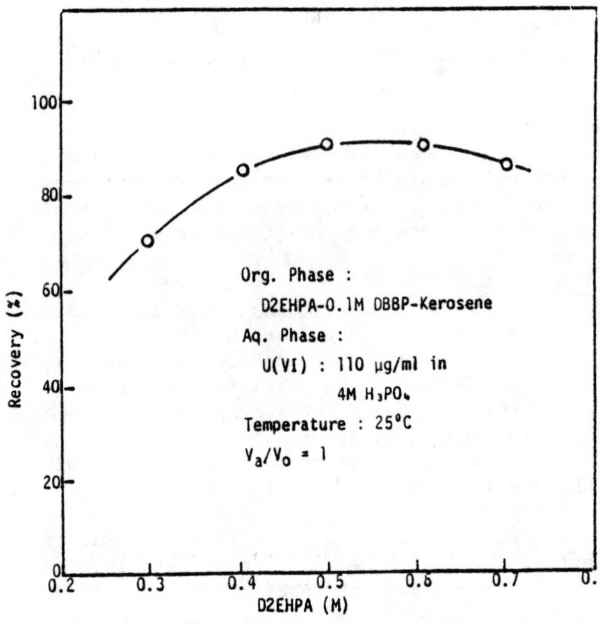

Fig. 2. Effect of D2EHPA:DBBP ratio on uranium(VI) extraction.

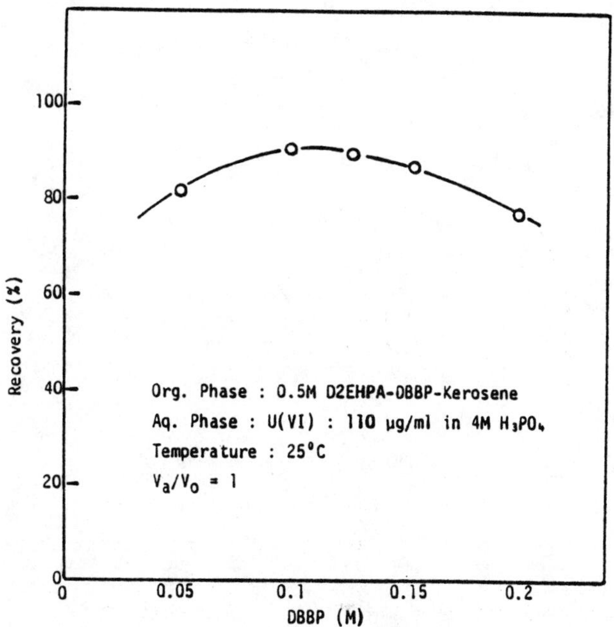

Fig. 3. Synergistic effect of DBBP on uranium(VI) extraction.

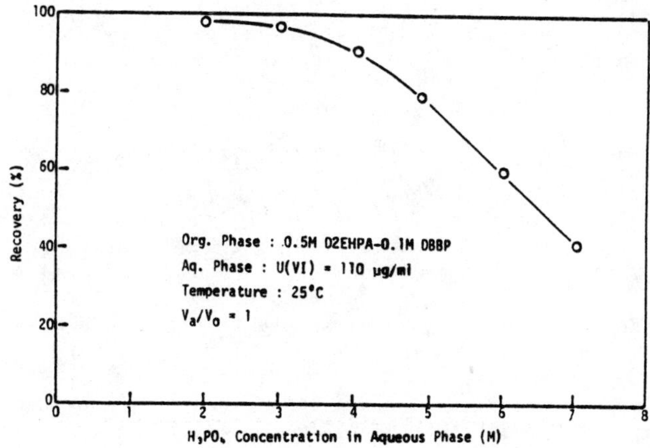

Fig. 4. Effect of H_3PO_4 concentration on uranium(VI) extraction with 0.5M D2EHPA-0.1M DBBP-kerosene.

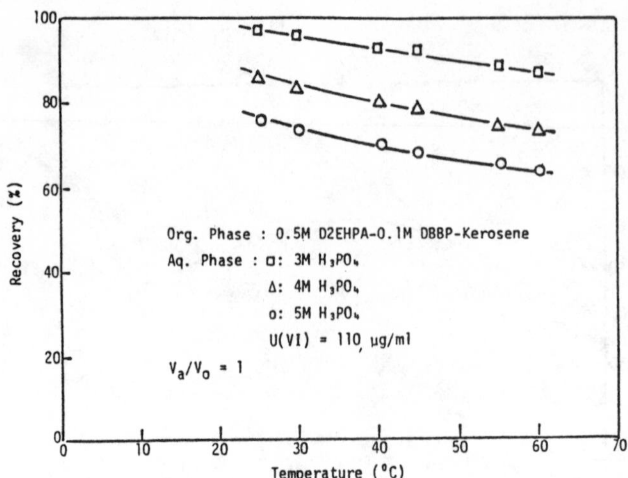

Fig. 5. Effect of temperature on uranium(VI) extraction with 0.5M D2EHPA-0.1M DBBP-kerosene.

The effect of temperature on uranium extraction is shown in Figure 5. Uranium recovery decreases as the temperature of the acid is increased. It is indicated that even at 60°C, the expected temperature of the wet-process phosphoric acid obtained from the plant, about 65% of uranium was recovered from acid having phosphate concentration of 5M.

Figure 6 shows the isotherms for extraction of uranium from 4M phosphoric acid and 5M phosphoric acid with 0.5M D2EHPA-0.1M DBBP at 25°C. McCabe-Thiele diagrams indicate that 99% of the uranium could be recovered from 4M phosphoric acid and 5M phosphoric acid, respectively, in five and thirteen ideal extraction stages by assuming that an aqueous/organic flow ratio of 4. These results are coincident with the predicated values by Fenske Equation.

Stripping of Uranium

The effect of the phosphate concentration of acid raffinate on uranium stripping is shown in Figure 7 from which it can be seen that the efficiency of reductive stripping of uranium decreases rapidly as the phosphoric acid concentration is decreased to below 4M. Further studies of the effect of Fe(II) concentration in acid raffinate indicate that there is only a slight increase in the uranium stripping efficiency when the Fe(II) concentration increases to more than 10 grams of Fe(II) per liter (Figure 8). Temperature effect has been examined with acid raffinate of different phosphate concentration at variant temperatures. It is suggested from the results shown in Figure 9 that operation at a temperature of 50°C or higher is needed to ensure a sufficient fraction of the U(IV) to be transferred to the aqueous phase.

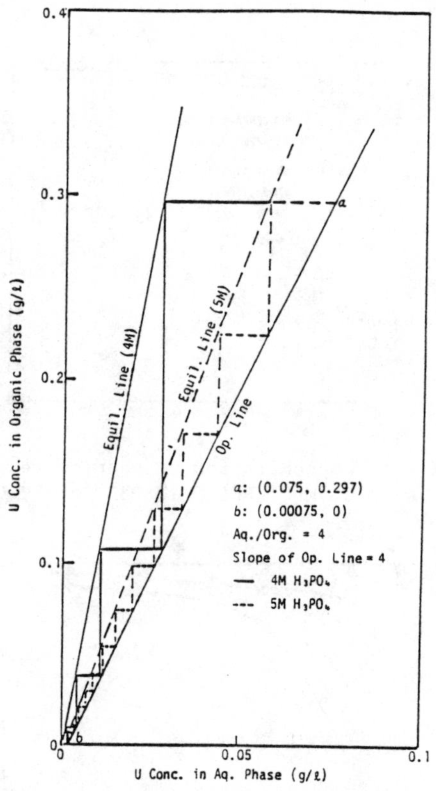

Fig. 6. McCabe-Thiele diagram of uranium(VI) extraction with 0.5M D2EHPA-0.1M DBBP at 25°C.

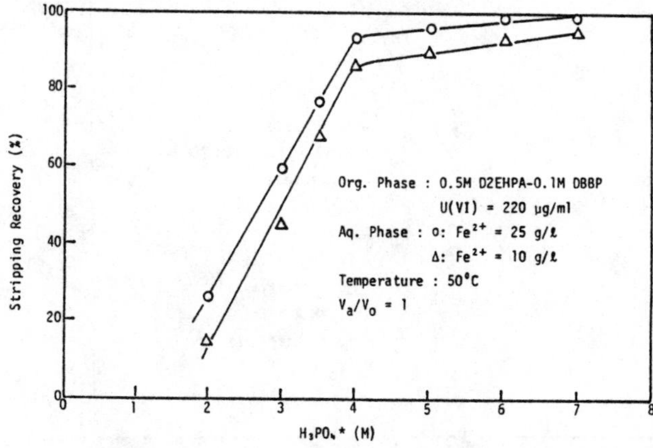

Fig. 7. Effect of H_3PO_4 concentration on uranium reductive stripping with 0.5M D2EHPA-0.1M DBBP at 50°C.
[*: H_3PO_4: extracted wet process phosphoric acid raffinate (U(VI) conc. less than 0.1ppm)]

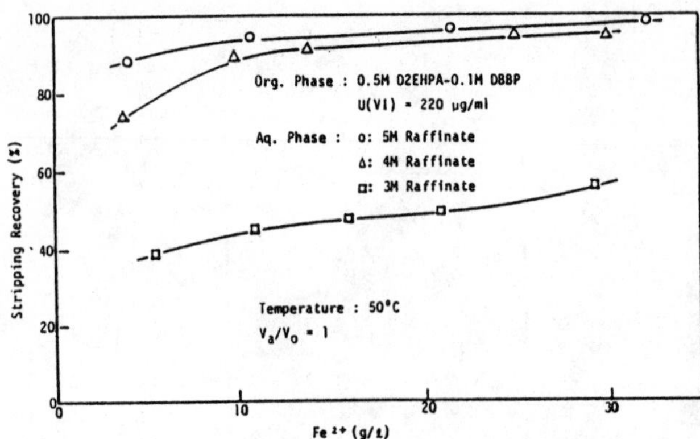

Fig. 8. Effect of Fe^{2+} concentration on uranium reductive stripping with 0.5M D2EHPA-0.1M DBBP at 50°C.

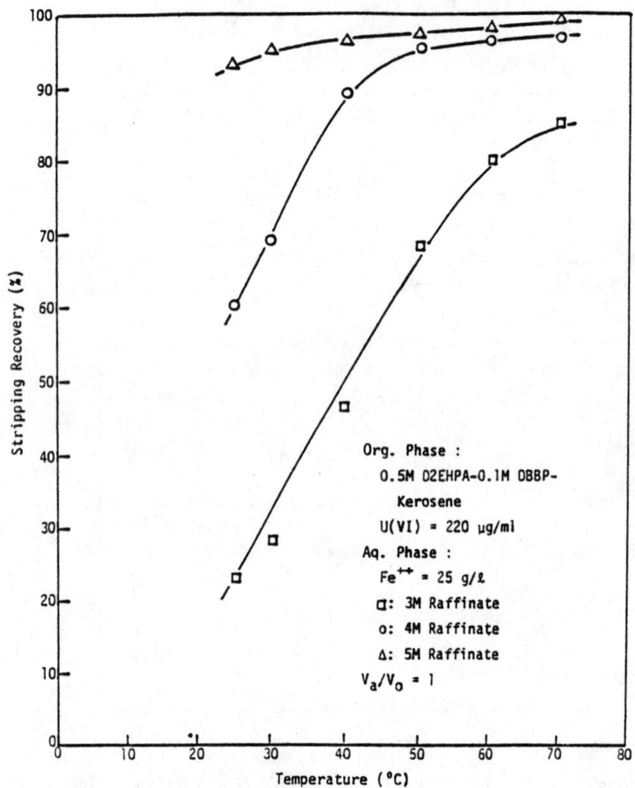

Fig. 9. Effect of temperature on the reductive stripping of uranium for D_2D process.

Carbonate Stripping

Uranium is effectively stripped from D2EHPA-DBBP by contact with ammonium carbonate solutions. A series of batch experiment indicated that operation with relatively concentrated ammonium carbonate solution of about 2~2.5 M resulted in direct precipitation of ammonium uranyl tricarbonate (AUT) in the stripping system. Calcination of the AUT precipitate yields U_3O_8 product.

Sulfuric Acid Treatment and Water Scrubbing of the Recycled Extractant

It was found in a bench-scale, continuous mixer-settler test and pilot plant test run that some kind of precipitated solid formed if the organic extractant D2EHPA-DBBP was recycled to the extraction steps directly from the ammonium carbonate stripping step. Analyses of the precipitated solid showed that its composition seemed primarily to be $Fe_3NH_4H_8(PO_4)_6 \cdot 6H_2O$. Lower efficiency of the extraction operations result from the fact that the precipitate formed has to be removed to prevent phase separation problems and scaling and clogging of the extraction equipment.

It was also found that the drawbacks mentioned above could be solved by subjecting the stripped organic extractant to sulfuric acid treatment, followed by a water scrubbing operation. The concentration of the feed sulfuric acid solution in a two-stage scrubbing section is preferably 6N to 16N and the sulfuric acid solution is recycled between the two stages to reduce the amount of the resulting liquid waste. The volumetric flow ratio of organic solvent to sulfuric acid solution is typically about 2/1 to 5/1 and operation temperature of 50~60 °C is preferred for efficient phase separation. The scrubbing efficiency is less then 90% when the H^+ concentration of the recycled sulfuric acid solution gradually decreased to below 0.6N. In process operation the sulfuric acid scrubbing section should be designed so that it can be by-passed and replaced with fresh solution when necessary. After sulfuric acid treatment, organic extractant contains about 10ppm of $SO_4^=$. The organic extractant is free of $SO_4^=$ when scrubbed with water. No precipitate such as $Fe_3NH_4H_8(PO_4)_6 \cdot 6H_2O$ is found after sulfuric acid treatment and water scrubbing are employed to remove the ammonia and water in the organic extractant. Also, loss of P_2O_5, contamination and dilution of wet-process phosphoric acid are also eliminated.

Process Demonstrations

We have performed a continuous test run of the D2EHPA-DBBP process in bench-scale mixer-settler units before and reported the material balance results for the two cycles to illustrate the advantages of the suggested process(19). The amount of ferrous iron used in the reductive strip solution was claimed to be 25~50 g/ℓ, which is apparently far beyond the expected stoichiometric amount. Hurst et al.(20) discovered that the presence of dissolved oxygen in the D2EHPA-TOPO extract led to an excessive consumption of ferrous iron in the stripping operation of the first cycle in D2EHPA-TOPO process.

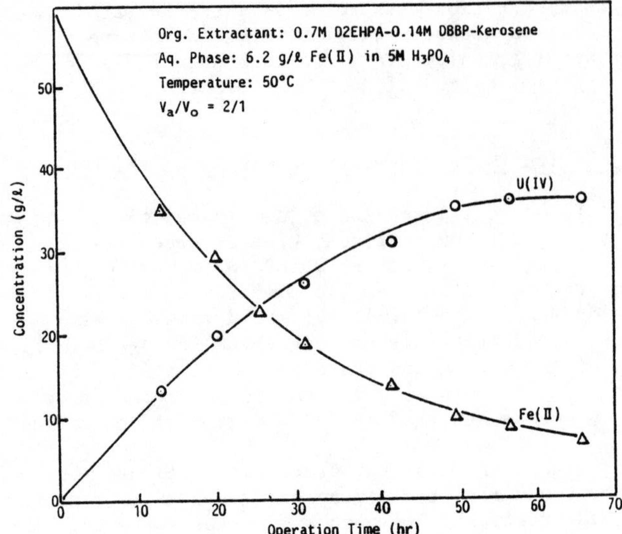

Fig. 10. The variation of U(IV) concentration and Fe(II) concentration as a function of the operation time. Dissolved oxygen in the organic extractant was not removed.

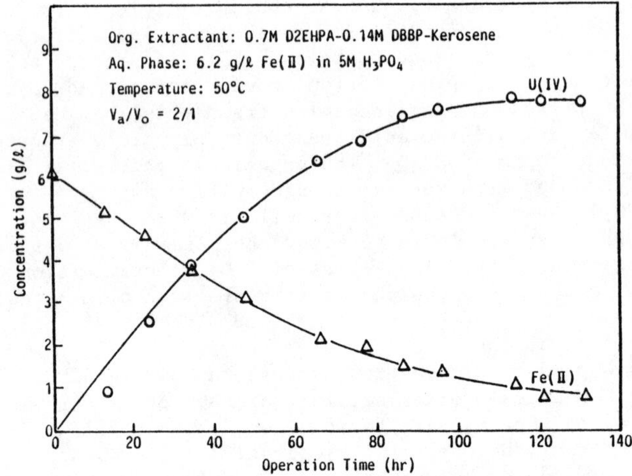

Fig. 11. The variation of U(IV) concentration and Fe(II) concentration as a function of the operation time. Dissolved oxygen in the organic extractant was removed with inert gas purging.

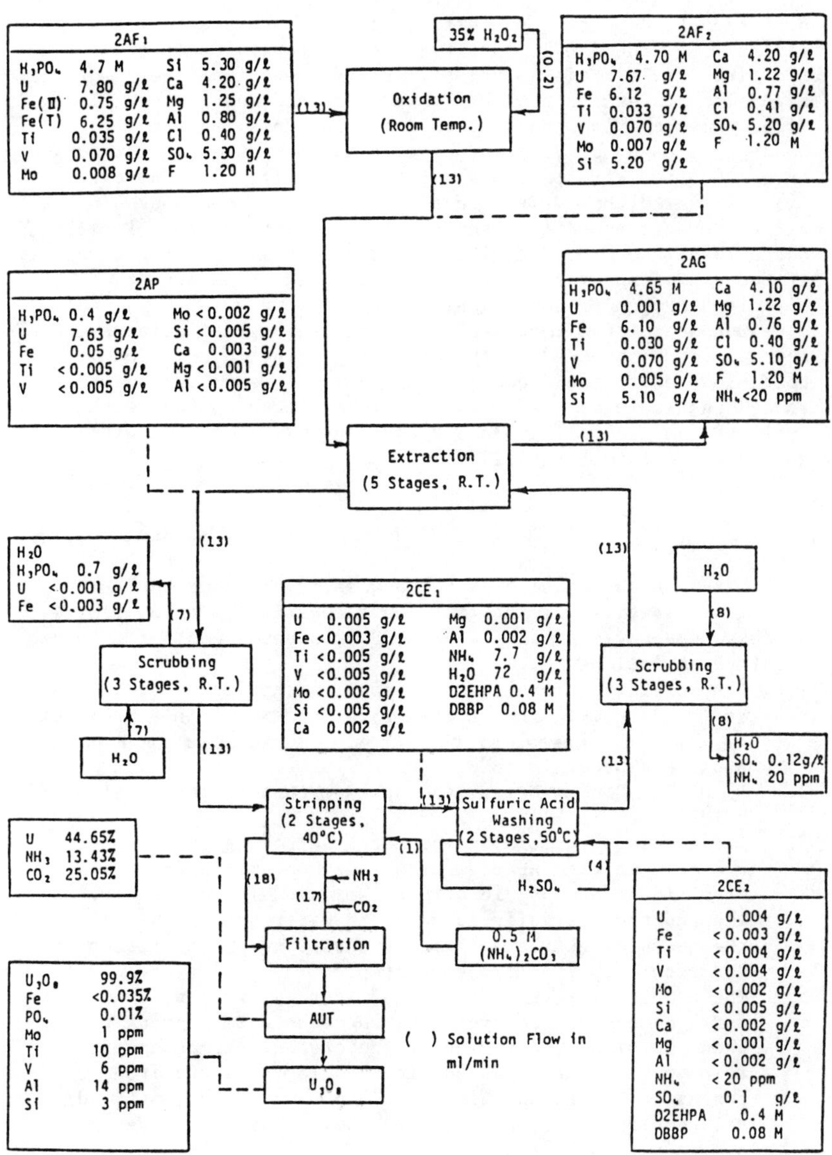

Fig. 12. Material balance diagram for the second purification cycle.

A continuous test run, in which the extractant was sparged with nitrogen before entering the stripping section, was further performed to determine the effect of the dissolved oxygen on the variation of the U(IV) concentration and Fe(II) concentration in the first cycle strip solution. It was shown from the experimental results that U(IV) was concentrated to about 3.5 g/ℓ after the operation had continued for 50 hours and the concentration of U(IV) never went up any more when the dissolved oxygen in the extractant was not removed (Fig. 10). However, Fig. 11 showed that U(IV) concentration was up to about 5.0 g/ℓ after 50 hours operation and it reached a steady state in 115 hours to give a product solution containing about 7.8 g/ℓ uranium if the extractant was sparged with nitrogen to remove the dissolved oxygen in the organic phase.

Because of the improvement of the process efficiency, less ferrous iron is needed in the stripping solution and iron contamination to the product is minimized. Demonstration test showed that calcination of the AUT precipitate in the second cycle yielded a final product containing 99.9% of U_3O_8 with less than 0.035% of Fe (Fig. 12).

CONCLUSION

Based on the results of batch-wise, continuous bench-scale and pilot plant experiments for the recovery of uranium from wet-process phosphoric acid by D2EHPA-DBBP, several conclusions are obtained as follows:

1. A synergistic extractant combination consisting of D2EHPA and DBBP in kerosene is effective and useful in a two-cycle separation process for the recovery of uranium from wet-process phosphoric acid.
2. In D2EHPA-DBBP process, the optimum concentration ratio of D2EHPA/DBBP is 5/1.
3. The novel sulfuric acid and water scrubbing steps for the recycled extractant eliminate the contamination and dilution of the phosphoric acid by ammonium ion and water.
4. The addition of sulfuric acid and water scrubbing steps for the recycled extractant also produces no precipitation in the second cycle extraction step.
5. The amount of ferrous iron used in the stripping unit of the first concentration cycle can be reduced to a great extent when the dissolved oxygen in the organic extractant is removed with inert gas purging. A lower ferric iron concentration in the first-cycle product solution (feed solution for the second-cycle) thus obtained can result in a more pure U_3O_8 product.
6. The advantages of the D2EHPA-DBBP process are lower chemical cost, higher product purity and better phase separation in comparison with the previous process.

REFERENCES

1. C. S. Cronan, Chem. Eng., 66(9), 108 (1959).

2. D. A. Ellis, The Recovery of Uranium from Industrial Phosphoric Acids by Solvent Extraction, Report DOW-81, Dow Chemical Company, Pittsburg, California, 1952.

3. B. F. Greek, O. W. Aller, and D. E. Tynan, Ind. Eng. Chem., 49, 628 (1957).

4. R. S. Long, D. A. Ellis, and R. H. Bailes, in Proc. Int. Conf. on the Peaceful Uses of Atomic Energy. Paper p.524, 1955. New York: United Nations, 1956, 8, p.77.

5. S. L. Reese and W. E. Schroeder, Recovering Uranium from Wet-Process Phosphoric Acid Using Alkyl Pyrophosphoric Acid Extractants, U. S. Patent 4,293,529 (1981).

6. W. W. Berry and A. V. Henrickson, Recovery of Uranium from Wet-Process Phosphoric Acid, U. S. Patent 4,302,427 (1981).

7. F. J. Hurst, D. J. Crouse, and K. B. Brown, Ind. Eng. Chem. Process Des. Dev., 11, 122 (1972).

8. F. J. Hurst and D. J. Crouse, Reductive Stripping Process for the Recovery of Uranium from Wet-Process Phosphoric Acid, U. S. Patent 3,711,591 (1973).

9. A. Sialino and A. Francois, Process for the Recovery of Uranium from Wet-Process Phosphoric Acid, U. S. Patent 4,238,457 (1980).

10. E. J. Steck, Barren Solvent Wash by Oxidized Raffinate Acid in the Process of Uranium Extraction from Phosphoric Acid, U. S. Patent 4,435,367 (1984).

11. W. R. York, Recovery of Uranium from Wet-Process Phosphoric Acid, Belg. Patent 894,090 (1983).

12. W. D. Arnold, Recent Studies of Uranium Recovery from Wet-Process Phosphoric Acid with Octylphenyl Acid Phosphate, paper presented at 1978 ACS Annual Meeting, Miami Beach, Florida, USA, Sept. 13, 1978.

13. W. D. Arnold, D. R. McKamey, and C. F. Baes, Progress Report on Uranium Recovery from Wet-Process Phosphoric Acid with Octyl Phenyl Acid Phosphate, Report ORNL-TM-7182, Oak Ridge National Laboratory, Oak Ridge, Tennessee, 1980.

14. F. J. Hurst and D. J. Crouse, Ind. Eng. Chem. Process Des. Dev., 13, 286 (1974).

15. F. J. Hurst and D. J. Crouse, Oxidative Stripping Process for the Recovery of Uranium from Wet-Process Phosphoric Acid, U. S. Patent 3,825,214 (1974).

16. F. J. Hurst, W. D. Arnold, and A. D. Ryon, Chem. Eng., 84, 56 (1977).

17. T. K. S. Murthy, V. N. Pai, and R. A. Nagle, in The Recovery of Uranium. Proc. Symp. Sao Paulo, Aug. 17-21, 1970. IAEA Proc. Series, Vienna, 1971, p.341.

18. C. A. Blake, C. F. Baes, Jr., and K. B. Brown, Ind. Eng. Chem., 50, 1763 (1958).

19. G. Ting, Y.-M. Tsai, H.-M. Chen, Separation Process for the Recovery of Uranium from Wet-Process Phosphoric Acid, U. S. Patent 4,544,530 (1985).

20. F. J. Hurst, G. M. Brown, and F. A. Posey, Lowering Oxygen Concentration While Isolating Uranium, Belg. Patent 894,858 (1983).

A SOLVENT ANODE FOR PLUTONIUM PURIFICATION *

by

David F. Bowersox, Keith W. Fife and Dana C. Christensen

Los Alamos National Laboratory
Los Alamos, NM 87545

The purpose of this study is to develop a technique to allow complete oxidation of plutonium from the anode during plutonium electrorefining. This will eliminate the generation of a "spent" anode heel which requires further treatment for recovery. Our approach is to employ a solvent metal in the anode to provide a liquid anode pool throughout electrorefining. We use molten salts and metals in ceramic crucibles at 700°C. Our goal is to produce plutonium metal at 99.9% purity with oxidation and transfer of more than 98% of the impure plutonium feed metal from the anode into the salt and product phases. We have met these criteria in experiments on the 100-1000g scale. We plan to scale our operations to 4 kg of feed plutonium and to optimize the process parameters.

I. INTRODUCTION

We define pyrochemical processes as nonaqueous processes which are conducted at elevated temperatures. In our plutonium reprocessing operations, we oxidize scrap metal to oxide, reduce oxide to impure metal, remove excess americium, cast to shape, and electrorefine to a pure product in a series of pyrochemical operations at temperatures between 700° and 900°C. Our research and development efforts are concentrated on improving the efficiencies of these processes, cutting costs, reducing residues and decreasing exposure of personnel to radiation. This paper will be restricted to a discussion of electrorefining and our efforts to develop a solvent anode to significantly increase our yields and reduce the residue volumes for our process.

Since the mid-forties, pyrochemical operations have been employed to reduce uranium and plutonium tetrafluorides to metal.[1,2] Baker's success with these early reductions led Los Alamos to study related pyrochemical process for recovering the actinides from spent reactor fuels. For metal-based fast

* Work done at the Materials Science and Technology Division of the Los Alamos National Laboratory under the auspices of the United States Department of Energy.

reactor systems, fuel could be reprocessed without altering the metal state. Studies of electrorefining were initiated to satisfy purity requirements for the Los Alamos Molten Plutonium Reactor Experiment.[3] This work was further expanded in the early sixties to an electrorefining operation in which impure plutonium metal was melted in a ceramic vessel and electrolytically oxidized into a molten salt, transported through the salt and reduced to high purity metal at a cathode.[4] With additional research, we developed an integrated method for producing metal from scrap by a series of pyrochemical operations.[5]

II. Plutonium Electrorefining at Los Alamos

A schematic diagram of the Los Alamos Electrorefining Cell is shown in Fig. 1.

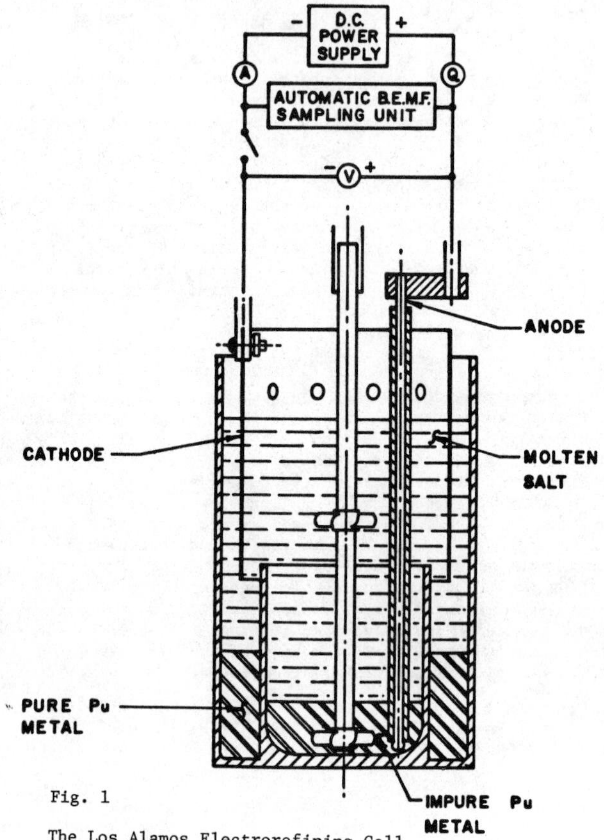

Fig. 1

The Los Alamos Electrorefining Cell

The cell is formed by two concentric magnesia crucibles joined at the base. In typical operations, 4-6 kg of impure plutonium metal is placed in the inner crucible and an equimolar mixture of sodium chloride-potassium chloride containing 5 w/o (weight per cent) magnesium chloride is added as an oxidant (to produce Pu (III) in the electrolyte prior to passing a current through the cell). The cell is assembled in a stainless steel furnace tube, evacuated and back-filled with argon. The crucibles are heated to 740°C and the stirrer, anode and cathode lowered into the positions shown in the figure. Current (dc) is passed through the cell, with stirring, until the plutonium is no longer the primary ion being formed at the anode. This is indicated by an increase in the static cell potential, which is measured by a back emf sampling unit. At a pre-set potential, the current is automatically terminated and an operator withdraws the hardware from the melt. The heat is turned off and the system cools to room temperature. The ceramic crucible is broken to recover the salt and metal product. The metal product ring, shown in Fig. 2, is mechanically separated from the salt, sampled for analysis and transferred for casting and future use.

Fig. 2 The Plutonium Product From Electrorefining

The purity of typical product is greater than 99.9% plutonium. The elements less active than plutonium remain in the anode while those more reactive concentrate in the salt. Impurities in the product are usually tungsten, which is introduced from the cell hardware, and americium. Other impurities are present at concentrations of less than 500 ppm (parts per million).

Two yields are important in considering the effectiveness of this process. The first, which we will call the oxidation yield, is defined as the per cent of the impure feed which is oxidized and transferred out of the anode. In our present gallium-plutonium system, this yield is about 90%; the remaining 10% of the plutonium forms a solid anode heel containing most of the initial impurities. The second yield, which we designate as the reduction yield, is the per cent of the plutonium in the initial feed recovered as product. In our present system, the reduction yield is typically 70-75%.

The three major residues are the solid anode heel, the electrolyte salt, and the crucible. The anode heel is treated by another pyrochemical process, pyroredox, to recover the remaining plutonium in a form suitable for additional electrorefining.[6] The salt is collected, melted, and contacted with excess calcium to recover residue plutonium.[7] Crucible residues are leached in acid and the plutonium recovered by aqueous methods.[8] The overall recovery of plutonium, considering the residue steps, is satisfactorily high.

Since the inception of the process in 1962, large quantities of extremely pure plutonium have been produced by this process. However, it has a history of somewhat low yields. As described above, the solid anode heel is a major residue. We are now investigating methods of oxidizing and transferring more of the plutonium from the anode compartment to eliminate processing of the spent heel. We propose using a low melting solvent metal in which plutonium is significantly soluble. This report will describe our experiments employing cadmium as the solvent metal in the anode pool along with a discussion of our preliminary results.

III. Approach to Solvent Anode Electrorefining

The behavior of impurities in the electrorefining can be predicted from

TABLE I.

ΔF Values At 1000°K For Impurity Elements In Electrorefining

Element	$-\Delta F^{o}$,kcal/g.atom Cl
Ni	18
Cu	21
Ta	22
Fe	27
Cr	32
Ga	32
Mn	41
Al	46
U	54
Th	59
Pu	59
Ce	66
Am	67

the free energies of formation of the chlorides. Some of these energies, at $1000^{o}K$, are listed in Table 1. For elements with more than one oxidation state, the $-\Delta F^{o}$ for the most stable form of chloride is given. Because plutonium has a quite high value for the energy of formation, most impurity elements remain in the anode pool while plutonium is oxidized. For example, for iron the reaction of interest is

$$3\ Fe(0)\ (\ell) + 2\ Pu(III) = 3\ Fe(II) + 2\ Pu(0)\ (\ell)\ . \qquad (1)$$

The calculated free energy for this reaction is 96 kcal/mol Pu at $1000^{o}K$, which is equivalent to an equilibrium constant of $10^{-21.9}$. The ratio of Fe(11) to Pu(111) is very low because

$$\frac{(Fe(II))^3}{(Fe(0))^3} = 10^{-21} \frac{(Pu(III))^2}{(Pu(0))^2} \qquad (2)$$

Assuming that the activity coefficients are within an order magnitude of being equal, and if the metal state of Pu(0) is taken as unity,

$$\frac{Fe(II))^3}{(Fe(0))^3} = 10^{-21}(Pu(III))^2 \qquad (3)$$

and the amount of ferrous ion in the electrolyte is negligible compared to Pu(III). From similar calculations, we predict that most impurities will remain in the anode. These predictions have been verified by our experimental data, as shown by the data for the behavior of impurity elements in electrorefining (Table II). The elements with higher free energy values, such as americium and cerium, will be preferentially oxidized and concentrated in the salt phase.

TABLE II.

BEHAVIOR OF IMPURITY ELEMENTS IN ELECTROREFINING

Element	Concentration (ppm) Feed	Concentration (ppm) Product	Element Concentrated in
Ni	750	<10	Anode
Cu	100	2	Anode
Ta	5 000	80	Anode
Fe	25 000	20	Anode
Cr	280	<10	Anode
Ga	10 000	<25	Anode
Mn	70	< 2	Anode
Al	2 700	< 5	Anode
U	275	<20	Anode
Th	160	<10	Anode
Pu	---	---	---
Ce	527	<25	Salt
Am	850	82	Salt

The major reaction at the anode, as long as Pu(0) reaches the metal electrode, will be

$$Pu(0),(\ell)_{impure} = Pu(III) + 3e \quad (4)$$

The Pu(III) will be carried through the salt phase to the cathode. The major reaction there, under our operating conditions, will be

$$Pu(III) + 3e = Pu(0),(\ell)_{pure} \quad (5)$$

and the overall reaction is

$$Pu(0),(\ell)_{impure} = Pu(0),(\ell)_{pure} \quad (6)$$

The plutonium collects on the cathode and then drips into the annulus between the outer and inner walls to form the product ring.

If a solvent anode pool is employed, reaction (1) is altered to become

$$Pu(0)_{impure}Cd(0),(\ell) = Pu(III) + Cd(0),(\ell) + 3\ e^{-}. \quad (7)$$

Transport and the cathode reaction will remain unchanged.

The amount of plutonium in the metal product ring can be estimated from the equilibrium constant for the reaction

$$Am(III) + Pu(0),(1) = Pu(III) + Am(0),(\ell) \quad (8)$$

and

$$\frac{(Am(0))}{(Am(III))} = \frac{1}{K_1}\ \frac{(Pu(0))}{(Pu(III))} \quad (9)$$

Under our operating conditions, about 10% of the americium is transferred into the product. Since the (Pu(0)) is approximately unity, we should be able to decrease this ratio by increasing the Pu(III) concentration in the electrolyte.

The composition of the salt phase should not change during electrorefining if the cell potential remains below its decomposition potential. In our regular process, we add $MgCl_2$ to the electrolyte to oxidize some of the feed plutonium prior to electrorefining. This introduces magnesium which can be vaporized out of the system or concentrated as an impurity in the anode. In our solvent anode experiments, substitution of tri-potassium plutonium hexachloride has eliminated this contaminant. If we recycle the electrolyte, we will already have sufficient Pu(III) in the salt phase to start reducing and collecting plutonium. However, the more active impurities, such as americium, will concentrate in the salt and eventually contaminate the product.

We believe that cadmium transferred into the product by vaporization and condensation in the experiments using NaCl.KCl at 750°C. We found that as much as 15% of the cadmium moved into the product. Lowering the operating temperature to 700°C and employing LiCl.KCl as the electrolyte eliminated most of this problem.

In most of our routine electrorefining, the plutonium has been alloyed with gallium to reduce the density and insure a critically safe operation. From the plutonium-gallium phase diagram, the alloy will be liquid as the electrolysis begins. As it continues, the plutonium concentration decreases, and the concentrations of gallium and the impurities in the anode increase.

Finally, the anode becomes pasty and solid as the gallium concentration approaches 16.8 a/o(atomic per cent). As the plutonium concentration continues to decrease, the stirring becomes less and less effective. At about 25 a/o gallium, the anode is completely solid. The principle desired reaction has become

$$Pu(0),(s) = Pu(III) + 3e^- \tag{10}$$

and the potential for plutonium oxidation is

$$E = 0.0291 \text{ ln } (Pu(0),\ (s))_{\text{anode surface}}\ . \tag{11}$$

The diffusion of plutonium through the solid anode material to the anode surface becomes much too slow to meet the conditions of high current density that prevail at the anode. Thus, the Pu(0) concentration at the anode approaches zero and competing reactions predominate. Impurity elements will be oxidized, transported to the cathode, reduced and contaminate the metal product. By monitoring the cell potential and terminating the run at a static cell potential of 0.35 to 0.4 v, we maintain high purity product. However, almost 10% of the plutonium feed remains in the solid anode heel.

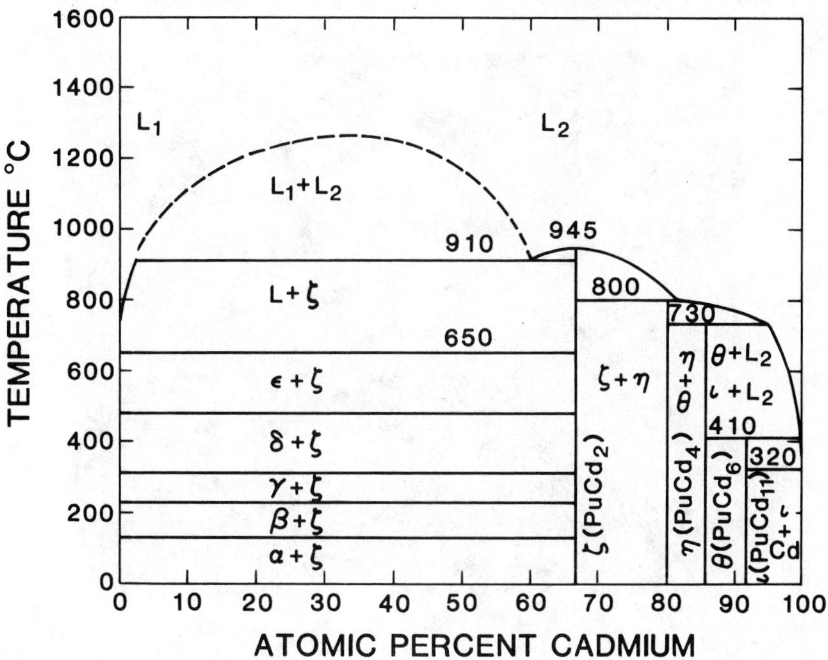

Fig. 3 The Plutonium-Cadmium Phase Diagram

By using a solvent anode pool, the anode material does not solidify, and plutonium should continue to be oxidized as long as it reaches the anode surface. Our experiments, which show very high oxidation yields, demonstrate the effectiveness of this approach. The high purity of the product indicates that even at very low plutonium concentrations (less than 1% of the initial feed value), impurity elements are not oxidized into the salt.

IV. The Solvent Anode Experimental Results

In a solvent anode system, a metal with a relatively low melting point and with no high melting compounds which form with plutonium is used to form an anode pool. We evaluated a number of possible candidate metals and selected cadmium for our initial experiments. Fig. 3 shows the plutonium-cadmium phase diagram.[10] As can be seen by the solidus lines, concentrations of cadmium greater than 85 a/o(atom per cent) will ensure some liquid throughout electrorefining at temperatures of 750°C. The cadmium-plutonium system seems quite promising, even though the vapor pressure of cadmium is quite high at the normal operating temperature of our electrorefining cells.[10] In our initial tests we modified a small-scale (500g) cell and tested it with high purity plutonium as the initial feed. The resulting product is compared in volume with the anode residue in Fig. 4. The ring was 99.9% pure metal, and

Fig. 4 Product From Small Scale Electrorefining

the oxidation yield was 98.8%. This shows that, with high purity metal in the initial feed, oxidation yields can be very high.

We then conducted experiments using cadmium metal and actual anode residues from the conventional electrorefining process. The product of an experiment is shown in Fig. 5. In the figure, the product metal (on the

Fig. 5

Product From a Small Scale Solvent Anode Experiment

bottom as a shiny ring), is 99.9% plutonium of 19.4 g/cc density. The salt, containing blue plutonium tri-chloride, lies on top of the product ring. Some salt has been broken away to show that the anode cup wall has not been wetted by the cell reagents or products. Over 98% of the plutonium was oxidized and transported from the anode cup. The high oxidation yields and high purity product encouraged us to continue with larger scale operations.

In 6 kg-scale electrorefining cells, we are limited to approximately 1 kg of plutonium as feed because of the volume of cadmium required for a solvent pool that contains no high melting components. Spent anodes from our electrorefining process were first contacted with calcium at 850°C to remove any surface oxides and then added to cadmium in the anode compartment. The Cd:Pu ratio was varied from 1.4:1 to 3.0:1. Oxidation yields increased with increasing ratios. Oxidation yields at 740°C varied from 86 to 90% and reduction yields averaged 80%. Results of a typical run are given in Table III. As

TABLE III.

SUMMARY OF EARLY ELECTROREFINING EXPERIMENT (ER 27R) AT 750°C

<u>Operating Conditions</u>

725-750°C for 24h with stirring
6 amps, 1:2v
End point potential: 0.4v

Initial Composition	
Pu	544g
NaCl.KCl	2350g
Cd	1500g
$MgCl_2$	75g

<u>Residues</u>	
Cathode	18g Pu
Black salt	23g Pu
White salt	10g Pu
Crucible shards	2g Pu
Anode heel	6g Pu

<u>Product</u>	
Cd (top ring)	15g Cd, 55 g Pu
Product ring	430g Pu, density 19.5g/cm^3

Oxidation Yield	98.9%
Reduction Yield	79.0%

shown, the white salt was very low in plutonium and, as a result, could be discarded. The other residues could be treated to recover the remaining plutonium. The product ring contained two phases. The upper, lighter phase contained 15g of cadmium and 55g of plutonium. This was easily separated from the high density, pure plutonium in the lower phase. We oxidized 98.9% of the feed plutonium and collected 79% as high purity product. An additional 18g remained on the cathode surface.

A spent cadmium anode and a product ring from a similar run are shown in Fig. 6. The lower metallic ring, which was 99.9% plutonium, contained 100 ppm

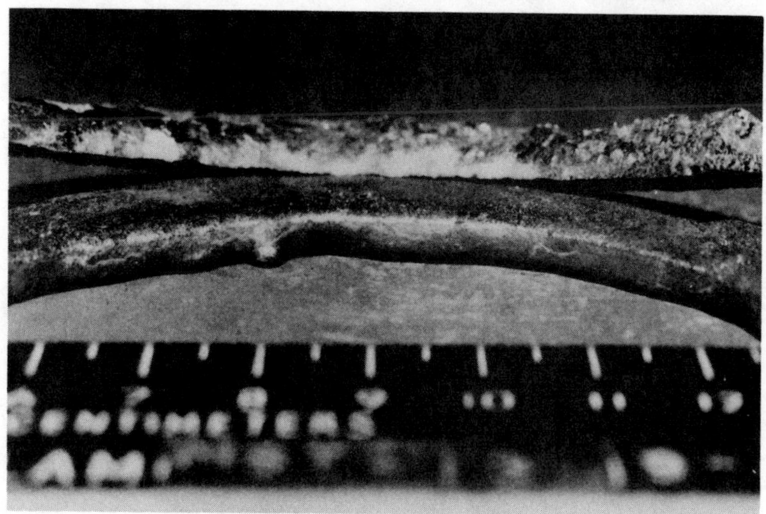

Fig. 6 Product Ring with Cadmium Transfer

cadmium. The upper ring was removed and recycled in a subsequent experiment. The electrolyte formed two phases; a lower black salt which contained almost all the plutonium in the salt, and a light-colored upper layer. The lighter salt and crucible could be discarded; the plutonium in the black phase was recovered pyrochemically.

Several steps were taken to improve our procedures and to further define the experimental parameters. We replaced the ceramic stirrer used on conventional electrorefining with an electrically-isolated tantalum stirrer designed for efficient mixing. We substituted a Lithium chloride-potassium chloride eutectic with a lower melting point for the sodium chloride - potassium chloride and lowered our operating temperature to 700°C. At this temperature, the cadmium vapor pressure is significantly lower, and the second layer of product did not form. Finally, we replaced magnesium chloride, which had been added as an oxidant, with tri-potassium plutonium hexachloride to provide initial Pu(III) in the electrolyte. This substitution reduces the Mg present in the anode compartment and the anode heel.

The results of three experiments with these modifications are summarized in Table IV. In each run, 400g of tri-potassium plutonium hexachloride was

TABLE IV.

SUMMARY OF MODIFIED ELECTROREFINING EXPERIMENTS, 700°C

Operation Conditions

Temperature 700°C

Stirring rate 200 - 250 rpm   400g K_3PuCl_6 in eutectic salt

Current 8 - 10 amps, 1.2v

End Point Potential 0.4v

ER	33	34	35
Initial Composition			
Pu, feed (g)	650	665	642
Cd (g)	1950	1950	1950
Eutectic (g)	2100[a]	2100[a]	2000[b]
Results			
Oxidation Yield	95	99	99
Reduction Yield	72	72	66
Product density, g/cc	19.2	19.3	19.7

[a]LiCl.KCl

[b]NaCl.KCl

added to the eutectic salt. The system was maintained at 700°C and the stirring rate adjusted to 200-250 rpm. The current was set at 8-10 amps, and the voltage was 1.2v. The runs terminated when the back emf reached 0.35-0.4v. A 1950g cast cylinder of cadmium was combined with the plutonium and placed in the anode compartment. The 2000g of eutectic salt was added to the crucible. In all three cases, the oxidation yield was greater than 95% and the metal product was of satisfactory purity (99.9%). The reduction yield, however, was only about 70%. Fig. 7 shows the product metal and the spent anode from ER

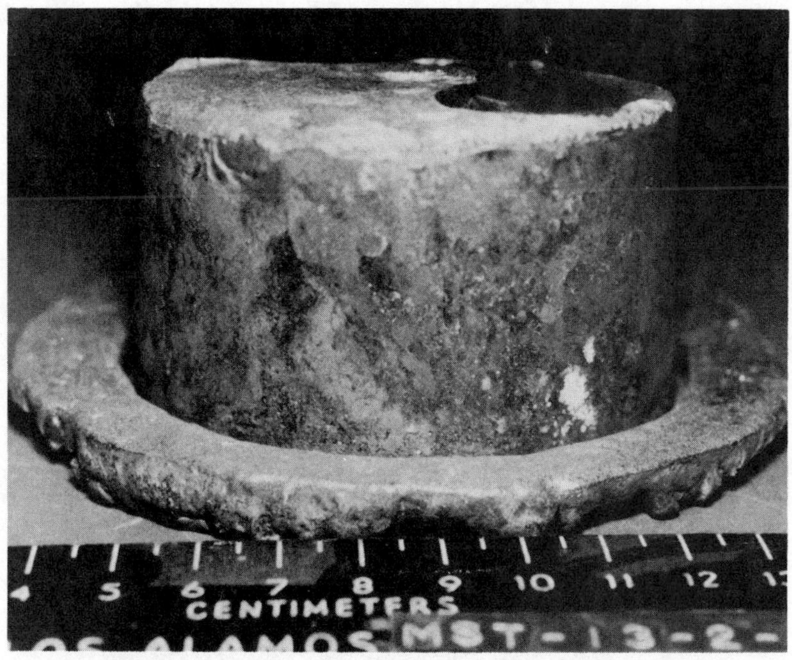

Fig. 7 Spent Anode and Product Ring

35. Fig. 8 is the product ring from a different angle. The spent anode contains less than 1.3% (8g) of the initial plutonium. The product is high purity plutonium with an unusually high cadmium concentration (3000 ppm).

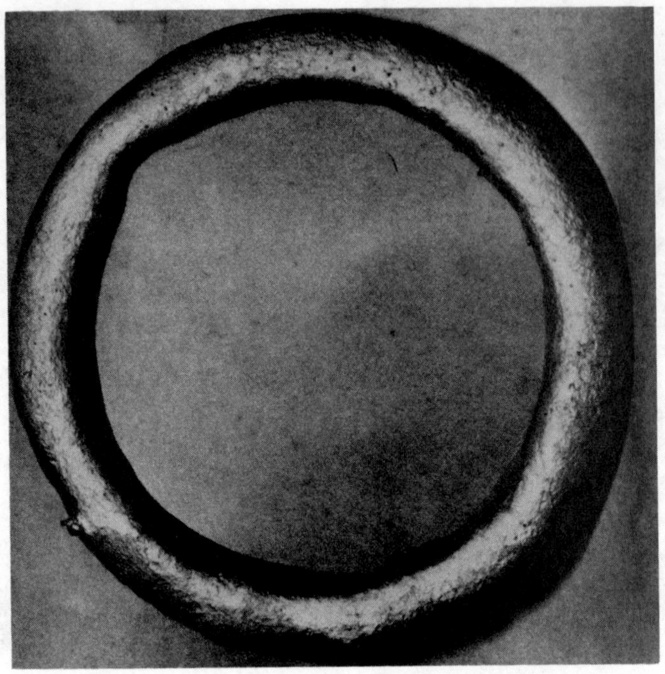

Fig. 8 Product Ring From 600g Solvent Anode Experiment

V. Discussion and Conclusions

The reduction yields of 70-80% are lower than we would like for this process. We believe that the reduction yield can be increased by increasing the initial Pu(III) concentration in the electrolyte. We also plan to vary some of the operating parameters. The cadmium-containing anode could be reused until the impurity buildup becomes high enough to contaminate the product or decrease the yields. At that point, most of the cadmium could be recovered by distillation, recast and recycled. The residue remaining after distillation would either be discarded or, if the plutonium content were above the discard level, treated by alternate means.

We do not recycle the electrolyte at this time; however, there are no major obstacles to instituting such a step in the process. Reuse would be limited by buildup of americium and its transfer into the product. When necessary, the plutonium and americium in the salt can be separated and recovered by pyrochemical reduction. The salt could then the reused or discarded.

We are successfully using cadmium as our solvent element, even though it exhibits a high vapor pressure at the temperatures of interest. Other metals with relatively low melting points may prove more effective, provided that plutonium is at least somewhat soluble and that no high-melting alloys or compounds are formed in the composition range of interest. The volume of cadmium required is too large for effective use of our present electrorefining cell with larger quantities of plutonium. We plan to redesign the cell. We may attempt to operate at lower temperatures and collect the plutonium product on a metal cathode.[11, 12] In such a system, the solid product could be withdrawn and separated. Additional electrodes and impure feed could be added to the cell for continuous processing. When impurity build-up occurred, the reagents could be withdrawn and recycled or discarded.

These experiments utilizing a cadmium solvent pool for the anode in electrorefining plutonium have been very successful. High oxidation yields, averaging 99%, demonstrate that the plutonium is indeed being oxidized to depletion in the anode under the operating conditions. The product metal is of quality at least equal to that obtained in our regular process. The spent anode no longer needs to be processed to recover 10% of the initial feed, and recycle of the reagents appears to be feasible. Reduction yields are lower than desirable, but we plan to change conditions to improve them.

More experiments are needed to define optimum stirring rates, ranges of effective initial Pu(III) concentrations and an optimum cell potential for current termination. We hope to develop reference electrodes suitable for examining electrode reactions and identifying cell potentials. Recycle of the electrolyte and the spent anode should be demonstrated. We plan to consider several cell designs and to test a semi-continuous cell. Other solvent metals and electrolytes may be tested. At lower temperatures, for example, plutonium could be collected as a solid on the cathode, removed, and treated separately. If mercury were the solvent, electrorefining could be carried out in Pyrex equipment at 300°C.

We are continuing this research program. The study is challenging, and the potential rewards are large. We plan to present more data as our study progresses.

VI. Acknowledgements

The authors wish to thank C. Brown and P. C. Lopez for the experimental work reported in this paper. We also acknowledge the ideas, advice and encouragement that we have received from J. G. Reavis of Los Alamos, L. J. Mullins of TRUECO, M. C. Coops and J. B. Knighton of Lawrence Livermore National Laboratory, and M. J. Steindler, L. Burris and W. E. Miller of Argonne National Laboratory.

REFERENCES

1. R. D. Baker, et. al., "Preparation of Uranium by the Bomb Method," Los Alamos Scientific Laboratory Report LADC-3012 (1946).

2. R. D. Baker, "Preparation of Plutonium by the Bomb Method," Los Alamos Scientific Laboratory Report LA-493 (1946).

3. J. A. Leary, R. Benz, D. F. Bowersox, C. W. Bjorklund, K. W. R. Johnson, W. J. Maraman, L. J. Mullins and J. G. Reavis, "Pyrometallurgical Purification of Plutonium Reactor Fuels," in Proceedings of the Second United Nations International Conference on the Peaceful Use of Atomic Energy, Vol. 17, United Nations, Geneva, 1958, pp. 376-382.

4. L. J. Mullins and A. N. Morgan, "A Review of Operating Experience at the Los Alamos Plutonium Electrorefining Facility, 1963-1977, Los Alamos National Laboratory Report LA-8943 (1981).

5. D. C. Christensen and L. J. Mullins, "Plutonium Metal Production and Purification at Los Alamos," in Plutonium Chemistry, W. T. Carnall and G. R. Choppin eds., ACS Symposium Series, American Chemical Society, (1983), pp. 409-432.

6. J. A. McNeese, D. F. Bowersox and D. C. Christensen, "The Recovery of Plutonium by Pyroredox Processing," Los Alamos National Laboratory Report LA-10457 (1985).

7. D. C. Christensen and L. J. Mullins, "Salt Stripping: A Pyrochemical Approach to the Recovery of Plutonium Electrorefining Salt Residues," Los Alamos National Laboratory Report LA-9464-MS (1982).

8. S. A. Apgar, "Monthly Report: Pyrochemical Research Project, January 1-February 15, 1985," MST-13:85-96 (1985).

9. M. S. Coops, J. B. Knighton and L. J. Mullins, "Pyrochemical Processing of Plutonium," in Plutonium Chemistry, W. T. Carnall and G. R. Choppin, eds., ACS Symposium Series, American Chemical Society, (1983) pp. 381-408.

10. D. E. Etter, D. B. Martin, D. L. Roesch, C. R. Hudgens and P. A. Tucker, "The Plutonium-Cadmium Binary System," Trans. Met. Soc. AIME, 233: 2011-2013 (1965).

11. L. Burris, M. Steindler and W. Miller, "A Proposed Pyrometallurgical Process for the Rapid Recycle of Discharged Fuel Materials from the Integral Fast Reactor," Proc. Fuel Reprocessing and Waste Management, Jackson, Wyoming, August 26-29, 1984, Vol. 2, p. 257, American Nuclear Society, La Grange Park, Illinois (1984).

12. Z. Tomczuk, D. Poa, W. E. Miller and R. K. Steunenberg, "Electrorefining of Uranium and Plutonium from Liquid Cadmium," Trans. Am. Nucl. Soc., 50, 204 (1985).

PLUTONIUM REMOVAL FROM NITRIC ACID WASTE STREAMS

A. C. Muscatello and J. D. Navratil
Rockwell International
Rocky Flats Plant
P.O. Box 464
Golden, CO 80401

Separations research at the Rocky Flats Plant (RFP) has found ways to significantly improve plutonium secondary recovery from nitric acid waste streams generated by plutonium purifications operations. Capacity and breakthrough studies show anion exchange with Dowex 1x4 (50-100 mesh) to be superior for secondary recovery of plutonium. Extraction chromatography with TOPO (tri-n-octylphosphine oxide) on XAD-4 removes the final traces of plutonium, including hydrolytic polymer.

INTRODUCTION

The plutonium content in RFP nitric acid waste streams ranges from 1 to 100 mg/l, arising mainly from the anion exchange purification of plutonium. One of our development tasks is to identify alternate candidates for secondary recovery of this plutonium, with a goal of reducing the concentration of plutonium to 0.01 mg/l.

The current secondary recovery process consists of evaporators to reduce the volume of ion column effluent (ICE) and recover nitric acid and Dowex 11 anion exchange columns to reduce the plutonium concentration of the ICE. Operating experience showed typical reduction of plutonium to about 0.5 mg/l (1), far short of the original goal of 0.01 mg/l. This report summarizes our efforts to further reduce plutonium concentrations in ICE. We chose alternative materials based on previous studies showing good results using extraction chromatography (2) and the alternate anion exchange resins (3) Dowex 1x4 and Amberlite IRA-938. The bifunctional extractants chosen for further investigation are the previously studied (2) DHDECMP (dihexyl-N,N-diethylcarbamoylmethylphosphonate) and OØD(IB)CMPO (octylphenyldiisobutylcarbamoylmethylphosphine oxide) and two monofunctional extractants TBP (tributyl phosphate) and TOPO (trioctylphosphine oxide). Mixtures of DHDECMP and TBP were also studied because our previous work (4,5) showed enhanced extraction with these materials. We used these materials in the extraction chromatography mode because of the relative simplicity of this technique, available equipment and column operating experience at RFP. Other studies (6,7) of secondary recovery have considered only anion exchange systems.

The goals of this study are (a) to evaluate the ability of the above named materials to lower the plutonium content of a simulated ICE to 0.01 mg/l and (b) to measure the capacity and elution characteristics of the best candidate materials.

EXPERIMENTAL

Materials

The extraction chromatography support, Amberlite XAD-4 (20 to 50 mesh), is a macroreticular, polystyrene-divinylbenzene cross-linked, non-ionic resin manufactured by Rohm and Haas Company, Philadelphia, Pennsylvania. Amberlite IRA-938 (20 to 50 mesh) macroporous anion exchange resin was also obtained from Rohm and Haas. Dowex 1x4 (50 to 100 mesh) gel-type anion exchange resin was received from Dow Chemical Company, Midland, Michigan.

DHDECMP was obtained from Bray Oil Company with a purity of 84% and free from acidic contaminants. OØD(IB)CMPO was purchased from M&T Chemical Co., Rahway, New Jersey as a 98% pure solid. TBP was obtained from Eastman Kodak Company, Rochester, New York, and is 98% pure. TOPO (mp. 50 to 54° C) was a product of Columbia Organic Chemical Company, Inc., Camden, South Carolina. All other chemicals were reagent grade or better.

The XAD-4 was conditioned by washing with distilled water, followed by methanol, and vacuum drying. The organic material to be sorbed was slurried with the prepared resin. The material stayed in contact with the resin for 7 days and was then suctioned off. The resin was washed with distilled water until no visible organic residue was observed in the wash liquid. The resin was air dried with suction and loaded into a column.

For loading the TOPO and the OØD(IB)CMPO, the resin was first conditioned as above, then a measured amount of solid organic extractant (0.36 g/ml XAD-4) was heated to 50° C in a beaker and slurried with the resin. The heating was discontinued after 2 hours, and the materials were allowed to stand overnight. The TOPO and OØD(IB)CMPO were completely sorbed by the resin using this treatment. The resin was washed with water before use.

The feed solutions were prepared by mixing the appropriate aliquots of a stock (36.8 g/l) plutonium nitrate solution (eluate from production anion exchange operations), giving 1 to 10 mg/l concentrations of plutonium in 7.2<u>M</u> nitric acid. For breakthrough capacity studies, the column effluent was restored to 10 mg/l plutonium by spiking with the stock solution and the effluent was rerun through the column in order to prevent the creation of large quantities of waste liquids.

Equipment and Procedures

Each type of resin was tested for its ability to decontaminate 1.0 mg/l plutonium solutions. The DHDECMP, TBP, OØD(IB)CMPO and TOPO were sorbed onto XAD-4 as described above. For DHDECMP+TBP, an equimolar mixture of the two materials was used on XAD-4. Dowex 1x4 and Amberlite IRA-938 were used as received. A total of 50 ml of resin was loaded into a 1.8 cm inside diameter (i.d.) glass column yielding a 20-cm bed height. Peristaltic pumps delivered a 7$\underline{M}$ HNO_3 conditioning solution and other feeds in an upflow mode at 10 ml/min. Four liters of feed were passed through the column. The plutonium content of the effluent was determined radiometrically every 250 ml. Corrections were made for the contribution of americium to the alpha count rate.

The candidate materials having effluents <0.1 mg/l were tested to 10% breakthrough capacity by using 25 ml of resin and a flow rate of 15 ml/min. The feed concentration was increased to 10 mg/l to avoid excessively long experiments. After each material was tested at least once under these conditions, higher than expected plutonium concentrations (0.1 to 1 mg/l) were observed in the effluents. The resin bed was increased to 50 ml to compensate for possible channeling problems. Plutonium concentrations in the effluent were determined every four liters.

After sorption, the materials were eluted with 0.1$\underline{M}$ ascorbic acid except where noted.

RESULTS AND DISCUSSION

Sorption Tests

The results of the study of the ability of the candidate materials to decontaminate 1.0 mg/l plutonium solutions in 7.2$\underline{M}$ HNO_3 are shown in Figures 1 and 2. The data are the average of duplicate runs for DHDECMP, OØD(IB)CMPO, TOPO, and DHDECMP+TBP. IRA-938 and TBP had triplicate runs.

From Figure 1, it is easy to see that TBP alone is not effective in removing plutonium and approaches breakthrough rather quickly. IRA-938 is also unsatisfactory since it sorbs only ~90% of the plutonium initially. The improved sorption by IRA-938 as the experiment progresses may be due to better packing of the resin. Figure 2 shows the results for OØD(IB)CMPO both with and without 0.12 g/l of iron(III) added to the feed. Without iron(III), the performance of OØD(IB)CMPO is better than any other material. However, iron(III) is a component of actual waste streams and OØD(IB)CMPO does extract it (8). With iron(III) present, OØD(IB)CMPO is better than only TBP.

The other materials each show similar sorption of plutonium to the 0.02 to 0.04 mg/l range, approaching the original goal of 0.01 mg/l. Table 1 summarizes the behavior of the materials in terms of the percent of the plutonium lost in the effluent and the percent of the sorbed plutonium recovered by elution with 250 ml of 0.1M ascorbic acid.

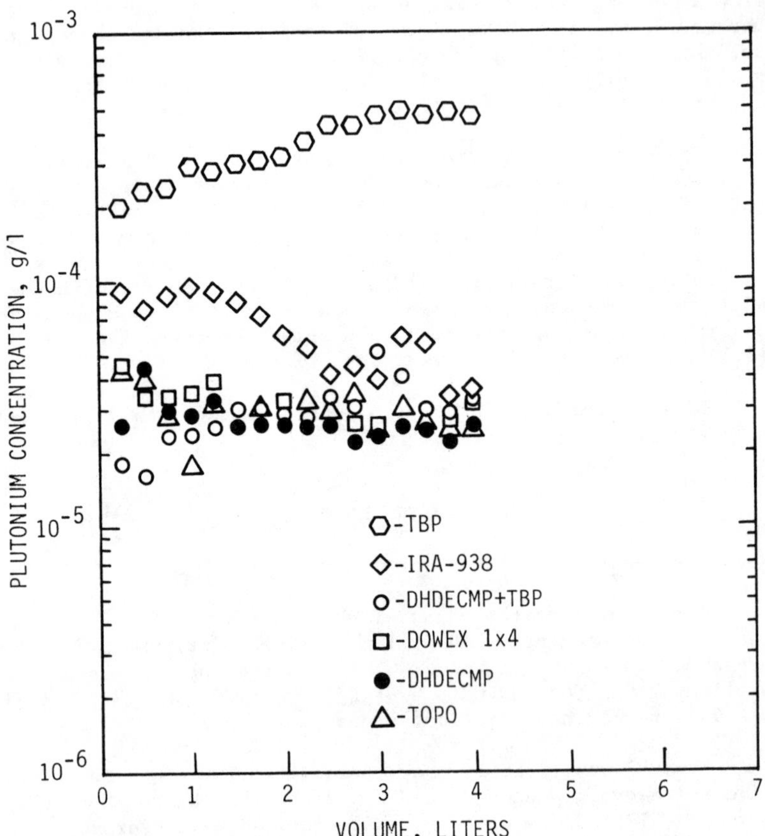

Figure 1. Sorption of 1.0 mg/l plutonium from 7.2M HNO_3 by candidate materials.

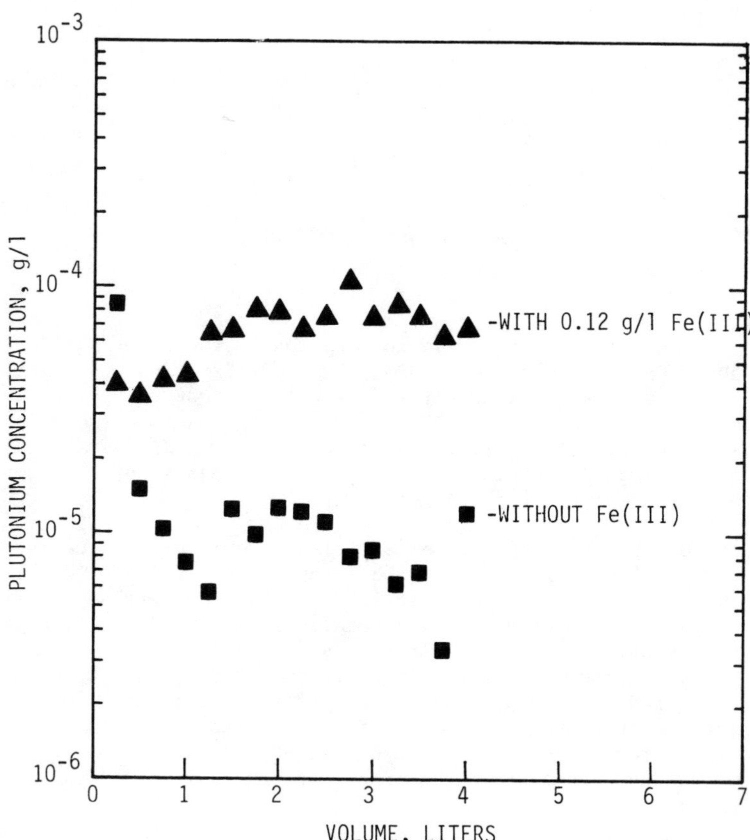

Figure 2. Sorption of 1.0 mg/l Plutonium from 7.2$\underline{M}$ HNO_3 by OØD(IB)CMPO on XAD-4.

Table 1. Sorption and elution of plutonium (1.0 mg/l)

Material	Pu %, in Effluent	Pu %, in Eluate*
DHDECMP	2.5	49.8
DHDECMP+TBP	4.0	75.6
TBP	32.8	46.1
TOPO	3.6	3.8
Dowex 1x4	3.5	80.0
IRA-938	6.3	50.8
OØD(IB)CMPO	1.5	23.0
OØD(IB)CMPO**	2.6	16.0

*Corrected for Pu not sorbed.
**With 0.12 g/l iron in the feed.

The materials decrease in sorption in the order DHDECMP > OØD(IB)CMPO > Dowex 1x4 > TOPO > DHDECMP+TBP > IRA-938 > TBP. The elution decreases in the order Dowex 1x4 > DHDECMP+TBP > IRA-938 > DHDECMP > TBP > OØD(IB)CMPO > TOPO. On the basis of sorption, TBP is an unsuitable candidate for further study. Similarly, IRA-938 was eliminated from further consideration because of its poor initial sorption of plutonium.

On the basis of elution efficiency, the rest of the materials are roughly equivalent with the exception of TOPO and OØD(IB)CMPO. However, we did not eliminate TOPO from further evaluation since it may be possible to use an incineration or oxidative acid digestion process to recover the plutonium (9,10). Acid digestion would completely decompose resin into gaseous products and allow precipitation and anion exchange recovery of the plutonium. TOPO on XAD-4 loaded with plutonium has been tested for acid digestion with satisfactory results (11). OØD(IB)CMPO was eliminated on the basis of its mediocre performance in the presence of iron(III) (Figure 2).

Capacity Tests

Additional breakthrough tests were performed using 10 mg/l plutonium in the feed for the final candidate materials. Table 2 summarizes the 10% breakthrough volumes. The materials decrease in capacity in the order TOPO > DHDECMP > Dowex 1x4 > DHDECMP+TBP. TOPO has a much greater capacity than the other three materials, twice as much as the nearest competitor material.

The theoretical capacities of the four materials are compared to the experimental capacities in Table 2, also. Again, TOPO shows the best loading, followed by DHDECMP, DHDECMP+TBP, and Dowex 1x4. The capacity of Dowex 1x4, 26.5 mg/l resin, compares favorably with

Table 2. 10% Breakthrough capacities with 30 mg/l plutonium feed

Material	10% Breakthrough Vol (l)	Pu Sorbed (g)	Capacity (mg/ml)	Theor Cap (mg/ml)	% Theor
TOPO	90	2.650	53.1	112	47.4
DHDECMP	48	1.550	31.0	115	27.0
Dowex 1x4	42	1.330	26.5	167	15.8
DHDECMP+TBP	32	0.956	19.1	115	16.6

the observed (14) capacity of IRA-938, 44 mg/ml resin, determined with 1 g/l plutonium in 7M HNO_3 feed and compares very favorably with 26 and 22 mg/ml resin for Dowex 1x4 determined with 0.5 g/l plutonium at 3.6 and 7.0 ml/min cm^2, respectively (3).

After loading, the materials were eluted with 250-350 ml of 0.1M ascorbic acid. TOPO columns were not eluted because of the low elution percentage observed in the sorption tests. In general, the columns were eluted until the blue color of plutonium(III) was no longer visible in the eluate.

For Dowex 1x4, one column was further eluted with 100 ml of 0.35M HNO_3 and 100 ml of 0.35M HNO_3 + 0.1M HF because the resin appeared to retain plutonium. For DHDECMP on XAD-4, the visual cut-off appears to be inadequate since the eluate plutonium concentrations remained around 1.0 g/l. The same is true for DHDECMP+TBP on XAD-4.

Plotting the percent of the plutonium eluted vs. the volume of the eluate shows that much greater volumes would be required to recover 90% of the sorbed plutonium. Table 3 summarizes the information. The percent plutonium eluted is based on the amount of plutonium sorbed in the complete run, beyond the 10% breakthrough. Unfortunately, there is a wide spread between the two runs except for DHDECMP+TBP. However, this mixture and Dowex 1x4 are on the average better than DHDECMP alone.

Table 3. Elution of plutonium sorbed in breakthrough tests

Material	% Eluted, Run 1	% Eluted, Run 2	% Eluted, Average
Dowex 1x4	61.8	34.0	47.9
DHDECMP	14.3	38.1	26.2
DHDECMP+TBP	54.8	45.7	50.3

CONCLUSIONS

Based on the preceding results, we can see that, unfortunately, none of the materials reached the original goal of <0.01 mg/l plutonium in the effluent. However, these studies do indicate that further reduction of the plutonium waste concentration is achievable, especially with more efficient, large-scale equipment and by operating at a lower linear flow rate. This could be very useful if the current discard limit of 1.0 mg/l is lowered below the capability of the current anion exchange secondary recovery system.

The conceptual process for such a further plutonium reduction in the ICE would entail a tertiary recovery after secondary recovery anion exchange. Based on the results given herein, the best candidate for tertiary recovery is TOPO on XAD-4 for the following reasons:

1) The TOPO has the greatest capacity for plutonium by far, with consistently lower effluent concentrations.

2) Although TOPO cannot be satisfactorily eluted, incineration or the acid digestion process can be used with a consequently large concentration factor of the plutonium recovered.

3) We have shown that TOPO extracts plutonium(IV) polymer from nitric acid (12), so the presence of polymer should present no problem, whereas anion exchange resins do not sorb polymer (Figure 3).

It is remarkable to note that solid TOPO sorbed on XAD-4 extracts plutonium at all, since diffusion through a solid would be very slow. Consequently, the distribution ratios of TOPO on XAD-4, TOPO dissolved in dodecane, and solid TOPO were compared. Table 4 shows the results of this study. We observed in the course of the experiment that the solid TOPO liquified upon contact with the nitric acid, probably due to the formation of the TOPO HNO_3 complex. Therefore, the TOPO in the resin becomes a liquid upon preconditioning with nitric acid and behaves as the other liquid extractants.

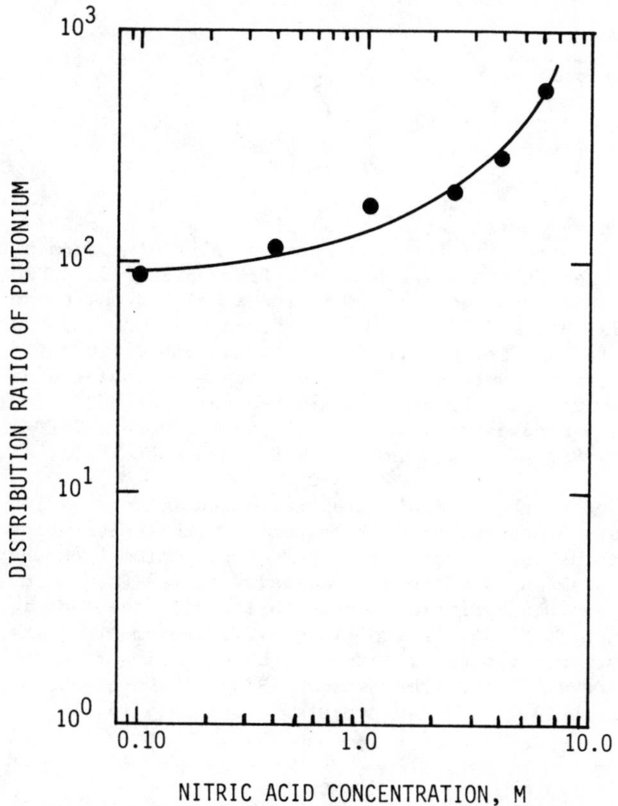

Figure 3. Distribution ratio of freshly prepared plutonium(IV) polymer as a function of nitric acid. 0.10 M TOPO in carbon tetrachloride; 0.5 g/l Pu.

Table 4. Extraction of plutonium(IV) from 7.75M HNO_3 using TOPO in different media*

Aqueous Phase	Organic Phase	[Pu] aqueous (g/l)	K_D
Feed	---	2.62×10^{-2}	--
Depleted Feed	0.5 g TOPO+1.38 ml XAD-4	1.92×10^{-4}	135
Depleted Feed	0.5 g TOPO+5 ml dodecane	1.52×10^{-4}	171
Depleted Feed	0.5 g TOPO	2.97×10^{-4}	87.2

*Aqueous Volume = 5.0 ml

SUMMARY

We have investigated the ability of several candidate materials to recover low concentration plutonium from nitric acid waste. The materials are DHDECMP, TBP, DHDECMP+TBP, OØD(IB)CMPO, and TOPO, each sorbed onto Amberlite XAD-4 and the anion exchange resins IRA-938 and Dowex 1x4. We found TBP, OØD(IB)CMPO, and IRA-938 to have insufficient ability to remove plutonium, and they were not tested further. The remaining materials reduce 1.0 mg/l plutonium to low 0.01 mg/l for 80 bed volumes (BV). Their 10% breakthrough capacities for 30 mg/l plutonium are TOPO - 1800 BV, DHDECMP - 960 BV, Dowex 1x4 - 840 BV, and DHDECMP+TBP - 640 BV.

Based on these results, we recommend TOPO on XAD-4 be used as a tertiary plutonium recovery material with incineration or acid digestion as the means to recover the plutonium from the TOPO. Further scale-up studies are necessary as well as an investigation to find an inexpensive support material with the same sorptive properties as XAD-4. Furthermore, an improvement in the secondary anion exchange process is proposed by switching from 20-50 mesh resins (Dowex 11 and Amberlite IRA-938) to the higher capacity 50-100 mesh Dowex 1x4, as reported previously (3).

ACKNOWLEDGEMENT

We gratefully acknowledge M. T. Saba for experimental assistance.

REFERENCES

1. S. C. Burkhardt, Rockwell International, Rocky Flats Plant, Golden, Colorado, unpublished data, October 14, 1985.

2. L. L. Martella, J. D. Navratil, and M. T. Saba, in Actinide Recovery from Waste and Low Grade Sources, J. D. Navratil and W. W. Schulz, Editors, Harwood Academic, New York, 27 (1982).

3. J. D. Navratil and L. L. Martella, Nucl. Technol., 46, 105 (1979).

4. J. D. Navratil and L. L. Martella, Sep. Sci. Technol., 16, 1147 (1981).

5. A. C. Muscatello, J. D. Navratil and M. E. Killion, Solv. Extr. Ion Exch., 1, 127 (1983).

6. S. G. Proctor and D. A. Burton, Rockwell International, Rocky Flats Plant, Golden, Colorado, unpublished data, August 7, 1972.

7. A. R. Kazanjian and M. E. Killion, Rockwell International, Rocky Flats Plant, Golden, Colorado, unpublished data, October 30, 1981.

8. G. F. Vandergrift, R. A. Leonard, M. J. Steindler, E. P. Horwitz, L. J. Basile, H. Diamond, D. G. Kalina, and L. Kaplan, Transuranic Decontamination of Nitric Acid Solutions by the TRUEX Solvent Extraction Process - Preliminary Development STUDIES, Report ANL-84-45, U.S. Dept. of Energy, Argonne, Illinois, 1984.

9. C. R. Allen, R. G. Cowan, M. D. Crippen, and G. L. Richardson, Radioactive Acid Digestion Test Unit (RADTU)--Final Status Report, HEDL-TME 82-23, Hanford Engineering Development Laboratory, Richland, Washington, 1982.

10. Y. Kobayashi, H. Matsuzuru, J. Akatsu, and N. Moriyama, J. Nucl. Sci. Tech., 17, 865 (1980).

11. A. R. Kazanjian and J. R. Stevens, Rockwell International, Rocky Flats Plant, Golden, Colorado, private communication, July 2, 1985.

12. A. C. Muscatello, J. D. Navratil, and M. E. Killion, Sep. Sci. Tech., 18, 1731 (1983).

AMERICIUM REMOVAL FROM NITRIC ACID WASTE STREAMS

A. C. Muscatello and J. D. Navratil
Rockwell International
Rocky Flats Plant
P.O. Box 464
Golden, CO 80401

Separations research at the Rocky Flats Plant (RFP) has found ways to significantly improve americium removal from nitric acid (7M) waste streams generated by plutonium purification operations. Partial neutralization of the acid waste followed by solid supported liquid membranes (SLM) are useful in transferring and concentrating americium from nitrate solutions. Specifically, DHDECMP (dihexyl-N,N-diethylcarbamoylmethylphosphonate) supported on Accurel polypropylene hollow fibers assembled in modular form transfers >95% of the americium from high nitrate (6.9M), low acid (0.1M) feeds into 0.25M oxalic acid stripping solution. Maximum permeabilities were observed to be 0.001 cm/sec, consistent with typical values for other systems. The feed:strip volume ratio shows an inverse relationship to the fraction of metal ion transferred. Cation exchangers may be used to concentrate americium from the strip solution.

Furthermore, OØD(iB)CMPO (or CMPO) (octylphenyl-N-,N-diisobutylcarbamoylmethylphosphine oxide) has been tested in an extraction chromatography mode. Preliminary results show CMPO to be effective in removing americium if the feed is neutralized to 1.0M acidity and iron(III) is complexed with 0.20M oxalic acid.

INTRODUCTION

Chemical processing activities at the RFP involve the recovery of plutonium from RFP scrap and residues. Nitric acid(7M) waste streams from aqueous plutonium processing operations contain 1-10 mg/l concentrations of plutonium and americium. The acid waste is sent through a secondary anion exchange process, and then to waste treatment where it is first neutralized with caustic and then processed through three stages of ferric hydroxide carrier precipitation. The resulting sludges are dried and drummed for offsite shipment and storage. The americium in the waste often times causes excessive radiation exposure to waste treatment operators. Furthermore, by reducing the actinide content of the waste and recovering additional amounts of plutonium, the stream might be able to bypass waste treatment and go directly to a spray dryer for conversion to non-TRU nitrate salts. Therefore, processes to selectively remove and concentrate plutonium and americium separately prior to waste treatment are being investigated. (The isolated americium could be placed in special casks for storage and the plutonium recycled in the process.) Work to date is described on the evaluation of solid supported liquid membranes (SLM) (1,2) and extraction chromatography techniques to remove americium from nitric acid waste streams.

EXPERIMENTAL

Materials

DHDECMP was obtained as an 84 vol% pure material free from acidic contaminants from Bray Oil Co. CMPO was obtained from M&T Chemical Co. at 98% purity. Dowex 50Wx8 resin(50-100 mesh) was obtained from Dow Chemical Co. Americium was obtained from RFP production operations. All other materials were reagent grade.

Procedures

Americium transfer behavior was studied using laboratory-built hollow-fiber membrane modules. A porous polypropylene hollow-fiber material was used: Armak Accurel fibers (2 per module), 11.5 cm long, 0.15 cm I.D., with a 0.1-cm wall thickness and 75% void volume. The hollow fibers were potted into a glass tube with RTV silicone adhesive.

The activity of americium in the feed as a function of time was followed by an on-line NaI(Tl) gamma detector in conjunction with a Canberra Series 35 multichannel analyzer operating in the multiscalar mode.

The feed and strip solutions were recirculated through the membrane module using Cole Parmer Master-Flex peristaltic pumps until no further change in feed americium concentration was observed. The initial and final americium concentrations were determined radiometrically.

Sorption of americium by inorganic ion exchangers was studied by mixing 1.0 g of the material with 10 ml of 0.10$\underline{M}$ nitric acid spiked with 0.1 mg/l americium on a rotating framework for 1 hour. The ion exchanger was filtered off and the liquid analyzed for americium concentration by gamma counting. Americium distribution ratios were calculated by difference.

For extraction chromatography sorption studies, the CMPO was sorbed by melting with Amberlite XAD-4 (0.36 g/ml resin) at 50° C. After cooling, the resin was loaded into 50 ml columns which were operated at 9-10 ml/min.

Permeability Calculations

Previous studies (3) have calculated the permeability (P_m) of the metal ion through the membrane from:

$$\mathrm{Ln}\ \frac{[M]_o}{[M]_t} = \frac{P_m A t}{V} \quad (1)$$

Where: $[M]_o$ = Feed concentration at time zero
$[M]_t$ = Feed concentration at time t
A = Surface area of membrane
t = Time
V = Volume of cell.

We may substitute

$$\mathrm{Ln}\ \frac{A_o}{A_t} = \frac{P_m A t}{V} \quad (2)$$

where A_o = Gamma activity at time zero
A_t = Gamma activity at time t

since the proportionality constant between gamma ray activity and concentration cancels out.

RESULTS AND DISCUSSIONS

Membrane Studies

Preliminary studies indicated incomplete transfer of americium from 7.0M HNO_3 to 0.25M $H_2C_2O_4$ at equilibrium through a membrane of undiluted DHDECMP. The reason for <100% transfer is the cotransfer and build-up of nitric acid in the strip solution and consequent back-transfer of americium (4). The transfer of nitric acid was demonstrated in a separate experiment. Therefore, the transfer of americium(III) was studied as a function of nitric acid concentration at a total nitrate concentration of 7.0M. Sodium nitrate was used to make up the balance of the nitrate. This procedure simulates a partial neutralization of a typical RFP 7.0M HNO_3 nitric acid feed stream.

Table 1 summarizes the percent americium transferred as a function of nitric acid concentration. The greatest transfer is observed at the lowest concentrations of nitric acid, because the driving force of a low nitrate concentration is maintained in the strip solution.

Only the experiment at 0.1M acidity has sufficient transfer to allow calculation of the permeability from a fit of Equation (2) to the straight line. This permeability is also listed in Table 1. For the Accurel fibers, the permeability is about twice that measured by Danesi et al. (4) in 1M HNO_3 for single hollow fiber modules, possibly because of the lower acidity and higher nitrate concentration of our experiments.

Table 1. Percentage transfer and selected permeabilities of americium(III) through a hollow fiber membrane of DHDECMP as a function of feed nitric acid concentration ([Am] = 0.1 mg/l).

$[HNO_3]$,M	$[NaNO_3]$,M	% Transfer	10^{-4} P_m $cm\ sec^{-1}$
7.0	0.0	61	-
5.0	2.0	64	-
3.0	4.0	76	-
1.0	6.0	90	-
0.1	6.9	94	7.2

These studies indicate the best conditions for americium transfer are 0.1M HNO_3 plus 6.9M $NaNO_3$. Actual waste streams may be brought to these conditions by adding the correct amount of sodium hydroxide.

Effect of Feed-to-Strip Ratio

Other experiments investigated the effect of the feed-to-strip(F:S) ratio using DHDECMP on Accurel fibers with a 7.0M HNO_3 feed and 0.25M $H_2C_2O_4$ strip. A roughly linear relation would be expected between the F:S ratio and the concentration factor. However, the observed concentration factors (Table 2) are less than the F:S ratio because a large fraction of the americium remains in the membrane at the conclusion of the experiment. A roughly inverse relationship between the ratio and the percent americium transferred is observed, because of the increased concentration of acid in the strip solution.

Table 2. Effect of feed:strip volume ratio on the transfer of americium from 7.0M HNO_3 to 0.25M $H_2C_2O_4$ through a DHDECMP SLM ([Am]=1 mg/l).

Volume Ratio	% Transfer	Concentration Factor
3.33:1	80	0.90
5:1	52	0.62
10:1	45	0.58

Cation Exchange of the Strip Solution

Table 3 shows the results of transfer experiments at 0.1M HNO_3 plus 6.9M $NaNO_3$ feed concentrations with 0.25M oxalic acid strip concentrations where the strip solution is circulated through a small column of Dowex 50Wx8 cation exchange resin. The cation exchange resin easily removes all the transferred americium from the strip solution, allowing a large concentration factor. The increased time for transfer in Run 2 is consistent with Equation (1) because of the increased feed volume.

Table 3. Results of the combined membrane/cation exchange system*

Component Run	Volume ml	Initial $[Am]gl^{-1}$	Final $[Am]gl^{-1}$	Transfer %	Equilibrium Time, min
Feed (1)	75	$8.0x10^{-4}$	$1.5x10^{-4}$	82	342
Strip (1)	15	0.0	$<4.1x10^{-6}$	-	-
Feed (2)	150	$8.0x10^{-4}$	$1.9x10^{-4}$	76	1510
Strip (2)	30	0.0	$<4.1x10^{-6}$	-	-

*0.89g resin for Run 1, 0.85g resin for Run 2

To avoid disposal of organic resins loaded with americium, we have also investigated inorganic ion exchangers for their ability to sorb americium. Table 4 shows the distribution ratio of americium with four promising candidates. These materials are sodium titanate, zirconium tungstate, zirconium phosphate, and bone char. At a pH of 1 and after 1 hour of contact, sodium titanate has the highest K_D.

Based on these results, it is possible to envision a useful membrane transfer system that removes americium from a nitrate stream and delivers it to a low acid strip solution. An inorganic cation exchange material could then be used to remove americium from the strip solution. The inorganic material could then be dried and shielded to become a safe material for long term waste storage.

Table 4. Americium distribution coefficients of various inorganic ion exchangers

Material	K_D
Sodium titanate	3500
Zirconium tungstate	800
Zirconium phosphate	1100
Bone char	600

Extraction Chromatography

The leading candidate for americium removal using liquid-liquid extraction techniques is OØD(iB)CMPO or simply CMPO (5), because it has the highest distribution ratio for americium from nitric acid combined with reasonable selectivity. We have tested CMPO using extraction chromatography to attempt to meet americium removal limits and the non-TRU limits.

DHDECMP was also considered. However, its breakthrough capacity was too small to be practical (6). Figure 1 shows much better results occur with CMPO in the absence of iron(III), with losses of only 2.6% of the americium and <10^{-4} g/l effluents for 120 bed volumes. However, elution with 0.1M oxalic acid was not effective, with only 1.6% recovery.

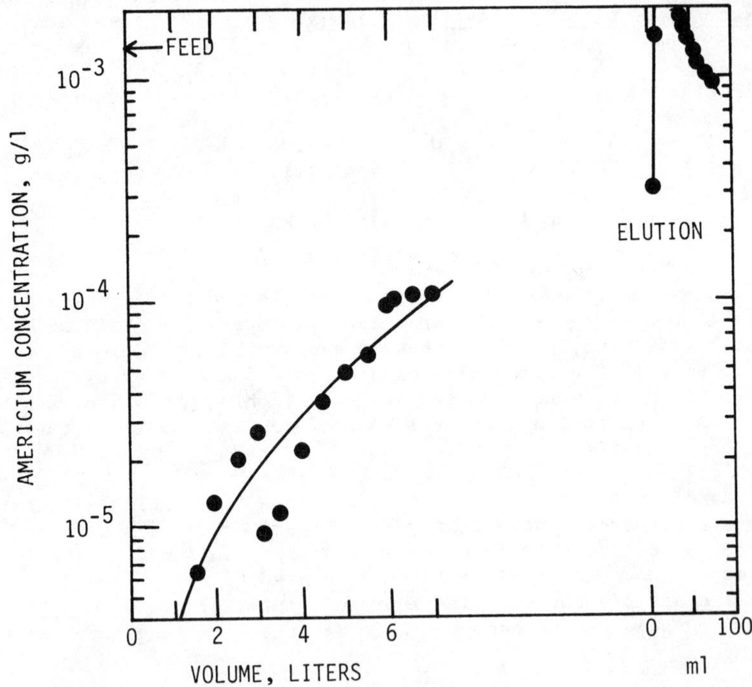

Figure 1. Sorption of americium from 7.4M nitric acid. Elution with 0.10M oxalic acid. Feed [Am] = 1.7 mg/l.

However, actual acid waste contains iron(III) in substantial amounts. Figure 2 shows that 0.25 g/l iron severely reduces the effectiveness of the CMPO. We observe 100% breakthrough in only 30 bed volumes. Again, only a few percent of the americium is eluted. Obviously, the iron saturates the column and prevents further americium loading.

Horwitz <u>et al.</u> (5) have shown that CMPO does not extract iron at lower nitric acid concentrations. It was not known whether simple neutralization would be effective in preventing iron extraction. Figure 3 shows the results of batch studies and that neutralization to 1.0 to 2.0M hydrogen ion concentration is very effective in removing iron interference. In addition, 0.2M oxalic acid was added to complex the iron(III). The distribution ratio of americium with CMPO on XAD-4 rises to 600-700 and that of iron decreases to 3, yielding a good separation factor, α, of >200.

The effectiveness of this neutralization has been tested on a 50 ml column of CMPO on XAD-4. Table 5 summarizes the results and shows that the removal of americium is very effective. The effluent contains <4.1 x 10^{-6} g/l americium, less than the detection limit for our analytical labs. We did not attempt to elute the column because of our previous poor results in this area. In retrospect, this elution may be possible because the CMPO would sorb less nitric acid that must be removed before americium would elute. This is a major reduction in the americium concentration and at least approaches a non-TRU limit for this isotope.

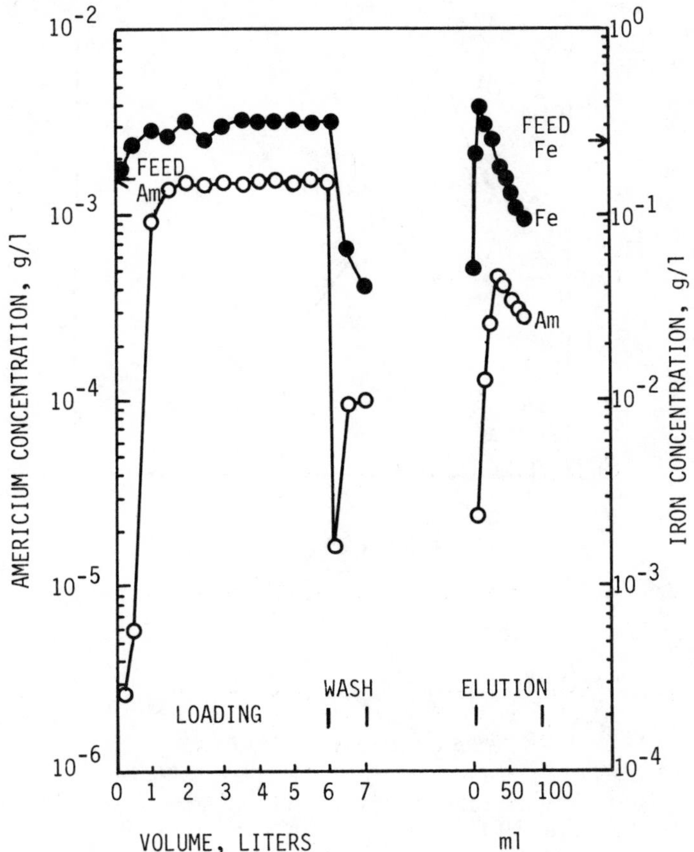

Figure 2. Sorption of americium from 7.4M nitric acid containing 0.25 g/l iron(III). Washed with 1.0M nitric acid. Eluted with 0.10M oxalic acid. Feed [Am] = 2.1 mg/l

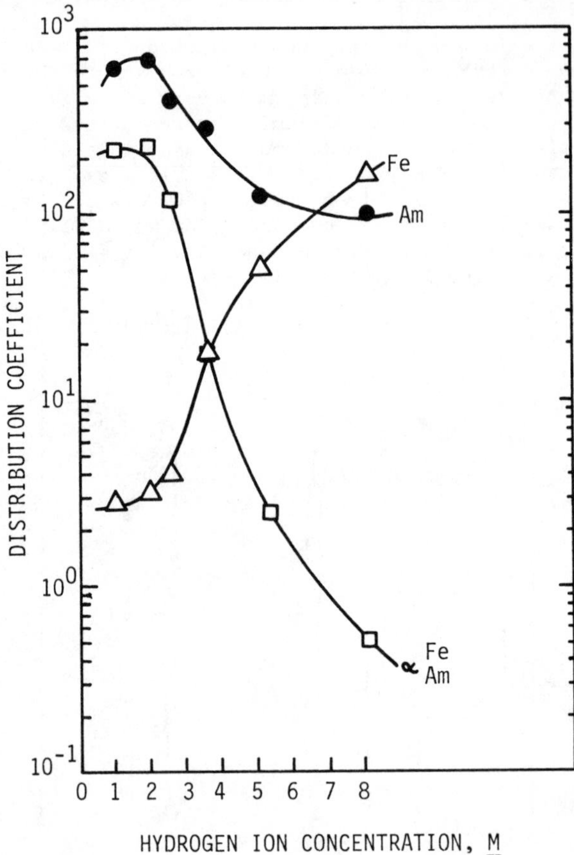

Figure 3. Volume distribution ratio of americium between neutralized nitric acid (10 ml) and CMPO on XAD-4 (0.50 g). Initial [Am] = 0.64 mg/l. Initial [Fe] = 74 mg/l.

Table 5. Extraction chromatography of americium by CMPO on XAD-4. Feed = 7.4M nitric acid neutralized to 1.0M acidity plus 0.20M oxalic acid. Initial [Fe] = 150 mg/l.

Solution	[Am],g/l	Vol.,l	% Am
Feed	1.2×10^{-3}	6.0	100%
Effluent	$<4.1 \times 10^{-6}$	6.0	<0.3%

SUMMARY

The planned treatment of the acid waste solutions at RFP is to remove as much of the plutonium as possible using secondary anion exchange. The effluent from anion exchange could then be reduced in acidity to 0.1M using sodium hydroxide and passed through SLM or extraction chromatography to remove the americium to 0.3 mg/l levels. The effluent should then produce non-TRU waste during waste treatment operations, which consist of iron carrier hydroxide precipitation (7).

These studies have defined some of the operating conditions for using SLM and extraction chromatography for the recovery and concentration of americium from nitric acid feeds. DHDECMP supported on Accurel hollow fibers shows promise in transferring americium from 6.9M $NaNO_3$-0.1M HNO_3 into oxalic acid strip solutions. Permeabilities of americium have been determined to be 0.001 cm/sec, in agreement with other membrane studies. Varying feed:strip ratios produce inverse percent americium transfer effects. Inorganic cation exchange materials can be used to give further concentration of the americium from the strip solution, and provide an excellent storage medium for the actinides. The strip solution could be recycled or sent to feed make-up. Results of extraction chromatography show CMPO to be very effective in removing americium at low acid, high nitrate concentrations, although work remains to be done to improve elution characteristics.

ACKNOWLEDGMENTS

This work was performed under U.S. Government Contract DE-AC04-76DP03533, and the U.S. Government retains a prior nonexclusive, royalty-free license to publish, translate, reproduce, use, or dispose of the published form of the work, or allow others to do so for U.S. Government purpose.

We also wish to thank M. E. Killion, K. S. Lubthisophon, and M. Y. Price for experimental assistance.

REFERENCES

1. P. R. Danesi, E. P. Horwitz, and P. G. Rickert, J. Phys. Chem., 87, 4708 (1983) and references therein.

2. P. R. Danesi, J. Membrane Sci., 20, 231 (1984) and references therein.

3. P. R. Danesi, E. P. Horwitz, G. F. Vandgrift, and R. Chiarizia, Sep. Sci. Technol., 16, 201 (1981).

4. P. R. Danesi, R. Chiarizia, P. Rickert, and E. P. Horwitz, Solv. Extr. Ion Exch., 3, 111 (1985).

5. E. P. Horwitz, H. Diamond, and D. G. Kalina, in Plutonium Chemistry, W. T. Carnall and G. R. Choppin, Eds., Am. Chem. Soc., Washington, D.C., 1983, p. 433.

6. L. L. Martella, J. D. Navratil, and M. T. Saba, in Actinide Recovery From Waste and Low Grade Sources, J. D. Navratil and W. W. Schulz, Eds., Harwood Academic, New York, 1982, p. 27.

7. T. E. Boyd, R. L. Kochen, J. D. Navratil, and M. Y. Price, Radioactive Waste Manag. Nucl. Fuel Cycle, 4, 195 (1983).

SEPARATION OF Sr IONS FROM DILUTE SOLUTIONS WITH MEMBRANE PROCESSES

C.FABIANI
Divisione Chimica
E.N.E.A. CRE-Casaccia
Rome Italy

INTRODUCTION

The separation of Sr-90 from dilute aqueous solutions(low activity nuclear wastes) represents a typical problem in nuclear liquid waste management: conditioning of water from fuel element storage ponds,concentration of tail-end solutions from ion-exchange treatments. Membrane processes are considered very promising methods(1)to solve the mentioned problems because of the continuity of the membrane based separation processes and the compactness of the membrane plants.
In this paper a comparison between three different membrane based separation processes (adsorption - ultrafiltration, electrodialysis and Donnan dialysis)is done to select the most useful approach.

EXPERIMENTAL

Two different electrolyte solutions were considered. They reproduce the main characteristics of two real nuclear liquid (low activity) wastes of a nuclear plant operating with magnox type fuel elements.In Table 1 typical compositions are reported.

Table 1.Composition of nuclear liquid wastes containing Sr-90.

	Storage Pond Sol. P	Tail-end Sol. S11
Temperature(C)	29-30	
Conductivity(uS)	1300	
pH	11.5	8.9
Sodium Carbonate(ppm)	250	---
Sodium Hydroxide(ppm)	110	---
Sodium Nitrate(w%)	---	3-4
Sr-90 (µCi/l)	2.2	0.3
Other radionuclides(µCi/l)	70	0.6

Solutions P reproduce the water of the fuel elements storage pond which are very dilute while solutions S11 represent tail-end solutions produced after ion exchange treatment of the previous one.
Separation experiments have been performed on Strontium nitrate 5 ppm solutions reproducing the ionic strength and the pH of the solutions reported in Table 1.
In the adsorption-ultrafiltration method the dissolved cations were adsorbed on Titanium hydroxide colloidal particles formed in situ from Titanium tetrachloride(2) by changing the solution pH after addition of a known amount of titanium tetrachloride(2).The adsorption mechanism for alkaline earth cations can be written(3)

$$TiOH + OH^- = TiO^- + H_2O \quad (1)$$
$$TiO^- + Sr(II) = TiOSr^+$$

and the adsorption kinetic is so rapid(2) that in less than 20 minutes a quasi-equilibrium is reached.
The obtained dispersion was cross-flow filtered by using an Amicon hollow-fiber cartridge with 0.06 sq.m of active surface and 5000 molecular weight cut-off.An initial volume of 2000 cu.cm of initial solution was circulated at a constant 600 cu.cm per minute under a 0.2 atm. inlet-outlet differential pressure.The oberved rate of volume reduction,for a maximum volume reduction factor of 20,was constant in any experimental run.Further details on the experimental study are reported elsewhere(2).

Electrodialysis were performed with a modified Bell II unit supplied by Berghof(Tubingen,D).
The unit uses a cell stack of six membrane couples with 37 sq.cm membrane area for each membrane.A standard 0.5M Sodium nitrate solution with 5 ppm of Sr as Strontium nitrate(2000 cu.cm) was used at a circulating rate of 1400 cu.cm per minute at a constant temperature

of 30+0.1 C (4) .The membrane couple was formed with a Nafion 125 cation exchange membrane(from DuPont) and a R-5035(from RAI Corp) anion exchange membrane whose characteristics are reported in Table 2.

Table 2.Membranes used in electrodialysis experiments

Membrane	Nafion 125	RAI R-5035
Thickness(cm)	0.025	0.020
Selectivity(mV)	KCl 1N/3N 70	KCl 0.5N/1N 82
Water content (%dry basis)	15-20	23
Resistance (ohm sq.cm)	(0.5m NaCl) 13-14	13-16

Current densities in the 0-50 mA/sq.cm range were used to test the dependence of the Na/Sr separation efficiency under different operating conditions.

Donnan dialysis experiments were performed by using a Nafion 125 membrane in tubular form in a cylindrical module of 264 sq.cm of total exchange surface.The feed solution(2000 cu.cm) Sodium nitrate 0.4M,Sr 5ppm as Strontium nitrate at pH 12 was circulated in the shell side compartment.The strip solution (200 cu.cm) Sodium nitrate 0.5M at pH=1 was circulated in the lumen side of the membrane tubes.

RESULTS AND DISCUSSION

Separation processes can be compared by correlating a concentration factor with the driving separation force(5).In isothermal conditions the molar flux referred to the membrane as a static phase can be written

$$J_i = -D_i\left[\nabla C_i + (C_i V_i/RT)\nabla P + (C_i z_i F/RT)\nabla\phi\right] + C_i J_v \quad (2)$$

By assuming a membrane control in the i-th species transport the parameters in the equation refer to the membrane inside:C(molar concentration),V(partial molar volume),P(pressure) and ϕ(electrical potential).J is the total volume flow.The driving force per unit volume (N/cu.cm) to move species i-th relative to the solution(excluding the volume convective effects) is given by

$$CRT\,d_i = C_i RT \nabla C_i + C_i V_i \nabla P + C_i z_i F \nabla \phi \qquad (3)$$

(C is the total molar concentration).This force will be compared for the different separation processes considered with an ideal concentration factor given by

$$\alpha = ([Sr] \text{ in conc.sol. } / \ [Sr] \text{ in feed}) \ . \qquad (4)$$

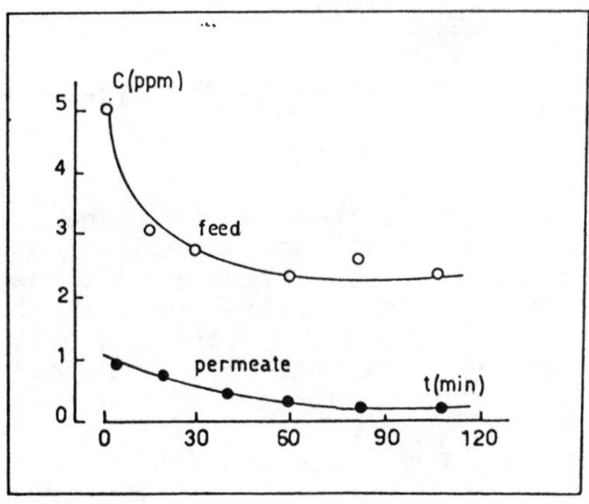

Figure 1.Sr(II) in feed and permeate during adsorption ultrafiltration;Ti/Sr=2,pH=10,P type solutions.From reference(2) with permission.

The typical experimental curve obtained in adsoption UF separation for the studied system is reported in Figure 1.From such plots the concentration factor can be calculated by considering the colloidal dispersion as the concentrated feed.
Several experiments have been reported in which the Ti/Sr ratio and the pH were varied(2) to define the best operating conditions according to the adsorption scheme equation 1 .From those data the concentration factor for P type solutions were calculated and are reported in Figure 2 as a function of pH and Ti/Sr ratio.The method is effective for concentrating a given

ion if no interfering ions are present or their concentration is very low as is the case of P type solutions.In fact in the presence of Sodium ions at 0.5M concentration, as in solutions S11, Sr is practically not adsorbed and therefore the concentration effect is very low:at pH=10 and for Ti/Sr ratio in the 1-4 range the Sr concentration factor varies between 1.14 and 1.70.In this separation process the concentration effect is simply the consequence of the decrease in volume of the colloidal dispersion as the solution flows through the membrane.The driving force is then just given by $\Delta P/l$ being l the membrane thickness.By assuming that practically the whole Sr is adsorbed on the colloid particles the concentration factor at time τ will be

$$\alpha = (V_o .k)/(V_o - J_v .S .\tau) \tag{5}$$

being Vo the feed initial volume ,k the colloid particles loading capacity(2)and S the membrane surface.The concentration effect is obtained ,on the basis of the experimental data reported,as a consequence of a separation force of 25 (N/cu.cm) as shown in Table 3.

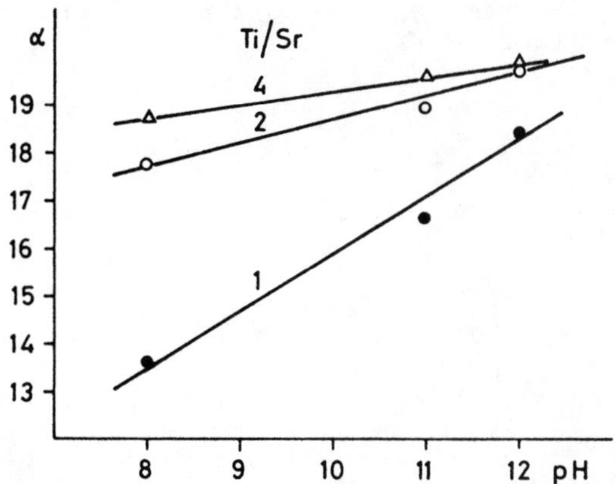

Figure 2.Concentration factors in the adsorption/UF experiments.

Table 3. Concentration factors and separation driving forces

Method	Conc. Factor	Minimum/Sr/feed (ppm)	Force (N/cu.cm)
Adsorption/UF		0.25	
Ti/Sr=2;pH=8	17.8		25
11	18.9		
12	19.8		
Electrodialysis			
(6 h ;pH=10)		0.15	
i= 8.1 (mA/sq.cm)	1.72		3.69
13.5	2.14		5.92
27.0	2.26		12.30
Donnan dialysis		0.50	
(6 h;pH=12)	8.6		350

In the case of S11 solutions to separate Na from Sr ions a selectivity effect should be introduced as in electrodialysis experiments. At the beginning of the electrodialysis experiments the same feed solution was introduced in both dilute and concentrate compartments and the changes of composition were followed for several hours. Current density was changed in a series of experiments (4). Examples of experimental curves are reported in Figure 3.

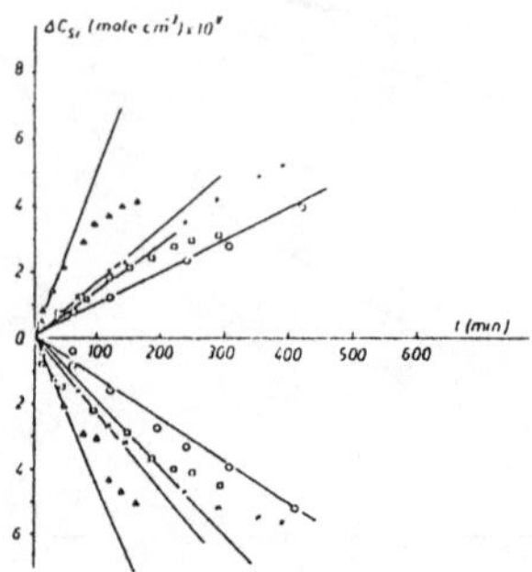

Figure 3. Sr concentration changes in dilute and concentrate compartment as a function of time and current density (mA/sq.cm): 8.1(o), 13.5(Δ) 27.0(x), 40.5(⊕). From reference (4) with permission.

Membrane selectivity defined as

$$P(Na/Sr) = (t_{Na}/C_{oNa})/(t_{Sr}/C_{oSr})$$

being the effective transport number(as difference between the cation transport number) across the cation and anionic membrane corrected by the electroosmotic effect(4),and Co the initial ionic concentration.The P(Na/Sr) values decrease for Sodium ions as the current increases (t=0.92 at 8.,1 mA/sq.cm;t=0.74 at 40.5 mA/sq.cm) while the Sr transport number remains practically constant around 2x10(-4).However the selectivity is always about 0.5 and it reflects the higher ion-exchange equilibrium constant for divalent alkali earth cations in Nafion membranes(K(Na/H)=1.22; K(Sr/H)=4.24)(6).Above 27.0 mA/sq.cm current density the polarization effect becomes important.However even at this low current density value the initial 0.5 M Na concentration decreases to 0.01 M after 6 hours while the Sr concentration decreases to 0.15ppm which means that about 97% of the initial Sr is transported.In electrodialysis at the beginning of the experimental run(ΔP=0,ΔC=0) and assuming no polarization,the separation driving force can be written RT(i/F)(t/D) (N/cu. cm) being i the current density,t the effective transport number and D the diffusion coefficient.The separation driving force reported in Table 3 was calculated by using a diffusion coefficient for sodium ions:9.44x10(-7) sq.cm/s (6).The concentration factor corresponding to the above driving force is given by

$$\alpha = 1 + (n.S/C_o V_o).t.(i/F).\tau \qquad (6)$$

being n the number of membrane couples in the stack,S the membrane surface and τ the time.The calculated values(4.02 at 8.1 mA/sq.cm,6.06 at 13.5 mA/sq.cm and 10.37 at 27.0 mA/sq.cm) are higher than the experimental data reported in Table 3 as obtained from the plots of Figure 3.This effect is probably due to the under estimation of polarization effects in the calculation of the ideal concentration.It is interesting to notice from the curves of Figure 3 that the enrichment of the concentrated solution is not symmetric with the decrease of Sr concentration in the dilute solution.This fact is due to a loading of the membranes.Therefore the concentration factors have been calculated from the data of the dilute solution taking into account the straight part of the curves.As a partial conclusion it can be said that the method seems very promising to separate different ions of similar concentrations.However the reduction in feed volume,which is an important aspect in treating nuclear liquid wastes, is very small.

A possible alternative to the electrodialysis method for Na/Sr separation(S11 type solutions) can be the Donnan dialysis.As reported in the experimental part a standard 0.4M Sodium nitrate solution containin 5 ppm Sr ions was used .The volume reduction was fixed at a factor equal 10 and a difference 12-1 in pH was used.The reduction in concentration in the feed solution is shown in Figure 4 together with the initial molar flow.The feed concentration can be reduced by a factor of 10 in a 6 hour experiment and 86% of the Sr transferred. In such experiments($\Delta P=0$; $\Delta C=0$) the separation driving force directly correlates with the concentration gradient of Sr in the membrane phase.At $\tau=0$ this force for a membrane in the Na-form and by considering Sr as the only diffusing species is given by $RT.(K(Sr/Na))^2.X^2.([Sr] / [Na]^2)$ being X the membrane site concentration(X=2.59 for Nafion 125 in NaCl 0.5M(7)) and K(Sr/Na)=3.4 (6) the exchange selectivity for the considered ionic couple.These data were used to calculate the separation force reported in Table 3.The corresponding concentration factor is then given by

$$\alpha = (1/V_{strip}).D_i.(K(Sr/Na))^2.(X^2/[Na]^2_{feed})(S/l)\tau \quad (7)$$

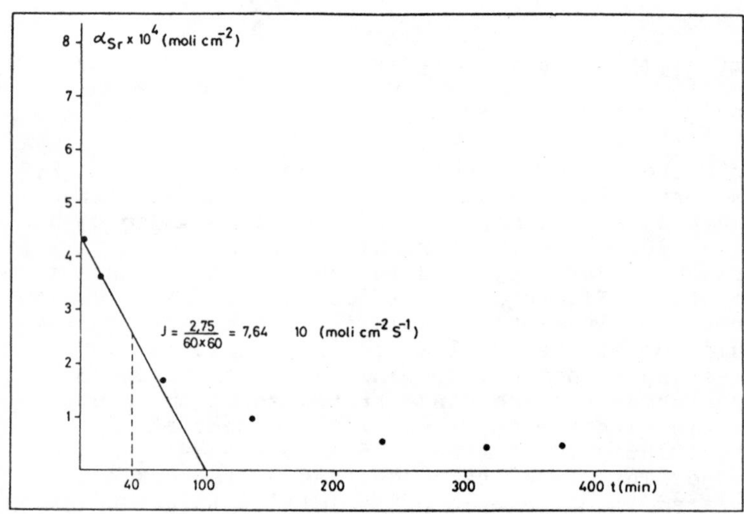

Figure 4.Donnan dialysis;rate of Sr concentration decrease in feed solution.

The method allows high concentration of the feed solution by reducing the volume of the strip solution.Its selectivity through the X and K(Sr/Na) parameters makes it possible to treat S11 waste solutions.
The data collected in Table 3 show that all the three methods can be of interest for treating the Sr containing solutions as they allow to reduce concentration below 0.5 ppm and,with the exclusion of the electrodialysis method,strong volume reductions are possible.On the contrary,electrodialysis needs a small separation driving force.

SUMMARY

From the above comparison(Table 3) it is shown that while the minimum concentration level that can be reached for Sr is similar in any of the three examined methods of separation,the driving forces needed are in the order Donnan dialysis UF Electrodialysis.
Donnan dialysis and Electrodialysis are selective methods useful to treat S11 type solution.On the contrary, with the adsorption-UF method only solutions with interferring ionic species at the same concentration level can be treated due to the lack of selectivity of the adsorption process.However this method allows for an effective volume reduction of the feed solution to be treated.

REFERENCES

1 . R.G.Gutman in"Radioactive waste:advanced management methods for medium active liquid waste",CEN-AERE Harwood Academic Pub.,New York,Ch.4,(1981).
2 . C.Fabiani,M.DeFrancesco,P.Galata,G.Bertoni, Sep.Science&Technology,2,3553(1986).
3 . D.W.Fueratnau.D.Manmohan,S.Raghavan in"Adsorption from Aqueous Solutions",P.H.Tewari (ed.),Plenum Press,New York,p.93,(1981).
4 . C.Fabiani,M.DeFrancesco in"Membranes and Membrane Processes",Plenum Press,New York,(1986).

5 . R.Krishna,The Inst.Chem.Eng:Symp Series 54,185(1978).
6 . A.Steck,H.L.Yeager,Anal.Chem.,52,1215(1980).
7 . L.Bimbi,M.DeFrancesco,C.Fabiani,Report ENEA RT/CHI(83)2Rome,Italy(1983).

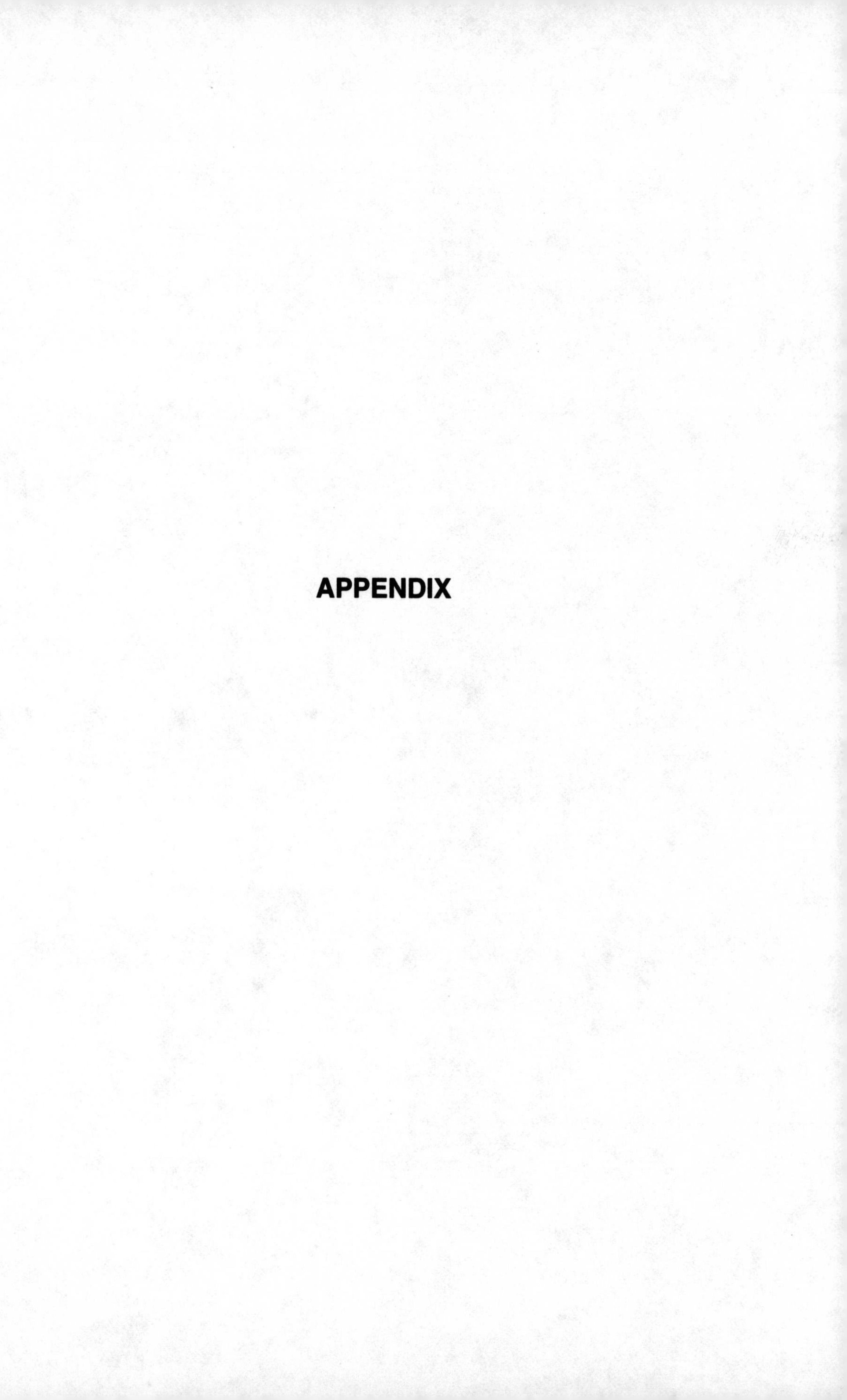

APPENDIX

CONTENTS FOR CHEMICAL SEPARATIONS
VOLUME I. PRINCIPLES

ELECTROPHORESIS

ION EXCHANGE AND ADSORPTION

SOLVENT EXTRACTION

INCLUSION

MEMBRANE PROCESSES

INFORMATION RETRIEVAL

APPENDIX

AUTHOR INDEX

SUBJECT INDEX